# GAS CHROMATOGRAPHY OF STEROIDS
IN BIOLOGICAL FLUIDS

# GAS CHROMATOGRAPHY OF STEROIDS IN BIOLOGICAL FLUIDS

Proceedings of the Workshop on Gas–Liquid
Chromatography of Steroids in Biological Fluids

Held February 25-27, 1965, at
Airlie House, Warrenton, Virginia

### Edited by
### Mortimer B. Lipsett, M. D.

PLENUM PRESS
NEW YORK
1965

The Workshop on
Gas-Liquid Chromatography of Steroids in Biological Fluids
was supported by Grant AM-9180 from the National Institute of Arthritis
and Metabolic Diseases to Dr. R. M. Dodson
University of Minnesota, Minneapolis, Minnesota

Library of Congress Catalog Card Number 65-25243

©1965 Plenum Press
A Division of Consultants Bureau Enterprises, Inc.
227 West 17th Street • New York, N.Y. 10011
All rights reserved

Printed in the United States of America

## CONTENTS

INTRODUCTION.................................................................. ix

EDITOR'S NOTE................................................................. xi

LIST OF PARTICIPANTS.......................................................... xiii

LIST OF ABBREVIATIONS......................................................... xv

STEROIDS - TRIVIAL AND CHEMICAL NAMES......................................... xvii

SESSION I: KETOSTEROIDS AND TESTOSTERONE

    THE SPECIFICITY OF A GAS CHROMATOGRAPHIC METHOD FOR THE DETERMINA-
    TION OF URINARY ANDROSTERONE, ETIOCHOLANOLONE AND DEHYDROEPIANDRO-
    STERONE IN URINE

        by B. S. Thomas................................................. 1

    THE INDIRECT ANALYSIS OF URINARY 17-OXYGENATED $C_{19}$ STEROID
    SULPHATES BY GAS-LIQUID CHROMATOGRAPHY

        by E. Menini..................................................... 11

    A TECHNIC FOR FRACTIONATION OF 17-KETOSTEROIDS FROM URINE

        by C. R. Berrett and C. McNeil................................... 17

    GAS CHROMATOGRAPHIC METHOD FOR THE DETERMINATION OF TESTOSTERONE
    IN HUMAN URINE

        by W. Futterweit, G. L. Siegel, R. Freeman, S. I. Griboff,
        M. Drosdowsky, N. Gibree, R. I. Dorfman and L. J. Soffer........ 19

    ESTIMATION OF TESTOSTERONE IN BLOOD BY GAS-LIQUID CHROMATOGRAPHY

        by A. C. Brownie................................................. 23

    THE ESTIMATION OF PLASMA ANDROGENS BY GAS-LIQUID CHROMATOGRAPHY

        by A. Goldfien, J. Jones, M. E. Yannone and B. White............ 35

    DISCUSSION....................................................... 43

SESSION II: CORTICOIDS

    THE ESTIMATION OF UNCONJUGATED CORTISOL METABOLITES IN URINE BY
    GAS CHROMATOGRAPHY

        by E. Bailey..................................................... 57

    GAS CHROMATOGRAPHY OF CORTOLS AND CORTOLONES

        by R. S. Rosenfeld............................................... 67

URINARY CORTICOIDS BY GAS CHROMATOGRAPHY

    by M. G. Crane and J. J. Harris............................... 73

A SPECIFIC METHOD FOR THE GAS CHROMATOGRAPHIC DETERMINATION OF ALDOSTERONE AND ADRENOCORTICAL STEROIDS: APPLICATION TO ADRENAL VENOUS BLOOD OF VERTEBRATES

    by H. Gottfried................................................ 89

ANALYSIS OF ALDOSTERONE IN URINE BY DOUBLE ISOTOPE DILUTION AND GAS-LIQUID CHROMATOGRAPHY

    by B. Kliman................................................. 101

    DISCUSSION................................................... 113

SESSION III: PROGESTERONE AND $C_{21}$-DIOLS AND TRIOLS

MEASUREMENT OF URINARY STEROIDS BY GAS CHROMATOGRAPHY. $3\alpha,17$-DIHYDROXYPREGNANE-20-ONE AND $C_{21}$ TRIOLS

    by R. S. Rosenfeld........................................... 127

ANALYSIS OF URINARY PREGNANEDIOL AND PREGNANETRIOL BY GAS-LIQUID CHROMATOGRAPHY

    by M. A. Kirschner and M. B. Lipsett......................... 135

THE APPLICATION OF GAS-LIQUID CHROMATOGRAPHY TO THE MEASUREMENT OF PLASMA PROGESTERONE

    by A. Goldfien, M. E. Yannone, D. B. McComas and C. Braga...... 143

DETERMINATION OF PROGESTERONE IN HUMAN PERIPHERAL BLOOD USING GAS-LIQUID CHROMATOGRAPHY WITH ELECTRON CAPTURE DETECTION

    by H. J. van der Molen and D. Groen.......................... 153

THE ANALYSIS OF PLASMA PROGESTINS BY GAS CHROMATOGRAPHY

    by I. H. Carlson, A. J. Blair and R. K. Meyer................. 169

    DISCUSSION................................................... 177

SESSION IV: ESTROGENS

GAS CHROMATOGRAPHY AND ITS ROLE IN THE VERSATILE ANALYSIS OF URINARY ESTROGENS

    by H. H. Wotiz and S. C. Chattoraj........................... 195

DETERMINATION OF URINARY ESTROGENS BY GAS CHROMATOGRAPHY

    by H. Adlercreutz and T. Luukkainen.......................... 215

GAS CHROMATOGRAPHY OF ESTRIOL IN PREGNANCY URINE

    by J. C. Touchstone.......................................... 229

ANALYSIS OF ESTROGENS IN THE URINE OF NON-PREGNANT WOMEN BY GAS-LIQUID CHROMATOGRAPHY

    by E. Menini................................................. 233

THE MEASUREMENT OF ESTRIOL IN URINE AND AMNIOTIC FLUID
BY GAS-LIQUID CHROMATOGRAPHY

      by A. E. Schindler, M. C. Lindberg and W. L. Herrmann.......... 237

GAS CHROMATOGRAPHY OF FREE AND CONJUGATED ESTROGENS IN BLOOD
PLASMA

      by J. C. Touchstone and T. Murawec........................... 243

ESTIMATION OF ESTRADIOL-17$\beta$ BY GAS-LIQUID CHROMATOGRAPHY WITH
ELECTRON CAPTURE DETECTION

      by K. B. Eik-Nes, A. Aakvaag and L. J. Grota................. 247

GAS CHROMATOGRAPHY OF ESTROGENS AND OTHER STEROIDS FROM ENDOCRINE
TISSUES

      by K. W. McKerns and E. Nordstrand.......................... 255

      DISCUSSION................................................... 263

SESSION V: SOME THEORETICAL AND PRACTICAL ASPECTS OF GAS-LIQUID CHROMATOGRAPHY

SOME ASPECTS OF THE CHEMISTRY OF GAS-LIQUID CHROMATOGRAPHY

      by W. J. A. VandenHeuvel..................................... 277

      DISCUSSION................................................... 292

SOME PRACTICAL PROBLEMS IN GAS-LIQUID CHROMATOGRAPHY

      by H. M. Fales............................................... 297

COMMENTS ON TECHNIQUES ON GAS CHROMATOGRAPHY

      by H. H. Wotiz............................................... 301

      DISCUSSION................................................... 307

THE UTILITY OF GLC FOR ANALYZING SPECIFIC STEROIDS IN BIOLOGICAL
FLUIDS

      by A. Karmen................................................. 309

INDEX................................................................ 313

# INTRODUCTION

During the past decade we have witnessed a revolution in analytical methods. The development of vapor phase chromatography for the separation and analysis of classes of substances ranging from metals and gases to a wide variety of organic materials has been one of the most exciting of these new techniques. Gas-liquid chromatography for the measurement of steroids is particularly significant for endocrinologists and reports during the past several years have demonstrated its usefulness.

Because of the growing interest in this method, a committee of the Endocrinology Study Section composed of Drs. R. M. Dodson, Seymour Lieberman, Hilton A. Salhanick, and Ralph E. Peterson, felt that the time was propitious to hold this Workshop and it is on their behalf that I welcome you. We hope to obtain enough data during these sessions so that those attending this conference and those who may read the proceedings will be able to make an informed judgement about the usefulness of gas-liquid chromatography for the analysis of steroids in biological fluids. Thus, I hope that there will be adequate documentation of the reliability of the methods as well as a comparison of the advantages and disadvantages of this analytical method with other classical methods. If we can do this, this Workshop will provide a significant base of practical considerations about gas chromatographic analytic techniques.

I would like to thank Drs. T. F. Gallagher, H. Wilson, H. Salhanick and L. Engel for agreeing to serve as Chairmen of the sessions.

## EDITOR'S NOTE

This book is divided into five sections: gas-liquid chromatography of ketosteroids, corticoids, estrogens, progestins, and some practical considerations in gas chromatography. Discussion follows each section. A list of abbreviations and of trivial names and abbreviations for steroids has been provided. An attempt has been made to index the discussion material and pertinent elements in the papers.

A word of caution seems appropriate here. Of the participants at this Workshop, only a few can be classified as professionals in the area of gas chromatography. The others are biologists or physicians who have attempted to use this technique with varying degrees of success. This book therefore is not a repository of wisdom about gas chromatography but is probably instead a fair reflection of the practice of the art as of 1965. Many of these methods will be revised, improved, or abandoned in the future and it is almost certain that none can be considered definitive at present. However, some of these methods have proved useful already and are worthy of consideration. Caveat emptor!

## LIST OF PARTICIPANTS

Bailey, E., Sheffield, England
Brownie, A. C., Buffalo, New York
Carlson, I. H., Madison, Wisconsin
Crane, M. G., Loma Linda, California
Dodson, R. M., Minneapolis, Minnesota
Eik-Nes, K. B., Salt Lake City, Utah
Engel, L., Boston, Massachusetts
Fales, H. M., Bethesda, Maryland
Futterweit, W., New York City
Gallagher, T. F., New York City
Goldfien, A., San Francisco, California
Gottfried, H., Sheffield, England
Karmen, A., Baltimore, Maryland
Kirschner, M. A., New York City
Kliman, B., Boston, Massachusetts
Lieberman, S., New York City
Lipsett, M. B., Bethesda, Maryland
Luukkainen, T., Helsinki, Finland
McKerns, K. W., Gainesville, Florida
McNeil, C., Salt Lake City, Utah
Menini, E., Edinburgh, Scotland
Patti, A., Philadelphia, Pennsylvania
Rosenfeld, R. S., New York City
Salhanick, H., Boston, Massachusetts
Schindler, A. E., Seattle, Washington
Thomas, B. S., London, England
Touchstone, J. C., Philadelphia, Pennsylvania
VandenHeuvel, W. J. A., Rahway, New Jersey
van der Molen, H. J., Utrecht, The Netherlands
Wilson, H., Bethesda, Maryland
Wotiz, H. H., Boston, Massachusetts

## LIST OF ABBREVIATIONS

| | |
|---|---|
| Centigrade | C |
| Centimeter (s) | cm |
| Counts per minute | cpm |
| Disintegrations per minute | dpm |
| Gas-liquid chromatography (ic) | GLC |
| Gram (s) | g |
| Hour (s) | hr |
| Inside diameter | id |
| Kilogram | kg |
| Meter | m |
| Microgram | µg |
| Microliter | µl |
| Microsecond (s) | µsec |
| Milligram | mg |
| Millimicrogram | mµg |
| Minute (s) | min |
| Normal (concentration) | N |
| Outside diameter | od |
| Pounds per square inch | psi |
| Relative retention time | $R_t$ |
| Retention time | $R_T$ |
| Specific activity | SA |
| Square centimeters | $cm^2$ |
| Standard deviation | SD |
| Standard error | SE |
| Thin-layer chromatography (ic) | TLC |
| Trimethylsilyl | TMS |
| Volume (s) | vol |

STEROIDS
TRIVIAL NAMES AND ABBREVIATIONS USED IN TEXT

| | |
|---|---|
| Aldosterone | 11β,21-dihydroxy-4-pregnene-3,20-dione-18-al |
| Allopregnanediol | 5α-pregnane-3α,20α-diol |
| Allotetrahydrocortisol | 3α,11β,17α,21-tetrahydroxy-5α-pregnane-20-one |
| Androstenediol | 5-androstene-3β,17β-diol |
| Androstenedione | 4-androstene-3,17-dione |
| 1-Androstenedione | 1,(5α)-androstene-3,17-dione |
| Androsterone, A | 3α-hydroxy-5α-androstane-17-one |
| Cholestane | 5α-cholestane |
| Cholestanol | 5α-cholestane-3β-ol |
| Cholesterol | 5-cholestene-3β-ol |
| Corticosterone, compound B | 11β,21-dihydroxy-4-pregnene-3,20-dione |
| Cortisol | 11β,17α,21-trihydroxy-4-pregnene-3,20-dione |
| Cortisone | 17α,21-dihydroxy-4-pregnene-3,11,20-trione |
| α-Cortol | 5β-pregnane-3α,11β,17α,20α,21-pentol |
| β-Cortol | 5β-pregnane-3α,11β,17α,20β,21-pentol |
| α-Cortolone | 3α,17α,20α,21-tetrahydroxy-5β-pregnane-11-one |
| β-Cortolone | 3α,17α,20β,21-tetrahydroxy-5β-pregnane-11-one |
| 11-Dehydrocorticosterone | 21-hydroxy-4-pregnene-3,11,20-trione |
| 11-Desoxycortisol | 17α,21-dihydroxy-4-pregnene-3,20-dione |
| Dehydroepiandrosterone, D, DHA | 3β-hydroxy-5-androstene-17-one |
| 11-Dehydroestradiol-17α | 1,3,5 (10),11-estratetraene-3,17α-diol |
| 11-Deoxycorticosterone | 21-hydroxy-4-pregnene-3,20-dione |
| 6β,20β-dihydroxy F | 6β,11β,17α,20β,21-pentahydroxy-4-pregnene-3-one |
| Epiandrosterone | 3β-hydroxy-5α-androstane-17-one |
| 16-Epiestriol | 1,3,5, (10)-estratriene-3,16β,17β-triol |
| Epitestosterone | 17α-hydroxy-4-androstene-3-one |

| | |
|---|---|
| Estradiol, estradiol-17β, E₁ | 1,3,5 (10)-estratriene-3,17β-diol |
| 17α-Estradiol | 1,3,5 (10)-estratriene-3,17α-diol |
| Estradiol-3-methyl | 3,17β-dihydroxy-1,3,5 (10)-estratriene-3-methyl ether |
| Estriol, E₃ | 1,3,5 (10)-estratriene-3,16α,17β-triol |
| Estriol-3-methyl | 3,16α,17β-trihydroxy-1,3,5 (10)-estratriene-3-methyl ether |
| Etiocholanolone, E | 3α-hydroxy-5β-androstane-17-one |
| 11β-Hydroxyandrostenedione | 11β-hydroxy-4-androstene-3,17-dione |
| 11-Hydroxyandrosterone | 3β,11β-dihydroxy-5α-androstane-17-one |
| 6β-Hydroxycortisol, 6β-OHF | 6β,11β,17α,21-tetrahydroxy-4-pregnene-3,20-dione |
| 20β-Hydroxycortisol, 20β-OHF | 11β,17α,20β,21-tetrahydroxy-4-pregnene-3,20-dione |
| 6β-Hydroxycortisone, 6β-OHE | 6β,17α,21-trihydroxy-4-pregnene-3,11,20-trione |
| 6α-Hydroxyestradiol-17β | 1,3,5 (10)-estratriene-3,6α,17β-triol |
| 6β-Hydroxyestradiol-17β | 1,3,5 (10)-estratriene,3,6β,17β-triol |
| 16α-Hydroxyestrone | 3,16α-dihydroxy-1,3,5 (10)-estratriene-17-one |
| 11-Hydroxyetiocholanolone | 3α,11β-dihydroxy-5β-androstane-17-one |
| 18-Hydroxyetiocholanolone | 3α,18-dihydroxy-5β-androstane-17-one |
| 17-Hydroxypregnanolone | 17α-hydroxy-4-pregnene-3,20-dione |
| 17-Hydroxypregnenolone | 3β,17α-dihydroxy-5-pregnene-20-one |
| 17α-Hydroxyprogesterone | 17α-hydroxy-4-pregnene-3,20-dione |
| 20β-Hydroxyprogesterone, 20-OHP | 20β-hydroxy-4-pregnene-3,20-dione |
| 11-Ketoetiocholanolone | 3α-hydroxy-5β-androstane-11,17-dione |
| 16-Ketoestradiol-17β, 16-ketoestriol | 3,17β-dihydroxy-1,3,5 (10)-estratriene-16-one |
| 2-Methoxyestrone | 2,3-dihydroxy-1,3,5 (10)-estratriene-17-one-2-methyl ether |
| 19-Nortestosterone | 17β-hydroxy-4-estrane-3-one |
| 11-Oxyandrosterone | 3α-hydroxy-5α-androstane-11,17-dione |
| Prednisone | 17α,21-dihydroxy-1,4-pregnadiene-3,11,20-trione |
| Pregnanediol | 5β-pregnane-3α,20α-diol |
| Pregnanetriol | 5β-pregnane-3α,17α,20α-triol |
| Pregnenediol | 5-pregnene-3β,20α-diol |
| Pregnenetriol | 5-pregnene-3β,17α,20α-triol |

| | |
|---|---|
| 5-Pregnenolone | 3β-hydroxy-5-pregnene-20-one |
| Testosterone | 17β-hydroxy-4-androstene-3-one |
| Tetrahydrodesoxycorticosterone, tetrahydro DOC, THDOC | 3α,21-dihydroxy-5β-pregnane-20-one |
| Tetrahydrocortisol, tetrahydro F, THF | 3α,11β,17α,21-tetrahydroxy-5β-pregnane-20-one |
| Tetrahydrocortisone, tetrahydro E, THE | 3α,17α,21-trihydroxy-5β-pregnane-11,20-dione |
| Tetrahydro-11-desoxycortisol, tetrahydro substance "S", THS | 3α,17α,21-trihydroxy-5β-pregnane-20-one |

# THE SPECIFICITY OF A GAS CHROMATOGRAPHIC METHOD FOR THE DETERMINATION OF URINARY ANDROSTERONE, ETIOCHOLANOLONE AND DEHYDROEPIANDROSTERONE IN URINE.

B. S. Thomas, Department of Clinical Chemistry, Division of Chemistry and Biochemistry, Imperial Cancer Research Fund, London, England.

## INTRODUCTION

A rapid and reliable method for the analysis of urinary androsterone (A), etiocholanolone (E) and dehydroepiandrosterone (D) was required for a forward study designed to investigate the relationship between these steroids and subsequent breast cancer in 5,000 normal women [1]. Gas-liquid chromatography seemed most likely to be the method of choice and this paper describes investigations into a suitable method.

## METHODS

1. Solvents and Reagents

    (a) Ethanol A.R. was stood over m-phenylenediamine in the dark for 7 days and then redistilled twice.

    (b) Ethyl acetate A.R. was redistilled over $K_2CO_3$ and stored at 4 C.

    (c) Heptane. Redistilled once before use.

    (d) Chloroform A.R. Redistilled once before use.

    (e) Hexamethyldisilazane, BDH Ltd. (Batches of this reagent from other sources sometimes contain impurities which have retention times that are similar to those of the 11-deoxy-17-oxosteroids.)

    (f) Trimethylchlorosilane, BDH Ltd.

2. Gas Chromatography Apparatus and Reagents

    (a) Pye Panchromatograph analyzers and electronic units.

    (b) Pye Macro-Argon detectors, changed at approximately 10 weekly intervals.

    (c) Pye 7 and 9 ft glass columns.

    (d) Stationary phases (Applied Science Laboratories, Inc.)

    (i) XE-60
    (ii) Hi-Eff 8B
    (iii) JXR

    (e) Column packings (Gaschrome P and Chromasorb W) were silanized by refluxing with 5% hexamethyldisilazane in toluene overnight and coated by an evaporation technique using a rotary evaporator.

    Packing was accomplished with the aid of a vibrator and the columns were then conditioned for 24 hr at 225 C under a stream of argon.

3. *Preliminary Purification of Urinary Extracts Before GLC*

Previous work had shown that the method of Thomas and Bulbrook (2) for the estimation of total 11-deoxy-17-oxosteroids (11-DOKS) in urine, gave results comparable to the sum of A, E and D estimated by gradient elution on alumina (3). It was therefore decided to investigate the use of the former method for the preparation of urinary extracts for subsequent GLC. A few minor modifications in the method were made as follows:

(a) 100 ml of a 24 hr urine was used.

(b) Heptane was substituted for benzene in replacing the ethyl acetate after the solvolysis stage.

(c) The final extract was made up to 2 ml, prior to the Zimmermann reaction for the estimation of the total 17-oxosteroids.

(d) The first two drops of the heptane eluate off the paper strips were discarded to remove pigments and traces of $5\alpha$-androstane-3, 17-dione, which is sometimes present.

The A, E and D were eluted from small strips of Whatman No. 4 paper (6.5 x 2.5 cm) in heptane using a special chromatographic apparatus (2). The heptane eluate was taken to dryness under a stream of air at 45 C and redissolved in 0.4 ml of ethanol; 0.2 ml of the ethanolic extract was analyzed by the Zimmermann reaction using D as a standard and results were expressed as total 11-DOKS in mg per 24 hr.

To the other half of the extract was added 10 µg of $5\alpha$-androstane-17-one as a GLC marker. The extract was then taken to dryness under a stream of air at 45 C and converted to TMS ethers by the method of Kirschner and Lipsett (4). The dried extract was then redissolved in 0.2 ml of chloroform and 20 µl was transferred to a Dixon gauze ring as described by Menini and Norymberski (5). Using this solid injection system with the Pye Panchromatograph column, the use of a flash heater was found to be unnecessary.

4. *Quantitation of Gas Chromatography Peaks*

In view of the agreement between the total 11-DOKS titer and the sum of the A, E and D titers obtained by gradient elution from alumina (2, 3), it seemed likely that the simplest method of measuring the amounts of A, E and D would be to calculate from the three peak areas the proportion of each of these compounds that was present in the total 11-DOKS fraction. This method might minimize errors likely to arise from adding small amounts of internal standards, injection variations, etc.

A correction of the peak areas was first necessary to compensate for individual differences in detector response. With the Macro-Argon detectors, these were found to be approximately as follows:

A = Peak area x 0.84         E = Peak area x 1.00         D = Peak area x 0.94

However, the detector response was checked daily with a series of standards and corrections were referred to this.

A second correction was necessary to allow for the difference in colorimetric response in the Zimmermann reaction since E gives 12% more color than either A or D when alcoholic KOH is used. This correction is as follows:

$$E = \frac{\text{Area of E} \times \text{total 11-DOKS}}{\text{Total area} + 0.12 \text{ (Area of E)}}$$

The corrected areas were summed and each area was expressed as a proportion of the total. From this the individual amounts of A, E and D were calculated directly from the total 11-DOKS, as mg per 24 hr. This method for measuring the amounts of D, A and E is referred to as the "proportionality method."

## RESULTS

1. *Comparison between the internal standard and the "proportionality methods."* A series of 15 urinary 11-DOKS extracts were analyzed and the amounts of A, E and D were determined by the "proportionality method" and by reference to an internal standard. There were 15 comparisons:

none of the differences was significant (Table 1). In view of the close agreement between the two methods of measuring the amounts of A, E and D, the "proportionality method" was used in the work reported below.

TABLE 1. Comparison between "proportionality method" and internal standard method.

Mean amounts (μg/24 hr)

|  | A | E | D |
|---|---|---|---|
| Proportionality method | 1454 | 1052 | 367 |
| Internal Standard method | 1487 | 1117 | 371 |
| Standard deviation of mean difference between methods | 24 | 17 | 4 |
| 95% of differences lie between: (of mean difference) | -1% to 5% | +3% to +9% | -1% to +4% |

2. <u>Comparison of results from GLC and gradient elution</u>. A series of comparisons were made between analyses carried out by the method described above and by gradient elution from alumina. The columns used in this comparison were 7 ft, 2% XE-60 on Gaschrome P (100-120 mesh). The results of 16 comparisons are shown in Fig. 1.

The titer of A was persistently higher by GLC, and that of E and D correspondingly lower. The mean difference in the A titer was 20% but in one instance a difference of 80% was encountered. The experiment was repeated using Hi-Eff 8B (6) and in four comparisons the results by GLC for A were still persistently higher than by gradient elution from alumina, although better separation was obtained with this stationary phase than with XE-60. At this stage it was not known whether the gradient elution method was at fault or the GLC, or both.

3. <u>Results from GLC after TLC of urinary extracts</u>. A pooled urinary neutral fraction was prepared (see methods section and 2) and three separate analyses by gradient elution on alumina by three separate workers are shown in Table 2. Aliquots from the pooled neutral fraction were then processed by the method of Thomas and Bulbrook (2) as described above and further aliquots by the method of Kirschner and Lipsett (4) using TLC. After conversion to TMS ethers the extracts were run on a 9 ft column of 1% Hi-Eff 8B on Chromasorb W (100-120 mesh). The results of these comparisons are shown in Fig. 2. The results were similar to those described in section 2 above, in that the titer of A by GLC was higher than that obtained by gradient elution, irrespective of the method of purifying the urinary extract.

TABLE 2. Results of 3 separate determinations of urinary A, E and D by gradient elution from alumina. Assays by 3 separate workers at 3 monthly intervals.

| Analysis No. | μg/ml of urinary extract | | |
|---|---|---|---|
|  | <u>A</u> | <u>E</u> | <u>D</u> |
| 1 | 85 | 87 | 67 |
| 2 | 88 | 88 | 72 |
| 3 | 90 | 90 | 66 |

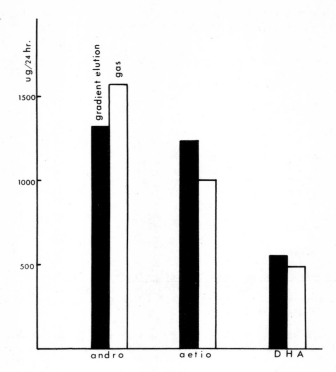

Fig. 1. Comparison of mean results of 16 determinations of urinary A, E and D by gradient elution, and by GLC on XE-60.

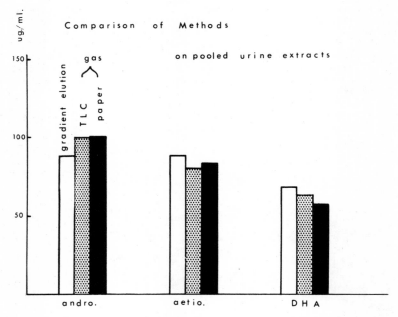

Fig. 2. Comparison of mean results of determinations of urinary A, E and D by gradient elution, by GLC and after TLC, and after paper chromatography (3, 5 and 5 experiments respectively). Stationary phase, Hi-Eff 8B.

4. _Investigation of purity of androsterone peak on GLC._ An interpretation of the above results was that there was an interfering compound in the peak which was not separable from A on either XE-60 or Hi-Eff 8B stationary phases. Accordingly, a Girard T separation of the pooled urine fraction was made. The non-ketonic fraction, after elution from the 6.5 cm paper strip, was treated with the TMS ether reagents and run on the 9 ft, 1% Hi-Eff 8B column. A peak was observed in the A region (Fig. 3).

The absence of peaks in both the E and D positions indicated that this compound was unlikely to have been carried over from the ketonic fraction. Furthermore, a blank of 100 ml of water carried through the extraction, hydrolysis, paper chromatography and silyl ether steps gave no peaks on the gas chromatogram. This indicated that the difference between the methods was due to an overestimation of A by the GLC method.

5. _Further resolution of the androsterone peak._ In an attempt to resolve the A peak a series of trials was performed, using varying combinations of a selective stationary phase (Hi-Eff 8B) and a non-selective phase (JXR). Using a coating of 0.6% of each on silanized Chromasorb W (100-120 mesh), a peak, not present when Hi-Eff 8B alone was used, was observed with a retention time greater than that of D. A corresponding drop in the ratio of A and E was also apparent as is shown in Fig. 4 and 5. A further comparison of 14 determinations of A, E and D from 14 different urine specimens by gradient elution from alumina, and by the use of the "hybrid" column described above, was carried out and the results are shown in Table 3. In four instances Hi-Eff 8B was used (as described in section 3). There is agreement between the results for A and D from the first two methods and, as expected, a large difference in the results obtained with Hi-Eff 8B in all three fractions.

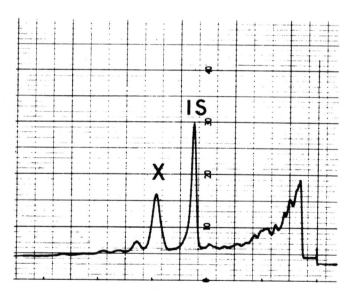

Fig. 3. GLC of TMS ethers of non-ketonic extract of urinary 11-deoxy-17-oxosteroid fraction. IS represents the internal standard (1 μg): X represents an impurity in the androsterone region. Conditions: 1% HiEff 8B, temp 205 C, flow 30 ml per min, 850 v, 9 ft column. Chart speed 12 in per hr.

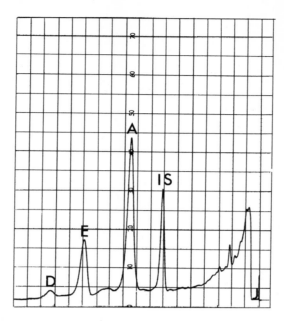

Fig. 4. GLC of urinary 11-DOKS fraction on Hi-EFF 8B 1%. Other conditions as in Fig. 3.

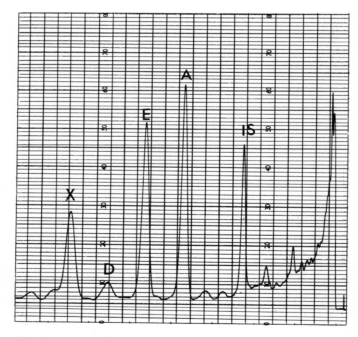

Fig. 5. GLC of urinary 11-DOKS fraction on "hybrid" column (JXR + Hi-Eff 8B). (See text for conditions). Extract identical to that used in Fig. 4 but more sensitive detector used. Note appearance of X and fall in A to E ratio compared with Fig. 4.

TABLE 3. Comparison of A, E and D values (μg per 24 hr) from urine extracts analyzed by (1) gradient elution from alumina, and (2) GLC on 0.6% Hi-Eff 8B + 0.6% JXR. Difference between means: Androsterone: $t = 0.04$, $p > 0.2$   Etiocholanolone: $t = 2.88$, $p < 0.01$   Dehydroepiandrosterone: $t = 0.04$, $p > 0.2$.

| No. | Gradient Elution | | | Hi-Eff 8B | | | Hi-Eff 8B + JXR | | |
|---|---|---|---|---|---|---|---|---|---|
|  | A | E | D | A | E | D | A | E | D |
| 1 | 1405 | 1766 | 251 | 1988 | 1037 | 152 | 1437 | 1490 | 216 |
| 2 | 1000 | 1042 | 176 | 1275 | 858 | 181 | 1121 | 1020 | 153 |
| 3 | 430 | 1008 | 674 | 830 | 876 | 433 | 586 | 1063 | 470 |
| 4 | 1389 | 908 | 121 | 1616 | 384 | 146 | 1345 | 744 | 56 |
| 5 | 2780 | 1572 | 419 | - | - | - | 2809 | 1481 | 425 |
| 6 | 1016 | 1311 | 470 | - | - | - | 976 | 1302 | 439 |
| 7 | 1134 | 704 | 196 | - | - | - | 1265 | 631 | 114 |
| 8 | 1109 | 1618 | 538 | - | - | - | 1113 | 1514 | 548 |
| 9 | 1643 | 1953 | 2730 | - | - | - | 1495 | 1588 | 2716 |
| 10 | 2368 | 2550 | 1080 | - | - | - | 2351 | 2101 | 1346 |
| 11 | 461 | 633 | 154 | - | - | - | 515 | 762 | 287 |
| 12 | 488 | 651 | 95 | - | - | - | 574 | 624 | 72 |
| 13 | 4434 | 1962 | 195 | - | - | - | 4117 | 1655 | 0 |
| 14 | 542 | 1015 | 1246 | - | - | - | 742 | 1068 | 1364 |
| Mean of Nos. 1-4 | 1056 | 1181 | 306 | 1427 | 789 | 228 | 1122 | 1079 | 224 |
| Mean of Nos. 1-14 | 1442 | 1335 | 596 | - | - | - | 1460 | 1217 | 586 |

A list of $R_t$ values for both columns is given in Table 4. These show that the unknown compound is not pregnanediol. The impurity in the A peak is still present when TLC is used for the preparation of the urinary extracts (Fig. 6).

TABLE 4. Retention times for TMS ethers of urinary steroids on Hi-Eff 8B and "hybrid" columns. The retention times are relative to that of 5α-androstane-17-one.

| Compound | 1% Hi-Eff 8B | 0.6% Hybrid |
|---|---|---|
| 5α-androstane-17-one | 1.00 | 1.00 |
| androsterone | 1.33 | 1.58 |
| etiocholanolone | 1.81 | 1.95 |
| dehydroepiandrosterone | 2.14 | 2.32 |
| epiandrosterone | 2.15 | 2.30 |
| X | 1.32 | 2.74 |
| pregnanediol | 1.53 | 2.94 |

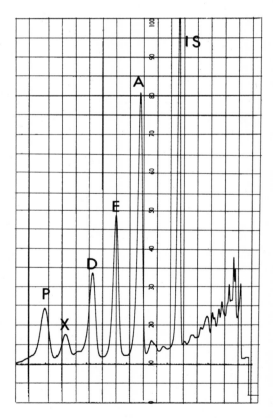

Fig. 6. GLC of urinary 11-DOKS following TLC of urinary extract. Conditions: "hybrid" column, 7 ft, 210 C, flow 30 ml per min, 850 v. Chart speed 12 in per hr. Note presence of ? pregnanediol (P) and Compound X.

## DISCUSSION

Emphasis has been placed on the comparison of results from the gradient elution method and that of GLC. This is necessary if continuity is to be observed: the previous results from A and D obtained by gradient elution on alumina in this laboratory are clearly still valid and are strictly comparable with those obtained with the new method. There is, however, a small but consistent difference between the values obtained for E. However, it should be borne in mind that absolute proof of specificity is virtually impossible to obtain for routine methods and that future work may very well demonstrate further errors in both of these methods. In fact, it has been shown (7) that estimation of D by gradient elution on alumina can be unreliable in some circumstances and this might apply to the gas method also. Furthermore, epiandrosterone, a urinary constituent, is not spearable from D on either XE-60, Hi-Eff 8B nor the 0.6% "hybrid" column, and would be measured by GLC as D. The small but significant difference in the slightly lower E values (9%) obtained by GLC may be due to some breakdown of the TMS ether, as recoveries of E when added to urine, are at least as high as those of A, up to and including the paper chromatography stage of the method (see 2). This point is now being investigated.

For routine determinations, the specificity of the method used is of primary importance since most urine estimations are "taken on trust" and only a few assays are rigorously investigated for specificity. At the present time parameter seems satisfactory and further investigations are now in progress to determine the accuracy, precision and sensitivity (as defined by Brown, Bulbrook and Greenwood (8) of the method.

## SUMMARY

1. A method for the estimation of urinary androsterone, etiocholanolone, and dehydroepiandrosterone has been investigated.

2. Comparison with results obtained by gradient elution on alumina gave higher androsterone values when either XE-60 or Hi-Eff 8B stationary phases were used.

3. The use of a dual phase ("hybrid") column containing equal amounts of Hi-Eff 8B and JXR resolved the androsterone peak into two fractions.

4. Results of androsterone assays using the "hybrid" column agreed closely with those obtained by gradient elution from alumina, but there were small differences in the values for etiocholanolone.

## ACKNOWLEDGEMENTS

The author wishes to express gratitude to Dr. R. D. Bulbrook for his encouragement and advice with this paper. Also, to Mrs. Sylvia Pearson, without whose technical assistance this work would not have been completed.

## REFERENCES

1. Bulbrook, R. D., J. L. Hayward, C. C. Spicer, and B. S. Thomas, Lancet, ii, 1238, 1962.

2. Thomas, B. S. and R. D. Bulbrook, J Chromatog 14: 28, 1964.

3. Kellie, A. E. and A. P. Wade, Biochem J 66: 196, 1957.

4. Kirschner, M. A. and M. B. Lipsett, J Clin Endocr 23: 255, 1963.

5. Menini, E. and J. K. Norymberski, Biochem J (in press).

6. Hartman, I. S. and H. H. Wotiz, Steroid 1: 33, 1963.

7. Bulbrook, R. D., F. C. Greenwood, and B. S. Thomas, Biochem J 69: 19, 1958.

8. Brown, J. B., R. D. Bulbrook, and F. C. Greenwood, J Endocr 16: 41, 1957.

# THE INDIRECT ANALYSIS OF URINARY 17-OXYGENATED $C_{19}$ STEROID SULPHATES BY GAS-LIQUID CHROMATOGRAPHY

E. Menini, Medical Research Council, Clinical Endocrinology Research Unit, Edinburgh, Scotland

The three principal urinary 17-oxosteroids, androsterone, etiocholanolone and dehydroepiandrosterone, cannot be conveniently separated by GLC on a single column. One approach to the analysis of these compounds by this technique is their conversion into suitable derivatives such as the TMS ethers and their resolution on a polar phase as QF-1 or XE-60 (1, 2).

Another approach is the oxidation of these compounds to a mixture of polyoxoandrostanes, which is readily resolvable into individual components on a non-polar phase such as SE-30.

Menini and Norymberski (3) reported the use of tertiary butyl chromate as an analytical reagent for the oxidation of hydroxysteroids and found that secondary alcohols are converted quantitatively to the corresponding ketones, and homoallylic alcohols of the $\Delta^5$-3β-ol type are converted in high yields into $\Delta^4$-3,6-diones. From the methodological point of view this conversion presents the following advantages:

1. The derivatives produced in the oxidation are chemically and thermally stable.

2. They can be easily measured by any of the colorimetric reactions based on properties of ketonic groups.

3. They can be purified by means of a Girard separation.

4. They can be conveniently analyzed by GLC.

The retention times of some polyoxoandrostanes, related to the compounds which are the object of this study, on an SE-30 column are listed in Table 1.

TABLE 1. Retention times ($R_T$) of some polyoxosteroids related to urinary 17-oxosteroids, on a 1% SE-30 column at 205 C.

| Main Urinary Precursor | Oxidation Product | $R_T$ (min) |
|---|---|---|
| Etiocholanolone | 5β-androstane-3,17-dione | 12.3 |
| Androsterone | 5α-androstane-3,17-dione | 13.7 |
| 11-Oxyetiocholanolone | 5β-androstane-3,11,17-trione | 15.6 |
| 11-Oxyandrosterone | 5α-androstane-3,11,17-trione | 18.0 |
| Dehydroepiandrosterone | 4-androstene-3,6,17-trione | 24.4 |

McKenna and Norymberski (4) found that 17-oxosteroid sulphates, in the presence of pyridinium ion, could be preferentially extracted into a relatively non-polar solvent such as chloroform and that this procedure afforded very clean extracts which could be solvolyzed by dioxan (5, 6).

It was expected that by combining these two principles, namely the preferential extraction of the steroid sulphates and the oxidation of the products of their solvolysis to a mixture of polyoxosteroids, the sulphates of androsterone, etiocholanolone, dehydroepiandrosterone, 11-oxyandrosterone and 11-oxyetiocholanolone could be determined respectively as 5α-androstane-3,17-dione, 5β-androstane-3,17-dione, 4-androstene-3,6,17-trione, 5α-androstane-3,11,17-trione and 5β-androstane-3,11,17-trione. To assess the potentialities of the proposed analytical method several

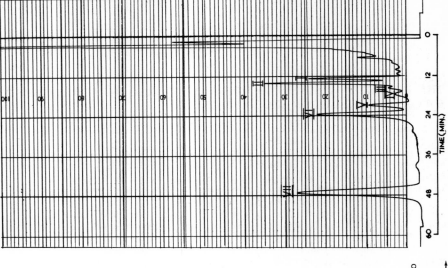

Fig. 1. Gas chromatogram of an extract containing the derivatives of the 17-oxosteroid sulphates of the urine of a normal female (E.K.). The portion chromatographed is equivalent to 1.5 ml of urine. The separated compounds have the same retention times as: (I) 5β-androstane-3,17-dione, (II) 5α-androstane-3,17-dione, (III) 5β-androstane-3,11,17-trione, (IV) 5α-androstane-3,11,17-trione, (V) 5β-pregnane-3,20-dione, (VI) 4-androstene-3,6,17-trione, (VII) pregnanediol diacetate, added as internal standard.

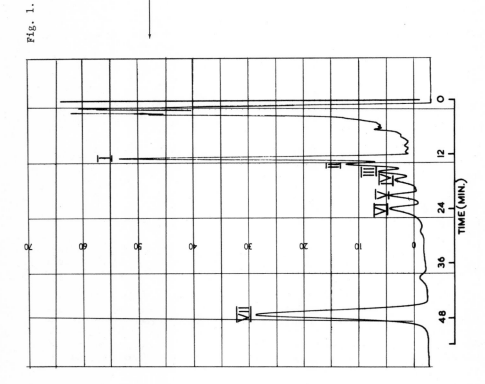

Fig. 2. Gas chromatogram of an extract containing the derivatives of the 17-oxosteroid sulphates of the urine of a normal female (D.I.H.). The portion chromatographed is equivalent to 2.0 ml of urine. Peaks correspond to steroids shown in legend to Fig. 1.

urine samples were submitted to the following procedure: urine (usually 1-20 ml) was made 0.5M in pyridinium sulphate and the pyridinium steroid sulphates were extracted into chloroform. A portion of this extract was evaporated to dryness and the residue was solvolyzed with dioxan. The solution containing the solvolyzed material was washed with aqueous sodium hydroxide and water and a portion of it was evaporated to dryness.

The dry residue was dissolved in pyridine and oxidized with t-butyl chromate according to Menini and Norymberski (3). A fraction of the final extract in ethylene dichloride containing the products of the oxidation was transferred onto a Dixon gauze ring by means of a teflon spotting plate (7) and analyzed by GLC on a 5 ft long glass column, packed with 1% SE-30 on Gas-Chrom P (100-120 mesh). Analyses were performed with a Panchromatograph (Pye Scientific Instruments, Cambridge, England) equipped with an argon ionization detector, fitted with a $^{90}$Sr source.

The gas chromatograms corresponding to samples consisting of pure sodium dehydroepiandrosterone sulphate and sodium androsterone sulphate added to water and submitted to the operations and reactions described, showed single peaks with the same retention times as 4-androstene-3,6,17-trione and 5α-androstane-3,17-dione respectively. 4-androstene-3,6,17-trione was not present in the gas chromatogram corresponding to a sample in which pure sodium dehydroepiandrosterone glucosiduronate has been added to water.

In all the urines analyzed, compounds with the same retention time as the 3,17-diketones and 3,11,17-triketones of the 5α and 5β androstane, and in many of them 4-androstene-3,6,17-trione and 5β-pregnane-3,20-dione (the expected derivative of pregnanediol) were found in the gas chromatograms of the final extracts. In addition these gas chromatograms showed that these compounds (the expected derivatives of the 17-oxosteroid sulphates) were the major, and often the sole, constituents of the samples analyzed. Further identification of the polyoxosteroids in the final fractions was carried out by paper chromatography in the systems $B_3$ (8) and $E_4$ (9).

Figures 1 and 2 show two typical chromatograms of extracts from urine of normal women.

This method in the form in which it has been described, should measure, not only 17-ketosteroid sulphates but also 17-hydroxysteroid sulphates, which, if present, will be oxidized to the corresponding 17-oxo compounds by the t-butyl chromate. In several cases the contribution from this group of compounds to the final measurements has been evaluated by analyzing, by the Zimmermann reaction, a portion of the sample before and after oxidation. It was found that the expected increase in chromogenicity due to the conversion of the 11-hydroxyl groups to the 11-oxo groups could account for the slightly higher values found in some cases after oxidation, indicating that 17-hydroxyandrostanes, if present at all, constituted a negligible fraction. However, should androstanediols or triols be expected in significant amounts, they could be analyzed separately by applying the described procedure to the non-ketonic fraction resulting from a Girard separation carried out after the solvolysis of the conjugates.

Although the reliability of the described method in terms of accuracy, specificificity, sensitivity and precision has not as yet been properly assessed, Table 2 shows that in general the values obtained fall within the range reported by other authors, for this group of steroids, using methods of proved reliability.

TABLE 2. Urinary excretion of individual 17-oxosteroid sulphates. Values are expressed in mg of parent steroid per 24 hr and were calculated from the results of gas-liquid chromatography.

| Urine No. | 55 | 58 | 66 | 74 | 74 | 77 | 78 | 87 | 87 |
|---|---|---|---|---|---|---|---|---|---|
| Sex | M | M | M | F | F | F | F | F | F |
| Treatment | - | ACTH | - | - | - | - | - | - | - |
| Remarks | (a) | (a) | | (b) | (c) | | | (d) | (e) |
| Compound | Measured as (*) | | | | | | | | |
| Etiocholanolone | 5β-A-3,17 | 0.58 | 1.21 | 0.89 | 2.65 | 2.40 | 0.36 | 0.31 | 0.35 | 0.34 |
| Androsterone | 5α-A-3,17 | 0.79 | 0.52 | 0.43 | 0.71 | 0.75 | 0.38 | 0.56 | 0.49 | 0.47 |
| 11-Oxyetiocholanolone | 5β-A-3,11,17 | 0.91 | 2.91 | 0.45 | 0.56 | 0.53 | 0.15 | + | + | + |
| 11-Oxyandrosterone | 5α-A-3,11,17 | + | 0.80 | 0.63 | 0.30 | 0.41 | + | 0.15 | + | + |
| Pregnanediol (+,-olone) | 5β-P-3,20 | + | 0.33 | + | 0.60 | 0.52 | 0.14 | 0.12 | 0.23 | 0.22 |
| Dehydroepiandrosterone | $\Delta^4$-A-3,6,17 | - | - | - | + | rec. | 0.22 | 0.47 | 0.52 | rec. |

(a) same subject; (b) and (c) duplicate analysis; second sample used for the recovery of DHA sulphate (rec. 66%); (d) and (e) duplicate analysis; second sample used for recovery of DHA sulphate (rec. 75%).
(*) 5β-A-3,17 = 5β-androstane-3,17-dione; 5α-A-3,17 = 5α-androstane-3,17-dione; 5β-A-3,11,17 = 5β-androstane-3,11,17-trione; 5α-A-3,11,17 = 5α-androstane-3,11,17-trione; $\Delta^4$-A-3,6,17 = 4-androstene-3,6,17-trione.

The following are shortcomings of the method known to affect, but not invalidate, the determination of dehydroepiandrosterone.

1. $5\alpha$-pregnane-3,20-dione cannot be separated from 4-androstene-3,6,17-trione, the expected derivative of dehydroepiandrosterone, on an SE-30 column. These compounds, however, can be resolved on a polar stationary phase.

2. Dehydroepiandrosterone is oxidized to 4-androstene-3,6,17-trione in a yield of only 70-75%, the rest being converted into more polar components, eliminated during the washing of the final extracts.

3. Using an argon ionization detector, and under the conditions used, the response obtained with 4-androstene-4,6,17-trione, is considerably lower than that obtained for the other normal constituents of the final sample.

Although the full value of this procedure remains to be established, the results obtained up till now seem to indicate that, provided that the implications discussed are realized, this method because of its simplicity could be particularly suited to the analysis by GLC on a single column, of the urinary fraction containing the derivatives of the sulphates of androsterone, etiocholanolone, dehydroepiandrosterone, 11-oxoandrosterone, 11-oxyetiocholanolone and pregnanediol.

## REFERENCES

1. VandenHeuvel, W. J. A., B. G. Creech, and E. C. Horning, Anal Biochem 4: 191, 1962.

2. Kirschner, M. A. and M. B. Lipsett, J Clin Endocr 23: 255, 1963.

3. Menini, E. and J. K. Norymberski, Biochem J 84: 195, 1962.

4. McKenna, J. and J. K. Norymberski, Biochem J 76: 60, 1960.

5. McKenna, J. and J. K. Norymberski, J Chem Soc p. 3889, 1957.

6. Cohen, S. L. and I. B. Oneson, J Biol Chem 204: 245, 1953.

7. Menini, E. and J. K. Norymberski, Biochem J, 1965 (in press).

8. Bush, I. E., Biochem J 50: 370, 1952.

9. Eberlein, W. R. and A. M. Bongiovanni, Arch Biochem Biophys 59: 90, 1955.

# A TECHNIC FOR FRACTIONATION OF 17-KETOSTEROIDS FROM URINE

Charles R. Berrett and Crichton McNeil, Holy Cross Hospital Research Foundation, Salt Lake City, Utah.

Supported by NIH Grant No. B1363-08.

An attempt has been made to shorten the extraction-hydrolysis procedure for fractionation of the urinary 17-ketosteroids to make this determination more attractive for the clinical laboratory use. The technic consists of extracting 20 ml of $(NH_4)_2SO_4$ saturated urine with tetrahydrofuran (THF), drying down the extract, and hydrolyzing according to the method of DePaoli, Nishizawa and Eik-Nes (1). The essential feature of this is the addition of 15 ml of THF containing $10^{-2}$ M perchloric acid with three hr incubation at 50 C. Ten percent aqueous KOH is added to neutralize the perchloric acid and remove phenols, the THF is evaporated, the solution is extracted three times with ether, and the ether pool washed with water. The ether extract is dried down and the residue chromatographed by TLC on a mixture of silica gel G and H (1:1). The solvent system is ethyl acetate, benzene (2:3). On a separate channel on the plate, dehydroepiandrosterone is chromatographed as a marker. When the solvent has reached the front of the chromatogram (approximately 14 cm), the plate is dried and exposed to iodine vapor. The area of sample chromatogram corresponding in mobility to the standard is marked and scraped off the plate and extracted with three, two, and one ml of ethanol. Fifty µg of estrone is added to this extract and the ethanol evaporated to dryness. The residue is then converted to TMS ether derivatives by adding hexamethyldisilazane and trimethylchlorosilane and allowed to stand overnight. The solution is then dried down.

The residue is taken up in 0.05 ml of toluene and a small amount injected into a six ft 2% XE-60 glass column, using a Barber Colman model 10 chromatograph with argon ionization detector. Column temperature is 195 C, flash heater 230 C, and the detector temperature is 230 C. Argon gas flow pressure is 12 psi.

Figure 1 shows the chromatographic tracing of dehydroepiandrosterone sulfate processed by the method. The steroid gives a symmetrical peak and the sample is free of impurities. (60% mean recovery, urines are corrected for this.)

Figure 2 depicts androsterone, etiocholanolone, and dehydroepiandrosterone as silylized standards not carried through the procedure.

Figure 3 shows a chromatogram of the steroids of male urine carried through the procedure and converted to TMS ethers. Quantitation is made by means of carrying 200 µg of dehydroepiandrosterone sulfate (132 µg free DHEA) through the method except for the final step of TMS ether derivative formation. The recovery of dehydroepiandrosterone gives the yield up to the point of silyl ether formation. The amount of estrone recovered acts as an internal standard and gives an indication of the losses occuring during the formation of the TMS ether derivatives (usually 100% yield) and GLC.

## REFERENCE

1. DePaoli, J., E. E. Nishizawa, and K. B. Eik-Nes, *J Clin Endocr* 23: 81, 1963.

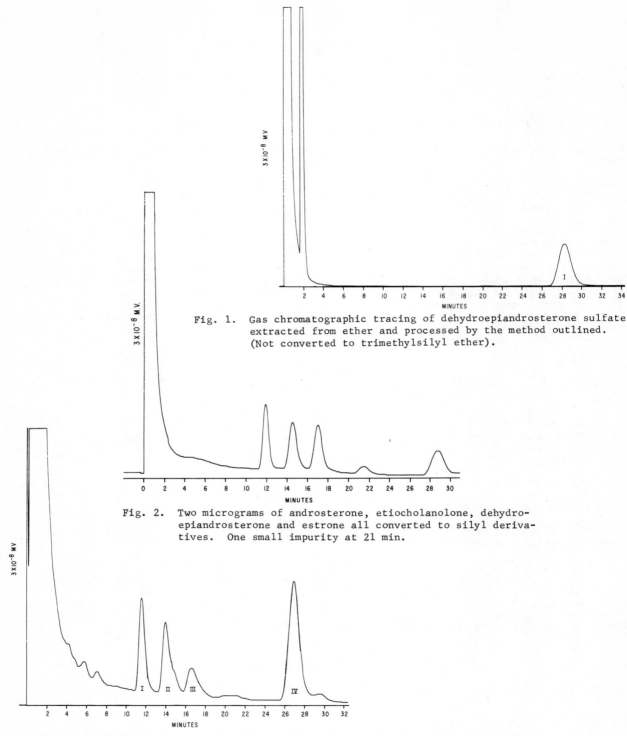

Fig. 1. Gas chromatographic tracing of dehydroepiandrosterone sulfate extracted from ether and processed by the method outlined. (Not converted to trimethylsilyl ether).

Fig. 2. Two micrograms of androsterone, etiocholanolone, dehydroepiandrosterone and estrone all converted to silyl derivatives. One small impurity at 21 min.

Fig. 3. Gas chromatographic tracing of an extract of normal male urine. The equivalent to 1.6 ml. Urine was applied to the column. I. Androsterone. II. Etiocholanolone. III. Dehydroepiandrosterone. IV. Estrone.

# GAS CHROMATOGRAPHIC METHOD FOR THE DETERMINATION OF TESTOSTERONE IN HUMAN URINE

Walter Futterweit, George L. Siegel, Ruth Freeman, Solomon I. Griboff, Michael Drosdowsky, Norman Gibree, Ralph I. Dorfman and Louis J. Soffer, Division of Endocrinology of the Department of Medicine, The Mount Sinai Hospital, New York City and the Worcester Foundation for Experimental Biology, Shrewsbury, Massachusetts.

This work was supported by Grant AM 08403-01 from the U.S. Public Health Service.

Clinical evaluation of hyperandrogenic states and gonadal function has been facilitated with the development of techniques for the microquantitative determination of testosterone in plasma and urine. Recently, we have devised a GLC method for analysis of testosterone glucuronide in urine and have utilized it as an ancillary test for the study of normal gonadal function and in patients with various endocrinopathies (1, 2).

The method is an isotope dilution procedure (1) employing testosterone-1,2,-$^3$H (SA 153 $\mu c/\mu g$). Approximately 30,000 cpm of the isotope is added to a 100 ml portion of a 24 hr collection of urine that is adjusted to pH 5.0 with 50% sulfuric acid and 5 ml of 0.1 M acetate buffer. Twelve ml of beef liver of b-glucuronidase (60,000 Fishman Units) are added and the mixture is incubated for 96 hr at 37 C.

Extraction. The urine is extracted twice with freshly distilled portions of 100 ml and once with 50 ml of anhydrous ether. The combined ether extracts are washed three times with 15 ml of 2N NaOH, and 7.5 ml of water. The ethereal extract is evaporated to dryness in a test tube under a gentle stream of $N_2$ at 40-50 C.

Girard separation. A modification of the Girard procedure was employed which allowed for recoveries of 70-90% in this step. The dried ether extract is warmed gently over steam after the addition of 1 ml of 80% methanol, 20 mg of Girard "T" reagent and 1 ml of glacial acetic acid. The mixture is then allowed to stand at room temperature overnight. The extract is then transferred to cold separatory funnels with five 10 ml portions of ice cold water. After neutralization with 15 ml of 1.0 N NaOH, the mixture is extracted with 5 portions of 10 ml cold methylene chloride. The methylene chloride extract is backwashed with 5 ml of water, and the non-ketonic methylene chloride portion is discarded. The aqueous portion is then adjusted to pH 1.0 by addition of 2 ml of concentrated HCL, and allowed to stand at room temperature for 2-3 hr. The latter mixture is then extracted with five portions of 20 ml methylene chloride and washed twice with 10 ml of water. The combined methylene chloride extracts are then evaporated to dryness over steam.

Thin-layer chromatography. The urinary extract as well as the testosterone standard were applied to a 20 x 20 cm glass plate containing the adsorbent Silica Gel G. Both were applied approximately 3.0 cm from the edge of the plate. The plate was placed in a tank containing a freshly prepared benzene:ethyl acetate (3:2) solvent system. The chromatogram was developed by the ascending technique. Upon completion of a run, the testosterone reference zones were identified with ultraviolet light, and the corresponding zone of the urine extract transferred and eluted with triple-distilled acetone. The eluate was dried under $N_2$ and a 10% aliquot taken for counting, employing a PPO-POPOP system, in a Packard Tri-Carb liquid scintillation spectrometer. The remainder of the urine extract (equivalent to 90 ml of urine) was transferred to a small conical tube, dried and refrigerated until GLC.

TMS ethers were prepared by reaction with hexamethyldisilazane and acetylation was carried out at room temperature with acetic anhydride and pyridine. (See specificity section.)

Apparatus. A Research Specialties Co. Model 600 gas chromatograph equipped with a Model 660-1B hydrogen flame ionization detector was used. The gas chromatographic curves were obtained on a 1 mv Sargent Model SR recorder operated at a chart speed of 1 inch per minute.

The samples were introduced using a solids injector. Briefly the sample is deposited on a metal insert 3/16 inch diameter x 3/4 inch long fastened to the end of a 1/4 inch diameter shaft. This insert is then introduced into a modified R.S. Co., Model 603-1B micropipette sample injector.

Collection of the peak material was carried out using a R.S. Co. Model 600 gas chromatograph equipped with an argon ionization detector Model B-600. The material was collected at room temperature in a glass or teflon capillary tube connected to the heated outlet of the chromatograph. The trapped material was eluted with methylene chloride and run as a micro KBr pellet on a Beckman IR7 spectrophotometer using scale expansion.

The area of the peak was determined with a Royson Model 210 planimeter recording 1200 counts per inch full scale. The use of a proportional temperature controller was found mandatory for proper maintenance of column temperature.

<u>Gas chromatographic technique</u>. A 3-4 ft stainless steel column (i.d., 1.9 mm) is employed containing 4-6.8% SE-30 (General Electric Co.) on Anakrom ABS (110-120 mesh). The precise percentage of the stationary phase is determined by extraction of a 3 gm aliquot of packing with methylene chloride in a Sohxlet extractor. The packing is prepared by the filtration technique, and screened before and after the support is coated.

The conditions found most suitable were as follows: column temperature, 240-245 C; vaporizer temperature, 310-325 C.

The dried residue of the neutral ketonic fraction is dissolved in 200-400 μl of acetone, and mixed with a Vortex Junior vibrator for 20-30 seconds. An aliquot of 5-20 μl of extract is transferred with a Hamilton lambda syringe to the solids injector, and placed in an oven at 75-80 C for 4-6 min. The injector is then held against the vaporizer block for 30 seconds. Three to four injections are carried out, and as an internal standard testosterone is added to one replicate injection of urine.

<u>Specificity of the method</u>. Thin-layer chromatography allows for separation of several steroids whose retention time on the non-selective SE-30 column are similar to that of standard testosterone (1). Aliquots of urine extract of the neutral ketonic fraction of adult male urine were subjected to acetylation and the formation of the TMS ether derivatives. In both instances, a complete peak shift occurred and a new peak appeared with a retention time similar to that of authentic testosterone derivative. Collection of peak material having a retention time of 1.00 relative to testosterone obtained from GLC of a urine extract of a normal adult male and the peak material from the gas chromatography of standard testosterone had identical infrared spectra. However, in both instances the peak material appeared to be a mixture of testosterone with the same carbonyl containing compound which arises as a result of the gas chromatographic process.

<u>Reproducibility</u>. The results of eight replicate analyses from a single extract of adult male urine indicated that testosterone was determined with an error of ± 7.0% ($P = 0.99$) (Table 1). Similarly, the limits of uncertainty for the retention time of testosterone was ± 1.3% ($P = 0.99$).

<u>Sensitivity</u>. The lower limit of sensitivity for the measurement of testosterone is 2 μg per 24 hr urine volume.

<u>Recovery</u>. All determinations were done in duplicate, and recoveries ranged from 55-80% after elution of the urine extract from the thin-layer plate.

Examples of gas chromatograms of urine extracts are shown in Fig. 1.

Results of analyses of subjects of both sexes are shown in Table 2. There is a distinct difference between the urinary testosterone excretion of normal adult male and female subject, and a suggestion of a decrease in testosterone excretion with age in men.

TABLE 1. Serial 10 μl/1000 μl injections of pooled male urine employing the solids injector technique. Integrator units were calculated with a Royson planimeter.

| Int. Units | μg Sample Injected | μg/l |
|---|---|---|
| 164 | 0.265 | 112 |
| 164 | 0.265 | 112 |
| 166 | 0.268 | 113 |
| 164 | 0.265 | 112 |
| 170 | 0.274 | 115 |
| 168 | 0.271 | 114 |
| 146 | 0.236 | 100 |
| 145 | 0.234 | 99 |

109 ± 7.8 μg/l

(± 7.0%)

n = 8   P = 0.99

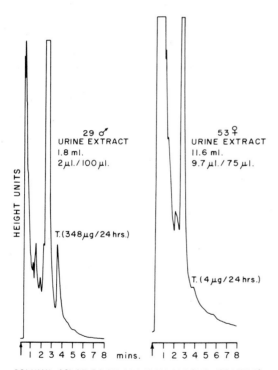

COLUMN: 4% SE-30 ON ANAKROM ABS (110-120 MESH)
4 FOOT, 1/8" O.D.
COL.TEMP: 245°C.
VAP.TEMP (SOLIDS INJ.): 320°C.
32 p.s.i.g., N₂
DETECTOR: FLAME IONIZATION

Fig. 1. Gas chromatograms of 2 urine extracts following acetone elution of the zone corresponding to testosterone of the silica gel G thin-layer plate. The peak preceding that of testosterone consists mostly of etiocholanolone.

TABLE 2. Urinary excretion of testosterone glucuronide in normal subjects.

| Subjects | (No.) | Age (Years) | Testosterone glucuronide ($\mu$g/24 hours) Mean | Range |
|---|---|---|---|---|
| Normal men | 15 | 19-29 | 170 | (65-348) |
|  | 13 | 30-40 | 94 | (45-178) |
|  | 20 | 42-50 | 76 | (25-159) |
|  | 7 | 55-67 | 70 | (19-154) |
| Normal women | 30 | 10-53 | 6 | (2-10) |

In summary, a GLC method for the detection of testosterone in urine has been developed and applied in normal subjects and patients with endocrinopathies. The advantages of the method lie in its relative simplicity, reproducibility, rapidity, and is therefore an easily available clinical endocrine research tool. Excluding incubation time, the procedure takes 3 days, and one technician can extract and analyze 12 urines a week.

Its disadvantages may be summarized in terms of specificity and applicability. Epitestosterone, the 17$\alpha$-hydroxy epimer of testosterone, may be present in human urine. Under such circumstances, our present procedure is incapable of separating the two epimers. Recent work has also demonstrated that testosterone glucuronide may not be an entirely unique metabolite of plasma testosterone (3), and therefore cannot be used in production rate studies. However, our experience to date with the GLC method for the determination of urinary testosterone has demonstrated a remarkable correlation with clinical androgenicity.

ACKNOWLEDGEMENT

We acknowledge the invaluable technical assistance of Mrs. Virginia Schapiro and Mrs. Marianne Berkovits. We also appreciate the cooperation of Dr. Aquiles J. Sobrero, Director of the Margaret Sanger Research Bureau, New York City, and Dr. Harold Weiner, Director of Medicine of the Veterans Administration Hospital, East Orange, N. J. for allowing us to study their patients with oligospermia and Klinefelter's syndrome.

REFERENCES

1. Futterweit, W., N. L. McNiven, L. Narcus, C. Lantos, M. Drosdowsky, and R. I. Dorfman, Steroids 1: 628, 1963.

2. Futterweit, W., N. L. McNiven, R. Guerra-Garcia, N. Gibree, M. Drosdowsky, G. L. Siegel, L. J. Soffer, and R. I. Dorfman, Steroids 4: 137, 1964.

3. Korenman, S. and M. B. Lipsett, J Clin Invest 43: 2125, 1964.

# ESTIMATION OF TESTOSTERONE IN BLOOD BY GAS-LIQUID CHROMATOGRAPHY

Alexander C. Brownie, Department of Pathology, State University of New York at Buffalo, Buffalo, N. Y.

This work was supported in part by research grants from the U.S. Public Health Service, Bethesda, Maryland.

Following the suggestion of Dorfman (1) that idiopathic hirsutism in the female may be due to excessive testosterone secretion, several methods have been developed over the past few years for the assay of testosterone in human plasma. Finkelstein et al. (2) converted testosterone to estradiol, which was then measured fluorometrically. Several laboratories (3, 4) have used this method for testosterone assays in normal males and females and in disease states. More recently the double isotope derivative procedure has been used by Riondel et al. (5), Hudson et al. (6), and Burger et al. (7) and these methods allow testosterone determinations to be carried out on 10 ml samples of normal female plasma. Plasma testosterone levels in normal women, determined by these methods, are shown in Table 1.

The method of Riondel et al. (5) employing $^{35}$S-thiosemicarbazide as used by Lobotsky et al. (8) gives the lowest values for plasma testosterone in the normal female. In addition it is the only method as yet which can show differences between the ovulatory and luteal phases of the menstrual cycle in normal women. The precision of this method is also very high at low testosterone concentrations. The mean testosterone level in the normal female reported more recently by Hudson et al. (9) is lower than a previous report (6) and this is apparently due to a more rigorous purification procedure. The non-specific water blanks of these methods (5, 9) are low and in the latter case, this has been due to the use of freshly prepared or redistilled $^{3}$H-acetic anhydride. The finding of a non-specific blank of zero by Burger et al. (7) after a less rigorous purification procedure than these other methods is difficult to understand.

It appears at this stage that there can be no short-cut to success with the double isotope derivative methods employing TLC and paper chromatography for purification. In the methods of Riondel et al. (5) and Hudson et al. (6), a derivative of the doubly-labeled testosterone derivative is prepared and this step achieves considerable purification of the testosterone derivative and much non-specific radioactive material is removed. The importance of this step is seen by comparing the results of Burger et al. (7) for normal female plasma testosterone with those of Lobotsky et al. (8) and Hudson et al. (9) (Table 1).

In spite of the undoubted success achieved by these workers (6, 8, 9) in applying these methods to the study of plasma testosterone in normal males and females and in clinical problems, many other workers have had considerable difficulty with the double isotope derivative method at these low steroid levels. The steroid thiosemicarbazones are difficult to prepare although this can be overcome. Frequently it is found that the non-specific blank is so high and variable that determination of testosterone in 10 ml of normal female plasma is impossible or highly inaccurate. However, the correlation which has been shown between testosterone in the plasma and hirsutism will no doubt increase the demand by clinicians for this assay; whereas, at this point it appears that very few laboratories could carry out these assays using the double isotope derivative method.

Recently, rapid and relatively simple GLC methods have been employed for the determination of steroids in urine. A GLC method was developed by Futterweit et al. (10) for the determination of testosterone in human urine; however, the sensitivity of that method in which an argon ionization detector was used, was not great enough to be used for the assay of testosterone in 10 ml of female plasma where one would be dealing with approximately 5 m$\mu$g of steroid initially.

TABLE 1. Plasma testosterone levels (μg/100 ml) in normal men and women

### Men

| Publication | No. of Subjects | Age Range | Mean Testosterone | Range | Water Blank | Ovariectomized Adrenalectomized Plasma |
|---|---|---|---|---|---|---|
| Fluorometric Method | | | | | | |
| Forchielli et al (3) | 9 | 23-74 | 0.56 | 0.1-0.98 | - | - |
| Lamb et al (4) | - | - | - | - | - | not detected |
| Double Isotope Methods | | | | | | |
| Riondel et al (5) | 11 | 21-36 | 0.80±0.25° | 0.5-1.1 | 0.0009 | 0.013±0.002* |
| Hudson et al (6) | 21 | 19-62 | 0.74 | 0.49-0.98 | 0.012 | 0.013$^x$ |
| Hudson et al (9) | 40 | 16-72 | 0.709±0.23* | - | 0.02±0.002* | - |
| Burger et al (7) | 7 | 19-33 | 0.70 | 0.32-1.07 | 0 | 0.08 |
| Lobotsky et al (8) | 14 | 19-38 | 0.64±0.18° | 0.36-0.94 | 0.02±0.005° | - |
| Gas Chromatography Method | | | | | | |
| Brownie et al (25) | 16** | 23-59 | 0.69 | 0.41-1.06 | 0 | - |

### Women

| Publication | No. of Subjects | Age Range | Mean Testosterone | Range | Water Blank | Ovariectomized Adrenalectomized Plasma |
|---|---|---|---|---|---|---|
| Fluorometric Method | | | | | | |
| Forchielli et al (3) | 10 | 22-35 | 0.12 | 0.02-0.26 | - | - |
| Lamb et al (4) | 20 | 17-38 | 0.11±0.06* | 0.05-0.29 | - | not detected |
| Double Isotope Methods | | | | | | |
| Riondel et al (5) | 2 | - | 0.059 / 0.079 | - | 0.0009 | 0.13±0.002* |
| Hudson et al (6) | 12 | 17-56 | 0.11 | <0.05-0.31 | 0.012 | 0.013$^x$ |
| Hudson et al (9) | 25 | 16-65 | 0.083±0.007* | - | 0.02±0.002* | - |
| Burger et al (7) | 8 | 19-40 | 0.18 | 0.06-0.31 | 0 | 0.08 |
| Lobotsky et al (8) | 9 | 19-38 | 0.054±0.015° | 0.033-0.10 | 0.02±0.005° | - |
| Gas Chromatography Method | | | | | | |
| Brownie et al (25) | 12** | 25-60 | 0.063 | 0.02-0.169 | 0 | - |
| | 10*** | 21-31 | 0.070 | 0.03-0.085 | 0 | 0 |

*S E  
$^x$Patient also hypophysectomized  
°S D  

**Carried out by Dr. H. J. van der Molen (Utrecht)  
***Carried out by Dr. J. Peaks (Salt Lake City)

The use of electron absorption for the identification of components in the effluent stream of GLC column was first reported by Lovelock in 1958 (11). Lovelock and Lipsky (12), Lovelock (13) and Landowne and Lipsky (14), have thoroughly examined many of the practical and theoretical considerations involved in electron capture analysis. The electron capture detector is an ionization chamber through which passes an inert gas (e.g., nitrogen), the source of ionizing radiation is commonly tritiated foil and the chamber is connected to a source of low potential just sufficient to collect all the electrons. In the electron capture detector the current produced is derived from free electrons and positive ions. In the presence of a gas containing neutral molecules ($A^o$) capable of capturing free electrons, negative ions are formed:

$$A^o + e^- \rightarrow A^- + \text{energy}$$

The negative ions so formed will tend to recombine with positive ions and whereas with most ionization methods this recombination interferes with the response, the electron capture detector is specifically designed to exploit recombination effects for the measurement of compounds having affinity for free electrons. The rate of recombination between positive and negative ions is much greater than between free electrons and positive ions, thus the capture of electrons by these neutral molecules is readily observed in terms of increased rate of recombination causing a decrease in current flow. The decrease is related to the concentration of compound with electron affinity and to the electron affinity of the compound.

The capture of slow electrons by compounds requires the presence within them of some element with an affinity for free electrons. Carbon and hydrogen have little affinity for free electrons but oxygen and halogens capture electrons readily to form stable negative ions. Hydrocarbons and steroids substituted with these atoms therefore capture electrons.

The potential applied to the electron capture detector is generally just sufficient to collect all the electrons set free in the chamber and this potential varies with the electron affinity of the electron absorbing vapor. One great convenience of the method is that the response to weakly-capturing compounds can be abolished by increasing the potential. It should be noted that the potential for maximum sensitivity depends on such other factors as the geometry of the detector, the conditions of temperature, pressure and carrier gas used and thus, calibration is required of any new system.

Many workers have used a d c potential to collect the electrons but Lovelock (13) has strongly advocated the use of a pulse sampling method to overcome many of the drawbacks such as contact potentials, electron energy changes, and space change development. In the pulsing method, no field is applied to the detector for most of the time and conditions are better for efficient recombination of positive and negative ions. The time for encounter between these ions can be extended to the point where the recombination between positive ions and electrons limits any further increase in sensitivity.

Landowne and Lipsky (14) suggested that derivatives should be formed to impart high electron affinity to compounds that ordinarily have little or no electron absorption capacity and thereby greatly augment their detection by this technique. In another publication (15) they showed that, by formation of simple haloacetates, cholesterol could be detected and measured in ultramicro quantities. They found that surprisingly the monochloroacetate was more sensitively detected than derivatives such as the trichloroacetate, dichloroacetate, bromoacetate and trifluoroacetate. In the method for blood testosterone assay to be described below, Brownie et al. (16) used electron capture of the monochloroacetate of testosterone to obtain the necessary sensitivity for work with small volumes of normal female plasma.

MATERIALS AND METHODS

All solvents were analytical reagent grade and were rigorously purified by standard procedures (16). All steroids, radioactive and non-radioactive, were thoroughly tested for purity and purified if necessary. Testosterone and other steroid chloroacetates were prepared as described by Brownie et al. (16) following the method of Landowne and Lipsky (15). The silica gel used was silica gel G (Merck, according to Stahl for TLC). Before use it was washed with boiling methanol after the addition of a phosphor (DuPont luminescent chemical, Index 609), 30 mg per 100 g silica gel.

*Apparatus.* The gas chromatograph was a standard Barber Colman model 10 instrument fitted with an "Aerograph" electron capture detector (Wilkins Instrument and Research Co.). A pulsating voltage was applied to the electron capture cell using a Model 214a pulse generator (Hewlett-

Packard Co.). The pulsating voltage was 50 v, having a pulse width of 10 μsec, a pulse position of 100 μsec with an internal repetition rate of 10 KC.

The column was 3 ft of 0.4 cm diameter borosilicate glass tubing (U-shaped) packed with 1% XE-60 stationary phase. The column support (Gas Chrom P) was washed and coated with stationary phase as described by DePaoli et al. (17). High purity nitrogen which was led through a tube filled with molecular sieve (Type 13X from Linde) was used as carrier gas at a pressure of 40 psi. The column temperature was kept at 210 C with the detector at 220 C and the flash heater at 250 C. Samples were introduced onto the column with a 10 μl calibrated Hamilton microliter syringe. Recorded peaks were quantitated by triangulation or with a planimeter.

TLC. This was carried out on silica gel plates prepared with a standard Desaga spreader to give layers of approximately 0.30 mm thickness. Plasma extracts were chromatographed on 2 cm lanes on the thin-layer plates and were separated from each other by a 1 cm lane and from standards by a 1.5 cm lane in order to avoid any possibility of contamination. Chromatograms were developed in an ascending manner and steroid standards on the chromatograms were detected using a Haines fluorescence scanner. Silica gel was usually removed from the plates by scraping with a spatula.

Assay of radioactivity. Tritium was assayed using a Packard Tri-Carb liquid scintillation spectrometer. Samples for assay were dried and dissolved in 10 ml of scintillation fluid (4 g PPO and 40 mg POPOP in 1 liter toluene) in 20 ml glass vials. Samples were counted for long enough to give an approximate accuracy of counting of 2%.

Estimation of testosterone in human plasma. The method used has been described in detail by Brownie et al. (16) and consists of the following steps.

1. Addition of about 2,500 cpm of high specific activity $1,2-^3H$-testosterone to 10 ml of female plasma or to 5 ml male plasma diluted with 5 ml of 0.9% aqueous sodium chloride. The plasma is made alkaline and extracted 3 times with 20 ml of di-ethyl ether.

2. After water washing, the ether extract is concentrated to dryness and chromatographed on silica gel thin-layer plates in the solvent system cyclohexane-ethyl acetate, 1:1 (v/v). At this chromatography, allowance has to be made for the slower running rate of testosterone in plasma extract and thus a generous area of silica gel is scraped off the plate. The silica gel is deactivated by the addition of a few drops of water and the testosterone is efficiently extracted from the silica gel using highly purified n-butanol:methanol, 1:3 (v/v).

3. The extracts from TLC are thoroughly dried in a vacuum desiccator and then monochloroacetylated by the addition of 0.5 ml of a solution of monochloroacetic anhydride in tetrahydrofuran (100 mg per 10 ml), and 0.1 ml pyridine, the reaction being carried out overnight in the dark in a desiccator. After addition of water, the testosterone chloroacetate is extracted with benzene and the benzene extract is subsequently washed with 6 N HCL and water and taken to dryness.

4. The extracts from monochloroacetylation are chromatographed on silica gel thin-layer plates using the solvent system benzene:ethyl acetate (4:1). The silica gel of the testosterone chloroacetate areas is scraped off, a few drops of water added and the testosterone chloroacetate extracted with benzene, high speed centrifugation being used to avoid silica particles in the benzene layer. The benzene extracts are transferred to 2.0 ml conical microcentrifuge tubes, drying down after each extraction.

5. To the dried extracts in the microcentrifuge tubes is added a solution of either cholesterol chloroacetate or 20β-hydroxy-4-pregnene-3-one chloroacetate. These compounds act as internal standards for the GLC and are used to check on losses occurring during the transference of material to the GLC column and during passage through the column. An aliquot of the extract is taken at this point for tritium assay, the tritium assay correcting for losses occurring up to this point in the procedure.

6. The mixture is taken to dryness, dissolved in 15 μl of toluene and as much as possible of this solution is injected into the column. Known amounts of internal standard and testosterone chloroacetate are chromatographed interspersed between the plasma extract samples.

7. The amount of testosterone in the plasma extracts is calculated from the areas of internal standard and testosterone chloroacetate, taking into account the losses occurring prior to GLC and losses occurring at the GLC step. Previously duplicate water blanks containing only

1,2-$^3$H-testosterone were processed through the method; however, using the high specific activity 1,2-$^3$H-testosterone this is unnecessary as this minute amount of testosterone is not detected by GLC.

## RESULTS

When the gas chromatograph and electron affinity detector are operated as described, testosterone and cholesterol chloroacetates or testosterone and pregnenolone chloroacetates are readily separated and testosterone chloroacetate can be detected at the nanogram level. Figure 1 shows the chromatographic tracing of 0.3 μg of pure cholesterone chloroacetate and 0.01 μg testosterone chloroacetate. Figure 2 shows a gas chromatogram after processing 10 ml of normal female plasma and Fig. 3 shows a gas chromatogram after processing 5 ml of male plasma through the method. In both these cases, cholesterol chloroacetate has been the internal standard employed and it can be seen that this is clearly separated from the solvent front and also from the testosterone chloroacetate. In some ways cholesterol chloroacetate is not an ideal internal standard and recently Dr. van der Molen (18) has been using 20β-hydroxyprogesterone chloroacetate as described above. The advantage of this chloroacetate is twofold; first of all it has a much higher electron affinity than does cholesterol chloroacetate and is in fact almost identical with testosterone chloroacetate and secondly, its retention time is much longer than cholesterol chloroacetate and thus one avoids the occasional contamination of the internal standard by interfering material in the extract. Figure 4 shows a GLC tracing of 0.01 μg of testosterone chloroacetate and 0.02 μg of the 20β-hydroxyprogesterone chloroacetate.

The accuracy and reproducibility of the method hinge on the accuracy of quantification at the GLC step. Figure 5 shows a typical calibration curve for quantitation of testosterone chloroacetate using the electron capture detector. As the sensitivity of the electron capture cell decreases slightly with use, several concentrations of pure testosterone chloroacetate are chromatographed each time with unknown plasma samples. The loss of sensitivity is overcome by cleaning the detector by the recommended procedure.

Recoveries in the order of 70% are now obtained up to the step before GLC. This high recovery has been brought about by the increased efficiency of extraction of free testosterone from silica gel and also by avoiding the use of hydrophilic solvents after the chloroacetylation step as the steroid chloroacetates have been found to be unstable in solvents such as ethyl acetate and ethanol.

The precision and accuracy of the method has been checked by carrying out replicate determinations on small quantities of testosterone added to water and also on a plasma pool of male plasma and another pool of female plasma. Table 2 summarizes the results obtained.

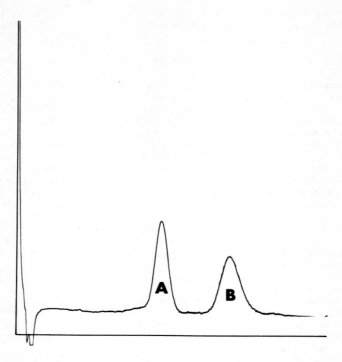

Fig. 1. GLC tracings of 0.3 μg of pure cholesterol chloroacetate (A) and 0.01 μg of pure testosterone chloroacetate (B).

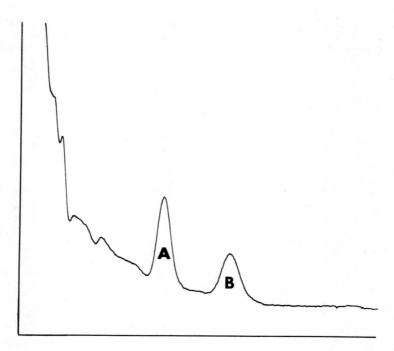

Fig. 2. GLC tracing after processing 10 female plasma. A = cholesterol chloroacetate. B = testosterone chloroacetate.

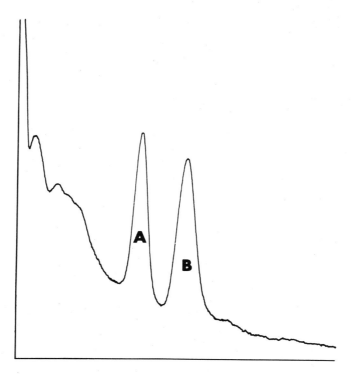

Fig. 3. GLC tracing after processing 5 ml male plasma. A = cholesterol chloroacetate. B = testosterone chloroacetate.

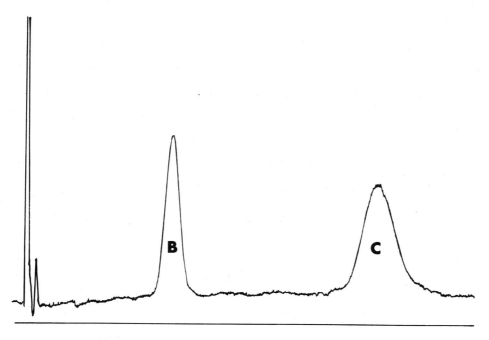

Fig. 4. GLC tracing of 0.01 µg pure testosterone chloroacetate (B) and 0.02 µg pure 20β-hydroxy-4-pregnene-3-one chloroacetate (C).

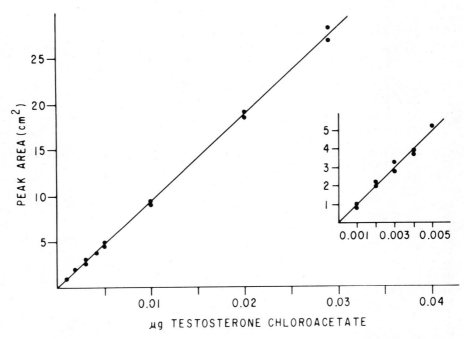

Fig. 5. Calibration curve for the quantitation of testosterone chloroacetate using the electron capture detector.

TABLE 2. Determination of testosterone added to water and in plasma.

|  | Testosterone Added μg | No. of Assays | Testosterone Found μg |
|---|---|---|---|
| Testosterone in 10 ml water | 0.010 | 7 | 0.010±0.005* |
|  | 0.020 | 8 | 0.021±0.005* |
|  | 0.040 | 10 | 0.041±0.005* |
| 5 ml samples of male plasma | - | 10 | 0.021±0.006* |
| 10 ml samples of female plasma | - | 15 | 0.006±0.003* |

\* = standard deviation

The specificity of the method was also examined by considering all steroids which might possibly be present in the plasma extracts. After the first TLC of the plasma extract an area of silica gel is scraped off and eluted corresponding to an $R_{testosterone}$ of 0.75 to 1.2. All steroids examined which would come within this range were chloroacetylated and chromatographed in the second TLC system. At this step an area of silica gel is scraped off and eluted corresponding to an $R_{testosterone}$ chloroacetate of 0.85 to 1.15. Of the steroids investigated, four were left as possible contaminants: - 17-nor testosterone, 17-epitestosterone, 20α-hydroxy-4-pregnene-3-one and 20β-hydroxy-4-pregnene-3-one and the chloroacetates of these compounds were exposed to GLC. Table 3 shows the chromatographic behavior of these compounds in the two TLC systems and in GLC. The progesterone chloroacetates have much longer retention times than testosterone chloroacetate, however, the other two chloroacetates were not completely separated. Although using the GLC conditions described her, 17-epitestosterone and testosterone chloroacetates are not completely separated, two distinct peaks are seen. If 17-epitestosterone is present it can be separated as the free steroid from testosterone by paper chromatography in the system hexane:benzene (2:1) - formamide, overrunning for two hours (19), prior to chloroacetylation and GLC. So far, no peak due to 17-epitestosterone chloroacetate has been seen on the chromatographic tracings from normal plasma samples.

TABLE 3. Thin-layer and gas-liquid chromatography of steroids.

|  | Thin-layer chromatography | | Gas chromatography on 1% XE-60 |
|---|---|---|---|
|  | $R_T$* in ethyl acetate: cyclohexane (1:1) | After chloroacetylation $R_{TCA}$* in ethyl acetate:benzene (1:4) | Retention time of chloroacetates (min) |
| 19-Nor-testosterone | 0.85 | 0.96 | 12 |
| 17-Epi-testosterone | 0.98 | 1.04 | 12 |
| Testosterone | 1.00 | 1.00 | 14 |
| 20α-Hydroxy-4-pregnene-3-one | 1.15 | 1.04 | 24 |
| 20β-Hydroxy-4-pregnene-3-one | 1.21 | 1.17 | 20 |

*Running rate relative to testosterone

\*Running rate relative to testosterone chloroacetate

## DISCUSSION

Outstanding success has been recently achieved in the application of GLC to the measurement of steroids in urine. Advantage has been taken of the high speed and high degree of resolution offered by this technique. For the assay of minute quantities of steroids in blood, perhaps the most important feature of GLC is the sensitive ionization detection methods which have become available. With the latest designs of flame ionization detectors it is possible to measure with accuracy about 15 m$\mu$g of steroid. In our testosterone assay method, advantage has been taken of the even greater sensitivity offered by the electron absorption detector when dealing with compounds with electron affinity.

It is possible with the instrument which we used to detect 1 m$\mu$g of pure testosterone chloroacetate. More recently, van der Molen (18) has set up this method in Utrecht using a standard F&M Model 400 series analytical gas chromatograph fitted with an F and M electron capture detector and using the pulsating voltage sampling method. With this particular apparatus, testosterone chloroacetate is more sensitively detected. Our method allows the measurement of testosterone levels in normal female plasma although the accuracy is less than desirable. With the greater sensitivity of his apparatus, van der Molen finds that the precision and accuracy of the method is much improved, the coefficient of variation being approximately 10% for the levels of testosterone found in the normal female. Thus the accuracy of this method can approach the accuracy of the double isotope derivative method of Riondel et al. (5), however, the speed and simplicity of the GLC method is much greater. In three working days, including one day for GLC, it is possible to process 12 plasma samples. Also, after the initial instrumentation the cost of this method is very low compared to the double isotope derivative methods. It is of interest to note that Kirschner et al. (19) have used a double isotope derivative method for plasma androgen assay in which following extraction and one TLC, GLC is used for the purification of labelled steroid derivatives. By using this high resolution chromatographic method, interfering radioactively labelled material is efficiently removed from the steroid fractions with a high degree of success and with considerably fewer manipulations than in the other double isotope derivative methods. The precision of measurement of this method is not as high as the method of Riondel et al. (5) although the plasma blank value for testosterone in two ovariectomized hypophysectomized women was relatively low.

The use of GLC as an analytical tool has certainly outstripped its use in qualitative analysis. Horning et al. (20) and Evans and Scott (21) have pointed out many of the sources of error associated with quantitative GLC and in particular Lovelock (13) has discussed this with regard to the use of the electron capture detector. A very important consideration in electron absorption spectrometry is the very small quantities of material normally being chromatographed. The effects of adsorption on the stationary phase support and other surfaces are important and in order to obtain accurate quantitative analyses some correction factor has to be used. In the method described here, an attempt has been made to obtain such a correction by the inclusion of an internal standard at the GLC step. In addition the use of a pulsating voltage sampling technique contributes to the uniformity of response seen and the apparent absence of anomalous results.

It is our impression that a single GLC step, without prior purification by TLC or paper chromatography, will be insufficient to guarantee the specific determination of testosterone in this method. It can be seen that apparently only 19-nortestosterone and 17-epitestosterone would interfere with the testosterone being present in biological fluids, however, it should be noted that 17-epitestosterone has been found in the urine of normal subject by Korenman et al. (22) and by Brooks and Giuliani (23) and in bovine spermatic vein blood by Lindner (24). Specificity is added to the method by applying to the electron capture detector a voltage at which other compounds, with much less electron affinity, do not capture slow electrons.

The values (Table 1) obtained using this method for plasma testosterone in normal men and women (25) agree well with those reported by other workers. Of interest is the finding that the plasma of ovariectomized-adrenalectomized women with breast carcinoma contains no testosterone.

As will be seen later in this conference, this method is readily applied to the determination of other steroids in blood. It is to be hoped that other simple derivatives will be found which will allow even more sensitive and accurate assays of blood steroid hormones by the electron absorption technique.

## ACKNOWLEDGEMENTS

The author would like to express his appreciation of the fruitful discussions and correspondence with Drs. K. B. Eik-Nes and H. J. van der Molen. He would also like to thank Professor B. Hudson for a preview of his paper on testosterone presented at the 2nd International Congress of Endocrinology, London, 1964.

## REFERENCES

1. Dorfman, R. I., Acta Endocr (Kbh) Supp. 50: 211, 1960.
2. Finkelstein, M., E. Forchielli, and R. I. Dorfman, J Clin Endocr 2: 98, 1961.
3. Forchielli, E., G. Sorcini, M. S. Nightingale, N. Brust, R. I. Dorfman, W. H. Perloff, and G. Jacobson, Anal Biochem 5: 416, 1963.
4. Lamb, E. J., J. W. Dignam, R. J. Pion, and H. H. Simmer, Acta Endocr (Kbh) 45: 243, 1964.
5. Riondel, A., J. F. Tait, M. Gut, S. A. S. Tait, E. Joachim, and B. Little, J Clin Endocr 23: 620, 1963.
6. Hudson, B., J. P. Coghlan, A. Dulmanis, M. Wintour, and L. Ekkel, Austral J Exp Biol 41: 235, 1963.
7. Burger, H. G., J. R. Kent, and A. E. Kellie, J Clin Endocr 24: 432, 1964.
8. Lobotsky, J., H. I. Wyss, E. J. Segre, and C. W. Lloyd, J Clin Endocr 24: 1261, 1964.
9. Hudson, B., J. P. Coghlan, A. Dulmanis, and M. Wintour, Proc 2nd Int Congr of Endocr, London, 1964.
10. Futterweit, W., N. L. McNiven, L. Narcus, C. Lantos, M. Drosdowsky, and R. I. Dorfman, Steroids 1: 628, 1963.
11. Lovelock, J. E., Institute of Petroleum, Gas Chromatography Discussion Group, Oxford, England, May 1957.
12. Lovelock, J. E. and S. R. Lipsky, J Am Chem Soc 82: 431, 1960.
13. Lovelock, J. E., Anal Chem 35: 474, 1963.
14. Landowne, R. A., and S. R. Lipsky, Anal Chem 34: 726, 1962.
15. Landowne, R. A. and S. R. Lipsky, Anal Chem 35: 532, 1963.
16. Brownie, A. C., H. J. van der Molen, E. E. Nishizawa, and K. B. Eik-Nes, J Clin Endocr 24: 1091, 1964.
17. DePaoli, J. C., E. Nishizawa, and K. B. Eik-Nes, J Clin Endocr 23: 81, 1963.
18. van der Molen, H. J., Personal communication, 1964.
19. Kirschner, M. A., M. B. Lipsett, and D. R. Collins, J Gas Chromatog 2: 360, 1964.
20. Horning, E. C., K. C. Maddock, K. V. Anthony, and W. J. A. VandenHeuvel, Anal Chem 35: 526, 1963.

21. Evans, R. S. and R. P. W. Scott, *Chimia* 17: 137, 1963.

22. Korenman, S. G., H. Wilson, and M. B. Lipsett, *J Clin Invest* 42: 1753, 1963.

23. Brooks, R. V. and G. Giuliani, *Steroids* 4: 101, 1964.

24. Lindner, H. R., *Nature* 183: 1605, 1959.

25. Brownie, A. C., H. J. van der Molen, E. E. Nishizawa, and K. B. Eik-Nes, *Proc 2nd Int Congr of Endocr*, London, 1964.

# THE ESTIMATION OF PLASMA ANDROGENS BY GAS-LIQUID CHROMATOGRAPHY

Alan Goldfien, James Jones*, M. Edward Yannone and Beverly White, Departments of Obstetrics and Gynecology, Medicine and the Cardiovascular Research Institute, University of California, School of Medicine, San Francisco.

*Part of this work was carried out while a U.S.P.H.S. trainee in the Department of Obstetrics and Gynecology.

This work was supported in part by NIH Research Grants HD 00640, H2685 and Training Grant 5TI RD-6.

During the past few years several methods have been developed for the measurement of plasma testosterone. Finkelstein et al. (1) reported a method based on the enzymatic conversion to estradiol of testosterone isolated by paper chromatography. The estradiol was estimated by fluorometry. Double isotope derivative techniques using either $^3$H-acetic anhydride (2) or $^{35}$S-thiosemicarbazide (3) have been described. Although remarkable in their sensitivity and reproducibility, these methods are extremely complicated and expensive.

GLC has been applied to the measurement of quantities of hormonal steroids of the order of magnitude required for the detection of testosterone in human plasma (4, 5, 6). The ability to measure such small amounts of steroids in these procedures was achieved by prior chromatographic purification and the use of sensitive detectors.

The following report describes the application of a method used for the measurement of progesterone (4, 7) modified to permit the estimation of plasma testosterone, androstenedione and dehydroepiandrosterone.

## MATERIALS AND METHODS

Reagent grade solvents and chromatographic grade methylene chloride were redistilled before use. Testosterone, androstenedione and dehydroepiandrosterone (Elite Chemical Corp.) were purified as required and used as standards. Testosterone-1,2-$^3$H (SA 135 μc per μg) was obtained as a gift from the National Institutes of Health. Androstenedione-1,2-$^3$H (SA 135 μc per μg) was prepared by oxidation of labeled testosterone with chromic acid followed by chromatographic purification. Dehydroepiandrosterone-7α-$^3$H (SA 12 μc per μg) was obtained from Nuclear-Chicago Corp. The purity of all tracers was checked periodically by paper chromatography. Phosphor solution was prepared by the dilution of Liquifluor® (Pilot Chemicals Inc.) with toluene. GLC was performed on a Model 60-10 gas chromatograph equipped with a hydrogen flame detector (Research Specialties Inc.). Radioactive steroids were located on paper chromatograms using an Autoscanner 880 (Vanguard Instrument Co.) and radioactivity was measured by scintillation counting in the liquid scintillation spectrometer (Packard Instrument Co.).

Plasma samples are thoroughly mixed with 15,000 cpm of each labeled steroid for which the sample is to be analyzed. Each sample is extracted twice with one volume of methylene chloride. The combined extract is then washed successively with 0.1 vol of 1N NaOH, water 0.3N acetic acid and water. The methylene chloride is then evaporated under filtered air and the residue redissolved in 5 ml of 70% methanol. The solution is then centrifuged at -15 C for one hr at a speed of 6,5000 rpm and the supernatant quickly drawn off leaving behind the precipitated solids. The solution is then washed twice with 2/5 vol of heptane. After the addition of 3 ml water the steroids are back extracted with 0.5 vol of methylene chloride which is then transferred to a 10x75 mm test tube and evaporated. The sample is then applied to 5 mm strips of Whatman #1 paper in the manner previously described (7). After equilibration, the strips are developed at room temperature in a cyclohexane:methanol:water (100:100:10) Bush-type system which affords excellent separation of the steroids concerned in approximately 15 hr.

Steroids are eluted from the sections of paper containing the peaks of radioactivity by the addition of a drop of water followed by 1 ml of 100% methanol. In order to minimize the amount of interfering material no attempt is made to recover the radioactivity at the leading and trailing ends of the peak. The eluates are then concentrated in the tips of conical vials and prepared for injection onto the column as previously described (7), including the removal of 1/11 aliquot (9 μl) for counting later.

After transfer to the solids injector, the eluted material is chromatographed on a 2 ft 8 in stainless steel column with a 2 mm id packed with Gas Chrom CLH 60/80 mesh coated with 1% XE-60 (Applied Sciences Lab., Inc.). The column is maintained at a temperature of 185 C and a carrier pressure of 20 psi of nitrogen. The mass of steroids is determined by comparing the area beneath the appropriate peak, which is estimated by multiplying the peak height by its width at half-height, with the area produced by the injection of known amounts of standards. The concentration of androgens in μg per 100 ml plasma is then calculated by the following formula:

$$\text{Concentration in } \mu g \text{ per } 100 \text{ ml} = \frac{100 \text{ SM}}{P(10C-V)}$$

Where S = cpm of added tracer

M = mass of steroid from GLC (μg)

P = volume of plasma (ml)

C = cpm in the nine μl (1/11) aliquot

V = cpm remaining in vial and syringe after transfer.

## RESULTS AND DISCUSSION

When testosterone in amounts varying from 3 to 110 mμg was injected into the GLC column, the area of the peaks was linearly related to the amount injected (Fig. 1). With less than 20 mμg, retention time increased slightly with decreasing samples size (Fig. 2). Steroid identification by retention time may be further complicated when the peak is superimposed on a steeply sloping baseline. This results in an apparent decrease in retention time. However, it is possible to estimate the degree of shift due to the changing baseline and determine the actual retention time.

When labeled testosterone, androstenedione and dehydroepiandrosterone are added to water and the sample carried through the entire procedure, peaks with a retention time similar to those of the above steroids are rarely observed if the solvents are sufficiently purified. Although we have not routinely run water blanks with each set of samples, random checks indicate that it is necessary to run such blanks with any change in materials.

In order to evaluate the specificity and reproducibility of the method, a pool of plasma was obtained by combining equal volume samples of plasma from 8 males. Seven 10 ml aliquots of this plasma were analyzed for testosterone and the values obtained appear in the first column of Table 1. Six additional 10 ml aliquots were extracted and chromatographed in cyclohexane-methanol-water and the testosterone area eluted in the usual manner. The eluates were then oxidized with chromic acid and rechromatographed in the same system. The radioactivity corresponding to the androstenedione standard was eluted and the eluates injected into the GLC column in the usual manner. The results of these determinations appear in the second column of Table 1. A few additional aliquots were analyzed by GLC after acetylating the material eluted from the testosterone area of the paper chromatogram. The mean of these determinations was 0.66 μg per 100 ml of testosterone measured as the acetate and all values were higher than the values obtained for testosterone. Concentrations of testosterone obtained from seven normal men ranged from 0.4 to 1.0 μg per 100 ml. The reproducibility of the method at these concentrations was assessed in two individuals in whom determinations were repeated (Table 2).

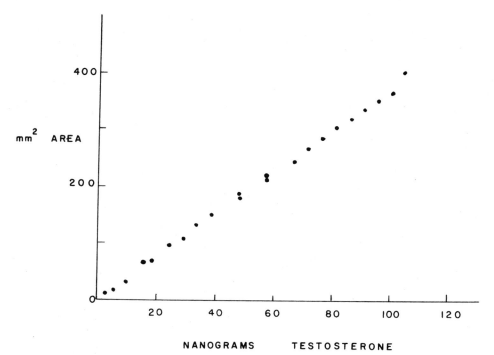

Fig. 1. Varying amounts of testosterone, subjected to GLC are compared to the resultant peaks observed. Note the linear relationship of mass (mµg) and area (mm$^2$).

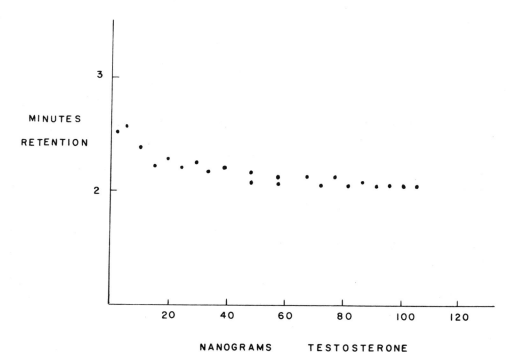

Fig. 2. The relationship of varying amounts of testosterone and resultant retention times is illustrated. Note the increase in retention time when less than 20 mµg was used.

TABLE 1. The first column represents the results of testosterone determinations performed on seven 10 ml aliquots of pooled, normal male plasma. In the second column are the results of six other 10 ml aliquots from the same pool in which testosterone was oxidized and measured as androstenedione.

| Aliquot No. | Testosterone micrograms/100 ml | Androstenedione micrograms/100 ml |
|---|---|---|
| 1 | 0.58 | 0.58 |
| 2 | 0.49 | 0.56 |
| 3 | 0.56 | 0.62 |
| 4 | 0.50 | 0.45 |
| 5 | 0.51 | 0.56 |
| 6 | 0.61 | 0.53 |
| 7 | 0.62 | |
| mean | 0.55 | 0.55 |
| S D | ±0.05 | ±0.06 |

TABLE 2. Results of repeated determinations in two normal male subjects. The values enclosed in brackets are dulpicate determinations. The asterisk (*) denotes that the analysis was performed on a different day by a different individual.

| Subject | Sample Date | Testosterone microgram/100 ml | % recovery |
|---|---|---|---|
| G | 1-24 | 0.40* | 33 |
| G | 2-17 | 0.43 | 26.4 |
| McC | 2-17 | 0.85 | 35 |
| | ⎡ 2-25 | 0.77 | 36 |
| | 2-25 | 0.79 | 33 |
| | 2-25 | 0.68 | 20 |
| | ⎣ 2-25 | 1.02 | 30 |
| | ⎡ 8-13 | 0.68 | 48 |
| | ⎣ 8-13 | 0.74 | 46 |

⎡⎣ Aliquots from same sample

* Different time and operator

When 20 to 30 ml specimens of plasma obtained from normal females were analyzed by this method, no peaks corresponding to testosterone were detected. Concentrations as low as 0.05 µg per 100 ml should have been detected if present. This finding is not in agreement with the observations of other investigators who have reported mean concentrations of testosterone in normal women to be between 0.05 and 0.1 µg per 100 ml (1, 2, 3, 8, 9). In order to determine whether or not these amounts of testosterone would be detected if present, 1 and 2 ml aliquots of the pooled male plasma were analyzed. Clearly discernible peaks were identified in each of the four aliquots and the values determined were 0.43, 0.52, 0.59, 0.61 µg per 100 ml. The mean of these values is 0.53 µg per 100 ml which is comparable to levels determined by the analysis of 10 ml aliquots.

Using larger plasma samples and increasing the final recoveries of testosterone it has been possible to find peaks corresponding to testosterone in normal women. Plasma (40-60 ml) obtained from 24 normal women was analyzed for testosterone, androstenedione and dehydroepiandrosterone. The results of these studies appear in Table 3. It was possible, to find a peak, in all but three of these samples. It is of interest that the mean of the concentrations of testosterone in these samples was 0.017 µg per 100 ml, a value lower than those previously reported. The values obtained for androstenedione and dehydroepiandrosterone do not differ significantly from those found in the normal males (Table 4).

During the past six months we have had the opportunity to perform approximately 600 analyses on 200 samples of human venous plasma. In order to evaluate the method, we have tabulated the difficulties encountered in the analyses performed on specimens obtained from normal and abnormal women. These are shown in Table 5.

Although further studies on the specificity of this method for androstenedione and dehydroepiandrosterone may indicate that additional chromatographic procedures are required before final GLC, the specificity of the testosterone measurements is supported by the studies reported above and by the concentrations observed in normal women which are lower than those reported using other methods. The sensitivity of this procedure does not appear to be as great as that of GLC method employing the electron-capture detector (6). However, the great advantages of this method are the requirements of only a single preliminary chromatographic procedure, the absence of the need for forming chemical derivatives and the use of a flame ionization detector which requires little attention.

TABLE 3. The results of testosterone, androstenedione and dehydroepiandrosterone determinations performed on 24 normal women using 50 ml samples of plasma.

| Subject | Testosterone | | Androstenedione | | DHEA | |
|---|---|---|---|---|---|---|
| | µg/100 ml | % recovery | µg/100 ml | % recovery | µg/100 ml | % recovery |
| 166 | 0.009 | 35.7 | 0.110 | 28.0 | 0.440 | 38.1 |
| 167 | 0.015 | 42.1 | 0.150 | 26.9 | 0.445 | 32.1 |
| 168 | 0.013 | 53.2 | 0.091 | 46.2 | 0.780 | 41.0 |
| 169 | 0.008 | 30.9 | 0.063 | 19.9 | 0.250 | 31.1 |
| 170 | 0.004 | 48.6 | 0.097 | 28.1 | 0.320 | 33.0 |
| 171 | 0.004 | 47.3 | 0.120 | 26.6 | 0.800 | 31.6 |
| 204 | 0.016 | 45.6 | 0.032 | 45.6 | 0.570 | 53.7 |
| 205 | 0.013 | 47.3 | 0.081 | 55.0 | 0.420 | 53.4 |
| 206 | 0.018 | 56.1 | 0.280 | 36.7 | 1.920 | 38.0 |
| 207 | 0.006 | 36.0 | 0.099 | 51.4 | 0.570 | 33.6 |
| 208 | 0.021 | 54.5 | 1.190 | 42.0 | 0.970 | 44.8 |
| 209 | 0.008 | 44.9 | 0.410 | 33.8 | 0.710 | 40.3 |
| 210 | 0.020 | 54.2 | 0.200 | 48.5 | 0.430 | 43.8 |
| 211 | 0.008 | 62.4 | 0.190 | 49.2 | 0.880 | 47.5 |
| 212 | 0.031 | 48.1 | 0.150 | 31.2 | 1.120 | 20.0 |
| 213 | 0.022 | 39.9 | 0.280 | 33.6 | 0.940 | 38.7 |
| 214 | 0.021 | 41.0 | 0.170 | 45.3 | 0.310 | 20.5 |
| 215 | 0.027 | 47.5 | 0.260 | 45.6 | 1.620 | 49.5 |
| 216 | 0.033 | 49.6 | 0.290 | 52.5 | 0.630 | 32.0 |
| 217 | < 0.033 | 35.0 | 0.190 | 40.0 | 1.070 | 26.7 |
| 178 | < 0.035 | 28.1 | 0.023 | 31.7 | 0.600 | 24.1 |
| 177 | not measurable | 42.2 | 0.100 | 25.1 | 0.180 | 21.1 |
| 175 | 0.036 | 34.9 | | | | |
| 176 | 0.018 | 31.5 | | | | |
| mean | 0.017 | | 0.163 | | 0.726 | |
| S D | ±0.009 | | ±0.096 | | ±0.432 | |

TABLE 4. The mean concentrations of plasma androgens from 10 ml aliquots of pooled normal male plasma are compared to concentrations found in 50 ml samples of female plasma.

| Source | Testosterone µg/100 ml | Androstenedione µg/100 ml | DHEA µg/100 ml |
|---|---|---|---|
| Pooled plasma from 8 "normal" males | 0.55 ± 0.05* | 0.15 ± 0.04* | 0.383 |
| Plasma from 21 "normal" females (not pooled) | 0.017 ± 0.009* | 0.163 ± 0.096* | 0.726 ± 0.432* |

*±S D

TABLE 5. Tabulation of the technical difficulties encountered in the analysis of human peripheral plasma for androgens. The four technical errors recorded resulted from breakage of the tube containing the sample.

|  | DHEA | Androstenedione | Testosterone |
|---|---|---|---|
| No. of samples | 65 | 183 | 179 |
| No. lost | 1 (1.5%) | 14 (7.7%) | 8 (4.5%) |
| Interfering peak | 1 | 10 | 5 |
| Technical error | - | 2 | 2 |
| Instrument malfunction | - | 2 | 1 |

## SUMMARY

A method is described for the simultaneous determination of testosterone, androstenedione and dehydroepiandrosterone. The method consists of a preliminary separation of the steroids to be measured by paper chromatography followed by GLC using an instrument equipped with a flame ionization detector. Studies supporting the specificity of the method for testosterone are reported. In normal men, plasma testosterone values ranged from 0.4 to 1.0 µg per 100 ml. A pool of plasma collected from eight normal men contained 0.55, 0.15 and 0.38 µg per 100 ml of testosterone, androstenedione and dehydroepiandrosterone respectively. The mean plasma concentrations (±SD) of these androgens found in normal women were 0.017 ± .009, 0.163 ± .096 and 0.726 ± .432 µg per 100 ml. The major advantages of the technic is the ease with which measurements can be made as compared to other available methods. The usefulness of the method is somewhat limited by the large plasma samples (50 ml) required for the measurement of testosterone in normal women.

## ACKNOWLEDGEMENT

We wish to acknowledge the valuable assistance of Shirley Gullixson, Barbara Borgers and Marie McCabe.

## REFERENCES

1. Finkelstein, M., E. Forchielli, and R. Dorfman, J Clin Endocr 21: 88, 1961.

2. Hudson, B., J. Coghlan, A. Dulmanis, M. Wintour, and I. Ekkel, Australian J Exp Biol Med Sci 41: 235, 1963.

3. Riondel, A., J. Tait, M. Gut, S. Tait, E. Joachim, and B. Little, J Clin Endocr 23: 620, 1963.

4. Yannone, M. E., D. B. McComas, and A. Goldfein, J Gas Chromatogr 2: 30, 1964.

5. Guerra-Garcia, R., S. Chattoraj, L. Gabrilove, and H. Wotiz, Steroids 2: 605, 1963.

6. Brownie, A., H. van der Molen, E. Nishizawa, and K. Eik-Nes, J Clin Endocr 24: 1091, 1964.

7. Goldfein, A., M. E. Yannone, D. B. McComas, and C. A. Braga, (This journal).

8. Lamb, E., W. Dignam, R. Pion, and H. Simmer, Acta Endocr 45: 243, 1964.

9. Lobotsky, J., H. Wyro, E. Segre, and C. Lloyd, J Clin Endocr 24: 1261, 1964.

# DISCUSSION ON 17-KETOSTEROIDS AND TESTOSTERONE

DR. WILSON: In regard to the quantitation of urinary testosterone from a peak on the GLC tracing, perhaps a note of caution is in order. We have come across a number of components which could interfere with this determination. Although most of them would usually be separated from testosterone before the final GLC, and some may not occur in all urines, it nevertheless seems important to keep these possibilities in mind.

The components listed below were not separated from testosterone by partition column chromatography, Girard fractionation, digitonin precipitation or TLC of either the free or acetylated fraction. All were detected by absorption of UV light, and all had $R_t$ values on an SE-30 column either identical with or within 10% of that of testosterone.

1) Epitestosterone, ($R_t$ = 1.0 of testosterone) amounts to about 30-50% as much as testosterone in normal men, and may be much higher in abnormal states. It can be separated by paper chromatography, where it is less polar than testosterone.

2) An unidentified UV absorbing compound ($R_t$ = 0.9 of testosterone) is much polar on a paper chromatogram, and occurs in most if not all urines in amounts similar to or greater than that of testosterone.

3) Several less abundant components are separated only by very careful paper chromatography.

Two other components have $R_t$ values identical with testosterone. One is UV absorbing, found in amounts roughly the same as testosterone in many urines, but does separate on TLC. Another is 16α-hydroxy-dehydroepiandrosterone, which may occur in large amounts in some abnormal states, but should have been separated by TLC or other means.

This list may not be complete, and moreover does not include other possible contaminants which do not absorb UV light. It would therefore seem that one must be very cautious about assigning a single peak on the GLC tracing to a single compound. Certainly any peak which is a little too wide, or asymmetrical, must be regarded with suspicion.

DR. LIEBERMAN: It was the principal objective of the initiators of this conference to evaluate GLC of steroids in naturally occurring mixtures, and specifically the concern was with the following points and not, for example, the question of whether or not GLC could distinguish a man from a woman or anything else.

The points that we hoped would be discussed were the accuracy of the method, the reproducibility of the method, sensitivity of the method, the specificity of the method, the convenience of the method and whether there were alternate methods available that did not involve GLC and how did this method compare with these alternate methods.

Specifically, as far as Mr. Thomas was concerned, I know he does more ketosteroid measurements than anybody in the world and the question I wanted to ask him was, what method does he use now? He has told us that the GLC method is adequate provided a variety of things are done, such as, TLC, paper chromatography, Girard separation, etc. What is the routine method that he uses presently in his laboratory?

MR. THOMAS: This is a question that I feared would come up because the answer is somewhat complicated. We are still running the gradient elution method of Kellie and Wade (1957) for the selection of patients with advanced breast cancer for endocrine ablation. Where possible, we run, any extract left over, on the gas chromatograph and another separate urine sample by the method I showed you (Thomas and Bulbrook, 1964).

The agreement between the gradient elution method and GLC, is plus or minus 10%. But the agreement within the gradient elution method is probably no better than that.

The difference between duplicates in the gas method is usually less than 5%. However, for this series of patients, we are still using the gradient elution method because we don't think it wise to change a method in the middle of an experiment. But for all other purposes we have now switched to the method that I described in my paper.

DR. DODSON: Mr. Thomas, how does the time required for your gradient elution method compare with the time required for the GLC method?

MR. THOMAS: The gradient elution assay takes five days. We can estimate the ketosteroids in six specimens by GLC, with a total working time of three days.

DR. LIPSETT: I would like to ask Dr. Futterweit to answer the question about the precision and accuracy of his urinary testosterone method in women. If I remember correctly you said this was plus or minus 10%. I am sure plus or minus 10% must apply to the higher levels of urinary testosterone. I also want to ask anybody else who presents a chromatogram showing a very small hump on a large solvent peak if it is legitimate to try to obtain any estimate at all?

DR. FUTTERWEIT: The reported values for precision were for the higher values in males. In view of the low peak heights in females, the precision is poor. When we use a larger aliquot we can, of course, have a larger peak. But the precision in the female range remains low.

DR. ROSENFELD: I would like to briefly change the subject somewhat and show this slide which illustrates an example of transformations which might take place on gas chromatograph column. Here are chromatograms of an eluate from a paper system in which 18-hydroxyetiocholanolone has been separated from other 17-ketosteroids. This is from the urine of a patient who received large amounts of etiocholanolone. On the left side (Fig. 1) are two principal peaks which are due to epimeric 18-nor-etiocholanolones. These compounds resulted from the elimination of formaldehyde from the original material injected on the column. When we chromatographed a pure sample of 13β-H-18-nor-etiocholanolone we again got the two peaks in the same ratio, so that the epimerization of the compound derived from loss of formaldehyde must have occurred on the column. However, when we prepared the TMS ether derivative of the 18-hydroxyetiocholanolone extracted from paper and chromatographed it under the same conditions, we obtained only one peak as you can see on the right. The peak on the far right is the internal standard. I just wanted to add this to the growing list of transformations that can occur on the gas chromatograph column.

DR. VAN DER MOLEN: I would like to reemphasize one of the comments that Dr. Wilson has made. We have used a method for the estimation of urinary testosterone, that is essentially similar to the method published by Drs. Wilson and Lipsett's group (J Clin Invest 42, 1453, 1963; Steroids 3, 203, 1964). This method includes an acetylation step followed by TLC and subsequent saponification.

On several occasions we have observed a large UV absorbing spot in the testosterone area during TLC before acetylation. This UV absorption was greatly diminished on the thin-layer place following acetylation and saponification. If these acetylation and saponification steps were left out, we have observed in the GLC tracings of the majority of our samples additional peaks that interfered with the testosterone peak. In such cases one has to go through all those quite tedious purification procedures in order to make sure that one really estimates testosterone using GLC.

Figure 2 shows examples of the GLC tracings obtained from standards and urine samples. Even following this extensive purification, the extracts finally obtained from low titer female urines, as shown in the second tracing, may give a small peak between the peaks of testosterone and the androstenedione. In order to overcome this, we often use the chloroacetylation and final purification of testosterone-chloroacetate as will be described by Dr. Brownie in the method for the estimation of plasma testosterone.

This figure is typical of the peaks obtained from a urinary extract. First, the testosterone peak. The left panel contains standards, the center panel picture is the female sample and the other one is a male.

If you will allow me one question, I would ask Dr. Menini about the results that he obtained with his method for the estimation of conjugates in urine. I wonder if he could elaborate a little more about the specificity of the technique, that he has been using.

DR. MENINI: The specificity of the method is based on the preferential extraction of the 17-oxosteroid sulphates and their solvolysis by dioxan. The conditions for the oxidation of the liberated steroids with *tert*-butyl chromate are taken from a procedure for the systematic analysis of urinary steroids (Menini and Norymberski, in press) in which 17-oxosteroid glucosiduronates are similarly measured as polyoxoandrostanes.

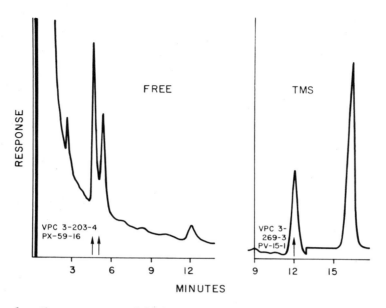

Fig. 1. Chromatograms of 18-hydroxy-etiocholanolone before and after TMS ether formation.

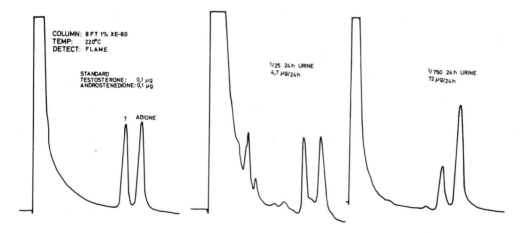

Fig. 2. Tracings obtained following GLC of extracts for the estimation of testosterone in urine.

DR. TOUCHSTONE: Before the subject gets off base, I notice that most of these methods for testosterone or ketosteroids were using SE-30. SE-30 is one of the poorest phases, if you want to get separation of steroids. Phases like QF-1, which is ketone selective may be more suitable for some of the 17-ketosteroid separations. Not only this, but in the past year or so we have used combinations of phases in one column. This may increase the resolution.

DR. LUUKKAINEN: The free 17-ketosteroids can be well separated on the Z-liquid phase. We use enzyme hydrolysis and no chromatography before GLC. If solvolysis in ethylacetate is used for hydrolysis and steroids analyzed as TMS in GLC, one finds a peak which has approximately the same retention time as androsterone TMS on many liquid phases. The same peak is found if only water alone is used as a blank in solvolysis. It is formed from ethylacetate during the procedure. This may be pertinent to Mr. Thomas' discussion.

MISS PATTI: With regard to quantitation with GLC, we have found that this is difficult to obtain with semipurified extracts. The concentration calculated may be too high because one can not ascertain the purity of the steroid peak on the chromatogram. In the event that TLC does not establish the purity of the steroid peaks on the original tracing, then the extract must be subjected to paper chromatography. Precise quantitation, as a rule, is possible after fractionation on paper. However, much can be learned from the rapid screening by GLC of the semi-purified extracts of biological specimens.

With reference to the transformation that is seen with GLC, we noted that the 17-hydroxycorticoids were pyrolyzed completely to their analogous 17-ketosteroids when chromatographed with SE-52 coated on Gaschrom-P. The transformation of steroids can not be attributed to the column. We must consider the effects of the flash heater temperature.

With regard to SE-30, I agree with Dr. Touchstone that there are other phases which are better and can separate steroids both in the free and conjugated states. We have obtained good separation of the sulfate esters of androsterone, etiocholanolone and DHEA on XE-60, QF-1, and SE-52. We were unable to separate these three sulfates on SE-30. Cholesterol-sulfate can be detected with the four liquid stationary phases. I was wondering if Dr. Menini has analyzed any sulfate esters of the 17-ketosteroids with GLC.

DR. MENINI: I have always solvolyzed the sulfates before oxidation and GLC.

DR. LIPSETT: To answer Dr. Lieberman's question, certainly, for clinical measurement of urinary ketosteroids one can use a combination of TLC and GLC with adequate precision and accuracy. We have used these methods for following urinary ketosteroids throughout the menstrual cycle.

DR. GOLDFIEN: I think it is important to understand the limitations of a particular method and to know when conclusions are justified. We might take, for example, any of the ketosteroid methods. If a normal pattern is obtained, then that's probably acceptable. When you find a very unusual pattern or peak which for some reason is unusual, it isn't safe to assume that you know what that peak represents and that patient, or that specimen, may require further study.

DR. LIEBERMAN: We have just heard Dr. Lipsett uses GLC for the routine measurement of urinary 17-ketosteroids. May I ask if there is anyone else in the audience who is using GLC for normal urinary 17-ketosteroids and how many have they done over the past years?

DR. KEUTMANN: We have performed, I suppose two or three hundred analyses of 24 hr urine specimens from normal individuals. Before discussing some of the results, I would like to mention several points.

It may be well to remember what we are trying to do when we prepare an extract for chromatography of any type. First, we would like the extraction to be complete for the compounds under consideration. While doing this we usually extract large amounts of unwanted materials which interfere with identification or quantitations. To remove non-steroidal contaminants, one usually uses a sodium hydroxide solution for washing. Then, one should remove from the organic solvent all of the water which holds contaminants. This is best accomplished by passing the organic solvent through a bed of sodium sulfate. Those steroids which can be pyrolyzed into 17-ketosteroids can be removed in several ways; we have used a solvent partition. The final problem in GLC is to produce a fraction or fractions which will give a reasonable number of steroid peaks. In the case of 17-ketosteroids we have depended on a Girard fractionation for this.

We feel strongly that an effective internal standard requires its addition at the very beginning of the procedure. After all, we are interested in what the patient excretes, not in what one finds in the extract. This makes things more difficult because an internal standard which will go through the entire procedure is required. One must submit to the whole procedure a series of known mixtures of internal standard and compounds to be measured, then determine the relative ratios of the peak areas, also whether linearity is achieved and over what range this holds. The large problem is always what sort of internal standard is suitable for which determination.

For identification of peaks we still use SE-30 as one of the phases because at the present time we like to chromatograph the steroids both as free compounds and as a derivative, in this case TMS ethers.

No one is more aware than we are of the fact that our method is not as good as it should be. It is too laborious and time consuming. The method which Drs. Lipsett and Kirschner described (J Clin Endocr 23: 255, 1963) is probably preferable at the present time. They accomplished the cleaning up and preliminary fractionation by TLC. The internal standard problem, however, is not entirely satisfactory.

DR. LIPSETT: Dr. Goldfien, how many nanograms of testosterone can you measure?

DR. GOLDFIEN: The area representing three or four nanograms can be measured. The error would be larger than with 15 nanograms and might be 20 to 30%. If our samples contained 0.05 µg per 100 ml of testosterone, we should end up with about 10 nanograms since our recovery is about 50% and we have used 50 ml of plasma. This large quantity would be easily measured and is half as much as we find in the 10 ml samples of male plasma.

DR. EIK-NES: We have used the testosterone method published by my laboratory for the measurement of testosterone in spermatic vein blood of the perfused rabbit testis (Ewing, L. L. and Eik-Nes, K. B., Fed Proc 24, 638, 1965). Better than 400 individual blood samples have been assayed and at the level of testosterone in such samples (.1-.3 µg) the precision of assay is 6-7%. Moreover, the rabbit testis has been infused with different labeled steroid precursors and labeled testosterone isolated and quantitated in spermatic vein blood. In these experiments the method has shown a high degree of specificity and no extra purification steps have been needed to get rid of excess radioactivity. I should, however, like to warn you that some trouble may be seen when infusing the testis with tritiated cholesterol and estimating tritiated testosterone in the spermatic vein blood by our method. On the last thin-layer plate, the one after chloracetylation, a measurable amount of tritium can be found in areas not corresponding in chromatographic mobility to testosterone chloroacetate. One should therefore advise that in experiments studying the conversion of tritiated cholesterol to tritiated testosterone and using our method for testosterone assay, precautions must be taken to remove excess tritiated cholesterol by either paper or column chromatography. Furthermore, freshly prepared, tritiated cholesterol is less apt to interfere in the experiments referred to above than tritiated cholesterol kept around for some time in the cold.

Finally, a comment on Dr. Goldfien's gas-chromatographic tracings. Have you ever tried to move the testosterone peak away from your solvent peaks? I, myself, would be somewhat worried about having to calculate the sample peak on an ascending curve. The sample peak of these tracings looks like an orange floating on the Atlantic Ocean in high wind.

DR. GOLDFIEN: We have reduced the temperature and increased the retention time for progesterone to bring it down off the solvent peak. Hirsute women have enough testosterone to allow us to make the measurement without difficulty on 20 to 30 ml of plasma.

DR. VANDENHEUVEL: It is difficult to measure peaks of this type. The amount and the type of background material that is going through the detector may well affect the response to a steroid. If your peak is on the solvent front, a great deal of material will be passing through the detector and this may affect the response to the component of interest.

So in addition to having a difficult time in measuring this peak, perhaps the peak area is meaningless, anyway.

DR. TOUCHSTONE: I think it would be much more simpler to develop a better column to move the peaks away from the solvent. Certainly quantitation would be improved.

DR. TAIT: As Dr. Brownie mentioned, it is encouraging that the $^{35}$S-thiosemicarbazide and GLC methods are beginning to give the same kind of value for testosterone in blood. The $^{35}$S values for testosterone in female blood are now of the same order of magnitude as measured by

Brownie and Goldfien. However, the values of Lobotsky et al. quoted by Dr. Brownie are uncorrected for blank. The net values have recently been published in J Clin Endocr (Dec. 1964) as 0.034 µg per 100 ml. There is, therefore, still some discrepancy. This could possibly be due to selection of subjects. Also the values obtained by Dr. R. Horton using a $^{35}S$ method for androstenedione in blood are three times higher in females than in males with no overlap of individual values so far. I believe these are in the reverse ratio to that found by Dr. Goldfien. The resolving of these discrepancies is important from the point of view of steroid dynamics.

I think the resolving of these discrepancies is also important in comparing the convenience of methods as suggested by Dr. Lieberman. This is rather difficult unless one has direct experience of them both which seems to be impossible.

If the GLC methods of Brownie, van der Molen and Eik-Nes give correct values which we take to be the lowest without adding further steps of chromatography then there seems to be little question that the GLC methods will prove to be more convenient than using $^{35}S$-thiocarbazide which involves three thin-layer and two paper steps. However, if more steps have to be added to the GLC methods then the matter becomes debatable.

DR. GOLDFIEN: I wonder if I could comment in regard to your question about the comparison of the male and female values for men and women. The level quoted for men is the mean of several determinations on a single pool, not a mean of male values. We don't have enough information on plasma androstenedione but in fact the mean level in women was higher than the mean of the men.

DR. KIRSCHNER: Our laboratory has been involved in developing methods to assay plasma androgens, and we also have explored the potential usefulness of GLC for such measurements. Since the quantity of steroid to be determined in the final aliquot here is often considerably less than 0.1 µg, I think it fair to say that such measurements using hydrogen flame or argon ionization press the limits of these detection systems. In addition, we were concerned that adsorption of the steroid to support or glass may be considerably more significant at these low levels. This might explain why Dr. Goldfien's values for testosterone in women are lower than previously reported non-GLC methods.

We have therefore utilized the double isotope method in connection with GLC for measuring the individual androgens in plasma. This approach used GLC as a high resolution chromatographic column to significantly reduce the number of purification steps, and did not press the limits of the detector system. Adding $^{14}C$-testosterone to measure recovery, the plasma is extracted, and after thin-layer and paper chromatographic separation, the testosterone fraction is acetylated with $^{3}H$-acetic anhydride of known specific activity. Unlabeled testosterone acetate is then added, and the sample is washed to remove unreacted $^{3}H$. For testosterone, a single TLC is performed prior to GLC. An aliquot of the sample is put through GLC (SE-30, 1% at 220 C) and the testosterone acetate fraction is collected on p̄ terphenyl cartridges using the Packard Fraction Collector. A second aliquot is converted to its dimethylhydrazone and this fraction is similarly collected. In each case the $^{3}H/^{14}C$ ratio of the dimethylhydrazone derivative was the same as that of the testosterone acetate. For the determination of testosterone in women, and for plasma androsterone, etiocholanolone and dehydroepiandrosterone, we found it necessary to use a paper chromatographic system after acetylation.

While on the subject of GLC and its use with double isotopic methods, I should like to present an interesting finding which may be of general concern, that is an isotopic effect in GLC systems. In previously described double isotopic methods, one way to assess the purity of the final derivatized product is to compare the $^{3}H/^{14}C$ ratios from several fractions of the chromatographic peak. If these ratios are constant, then the product is considered free of contaminants. When highly purified testosterone-4-$^{14}C$-acetate-1-$^{3}H$ was chromatographed by GLC and serial fractions collected throughout the peak, an orderly discrepancy of $^{3}H/^{14}C$ ratios were noted, as in Fig. 3. The $^{3}H/^{14}C$ ratios were higher at the beginning and end of the peak, and $^{3}H$ counts reached their maximal elution before the $^{14}C$. This phenomenon was observed using both polar and non-polar phases, second derivatives, and was also seen when the labels were reversed (testosterone-7α-$^{3}H$-acetate-1-$^{14}C$). Although this isotopic effect can be eliminated by collecting the entire peak, it would seem that GLC systems in common use have sufficient resolution properties to affect partial separations of $^{3}H$ and $^{14}C$ labeled isotopic species.

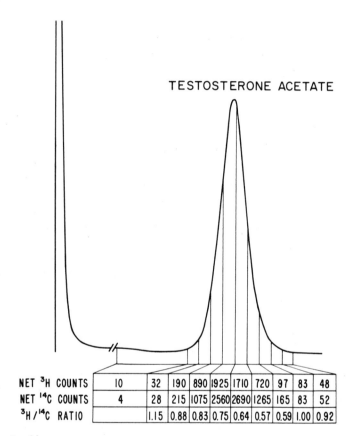

Fig. 3. $^3H/^{14}C$ ratios of fractions collected from GLC of testosterone-4-$^{14}$C-acetate-1-$^3$H. Reproduced by permission J Lipid Research.

DR. WOTIZ: I would like to inject a word of caution here. One of the problems we have always run into is the adsorption of minute amounts of material on the GLC column. This adsorption varies considerably between different types of derivatives and types of stationary phases.

We have recently obtained a flow through proportional counter which was attached to a gas chromatograph. Testosterone acetate, having approximately twenty thousand dpm and a very low mass, less than .002 μg, was chromatographed on a 3% SE-30 column and no radioactivity was detected in the effluent stream. However, the background count jumped from 200 cpm to 600 cpm after this, followed by a gradual decrease.

We felt this was a pretty good demonstration of adsorption of the radioactive steroid onto the column followed by a very slow, gradual elution of the material or possibly the elution of small pyrolytic fragments.

If this were the case, we felt that injection of cold material on the column might desorb some of the radioactivity. We injected one or two micrograms of steroid on the column and obtained peaks of radioactivity of several hundred cpm for each injection of cold testosterone acetate until after about six or eight injections there was a virtually complete return to the original background count.

I don't know that this necessarily needs to interfere with the kind of analysis just described, however, if one chromatographs radioactive substances one must be cognizant of the existence of the adsorption and equilibration effect.

Thus, if one injects a radioactive sample, one must be certain that the residual radioactive material is completely desorbed or one may record radioactivity in a sample for entirely spurious reasons.

DR. VAN DER MOLEN: I would like to discuss further this point of adsorption. It is absolutely necessary to use an internal standard in the technique that we use for the quantitation of testosterone-chloroacetate by electron capture detection. The results obtained by injecting a mixture containing equal amounts of testosterone-chloroacetate and 20β-hydroxy-4-pregnene-3-one-chloroacetate under supposedly identical conditions within a period of several months, clearly showed large differences in the recorded peak areas following electron capture detection. As the most important factors, that might contribute to this variation we considered the accuracy of injection, possible losses due to absorption onto the columns and variation in detector sensitivity. To overcome the factor of injection accuracy, we standardized our injection volume to 5 μl, a volume, that using a 10 μl graduated Hamilton syringe, should be injected with an accuracy better than that obtained for the absolute peak areas. We know, that the sensitivity of the electron capture detector may be severely affected by outside influences. We have every reason to believe, that the variation in detector sensitivity causes the large variation in this series of experiments because the ratios of peak areas remained constant within reasonable limits. Occasionally, we have noted that even the ratios of peak areas would not remain constant; in these instances we thought we had adsorption onto the columns. If we injected a mixture containing a different amount of steroids (for example: 0.005 μg testosterone-chloroacetate and 0.100 μg cholesterol-chloroacetate) that resulted in all but equal peak areas, then the peak area of the amount present in the largest amount was hardly influenced, but the peak area of the compound present in the smaller amount varied greatly between different injections. Whereas good columns will invariably give nice sharp peaks with amounts as small as nanograms, with bad columns we observe the same trouble as Dr. Wotiz describes with his small amounts of radioactive steroids: a continuous wide increased background signal. Consequently such columns are rejected.

Though admittedly we may unknowingly have primed our columns by the first injections that were done to estimate the retention behavior of the compounds, it would still surprise me if such a column would remain good over a period of months and still have, if there is adsorption, the same adsorption affinity for any pair of steroids that we inject. We have examined several other mixtures of steroids purposely combining polar with non-polar steroids, and always found the ratios of peak areas to be constant within 4 to 5%.

DR. WOTIZ: The fact remains that column priming is necessary under certain circumstances. If you take a new column or one which has not been used for a day or more and chromatograph a small smount of free steroid, especially one with several hydroxyl functions, you will not see a normal mass response. This lowered response is not a function of the detector. In fact, the detector never receives this material. The steroid stays on the column. Once the binding sites are saturated, further chromatography of small quantities yields the proper detector derivative.

It appears that hydroxy compounds are strongly adsorbed, acetates are less adsorbed, followed by TMS ethers. Fluoroalkyl derivatives appear to pass through the column in concentrations as low as a few picagrams.

Adsorption seems to occur irrespective of the method of coating and irrespective of the amount of stationary phase. Once the column is primed with standard, injection of the sample to be analyzed results in an equilibration. Thus, if 10% of testosterone is adsorbed, an equal amount is desorbed from the column. Quantitatively no difference can be noted. Problems occur only if one uses radioactive material and are selectively adsorbing some radioactive steroid.

The amount of adsorption we are talking about is relatively small. Thus, if you are injecting one or more micrograms, you will not find any significant loss. If, on the other hand, one injects very small mass, loss from adsorption might become very significant.

DR. VAN DER MOLEN: If you prime your column before a series of testosterone injections with, say, estrone, do you observe the same results? If "priming" would only serve to saturate the irreversible binding sites on the column, any steroid of comparable polarity would be expected to do that job. But if you consider a dynamic equilibrium, with a continuous exchange between this "irreversibly" bound testosterone and the injected testosterone, this certainly would explain why you need to prime with the particular steroid under investigation. So, I wonder if you take estrone or some other steroid, whether this would prime your column for testosterone or not.

DR. WOTIZ: Testosterone is essentially irreversibly adsorbed. The process is reversible over a period of several hours. For a more labile steroid like aldosterone or cortisone the priming effect will only last one or two hours, probably due to the elution of the pyrolyzed fragments.

If you are planning to analyze estrone, then you are going to have to prime the column with estrone. You have to do the same with testosterone. We have not noticed any cross-priming effects.

DR. SMITH: How many different halogen-containing acids have been utilized in evaluating the electron capture detector?

How much of a comparison of materials had been carried out in the search for acids which would exhibit increased specificity, sensitivity or stability? Can someone review for us what has been done?

DR. WOTIZ: I believe Dr. van der Molen might have some information on this. By and large trifluoroacetate is not a good derivative to use in electron capture detection. Fluorine is not too good. Chlorine is better. However, if there is enough fluorine substitution you can make a derivative with a high electron capturing coefficient.

We have been working with the heptafluorobutyrates. There were problems at first with the yields but we are now able to obtain approximately 90% yields of the derivatives of several estrogens and androgens with commercially obtained heptafluorobutyric anhydride. We are able to measure at 20% chart deflection and the most sensitive useful setting of the electrometer, concentrations of $10^{-11}$ gram.

DR. GALLAGHER: Dr. Wotiz, if you have a well primed column for testosterone, and you inject a suitable weight of testosterone with 10,000 counts of radioactivity in that, approximately how many counts do you recover?

DR. WOTIZ: I cannot really answer this because the instrument I mentioned before was received only a few weeks ago and was just put into operation. However, if you will accept a very rough estimate, I will try to answer you. You can recover of the order of 90% on a well-primed column. One thing however, needs comment, with tritiated substances, at the high temperatures used, tritium exchange with the stationary phase occurs.

DR. KLIMAN: I would like to supply some information on this capture mechanism. When 2 µg testosterone acetate were injected into a column which had been idle for several days, successive injections of testosterone acetate cause a gradual increase in the mass peak until it finally became stable in terms of area.

We have also studied this phenomenon using radioactive steroid to answer the question that was recently posed. If the carrier is injected with every isotope analysis the column appears to be clean when labeled testosterone acetate is reinjected.

However, if the column has not been primed, there is significant loss of labeled testosterone acetate when it is injected in millimicrogram quantities.

I would like to call attention to the fact that the procedures that have been described by the two preceding authors are not isotope dilution procedures. It is usually true that you cannot underestimate with isotope dilution. This is only the case if the tracer is carried through the entire procedure. In the GLC methods this is not done, since the isotope is only carried up to the point of GLC. If losses occur in the instrument such as by adsorption onto column, there will be an underestimation of the compound. The fact that low testosterone levels are found in females is not a satisfactory proof of specificity and sensitivity of analysis at millimicrogram levels.

DR. VANDENHEUVEL: I would like to make one comment on the heptafluorobutyrate and other derivatives. You might think by adding more fluorines to the derivative that this would increase the retention time considerably. However, heptafluorobutyrates have retention times similar to those of the trifluoroacetates. As you increase the amount of fluorine and ability to capture, you don't necessarily add to the retention time.

DR. VAN DER MOLEN: I think, that it might be fair to give Dr. Smith who had a specific question about the compounds that have been used for electron capture detection, a more detailed answer. Following the initial reports about the extreme sensitivity of electron capture detection of halogen containing pesticides, we started thinking about the introduction of halogen as such in the steroid molecule. At that time Landowne and Lipsky reported the sensitivity of electron capture detection of cholesterol-chloroacetate and we have followed their approach for other steroids. We have also observed that di- and trichloroacetates of steroids could not be detected in smaller amounts than the monochloroacetates. Recently we have tried to prepare trichloro- and pentachlorophenylhydrazones of steroids, but have thus far been unsuccessful to obtain these derivatives.

Note added in proof: As a result of the Workshop on Gas Chromatography, we obtained from Dr. Menini (Edinburgh) samples of 4-androstene-3,6,17-trione and 4-pregnene-3,6,20-trione. It appears, that the $\Delta^4$-3,6-diketo structure is extremely sensitive in the electron capture detector. Amounts in the order of 0.1 nanogram of these steroids could easily be detected and resulted in nice peaks on our columns. Detection of these compounds might be applied to the estimation of small amounts of dehydroepiandrosterone, 17α-hydroxypregnenolone and pregnenolone, which may easily be oxidized to the $\Delta^4$-3-6-diketo structures (Menini and Norymberski, Biochem J 84: 195, 1962).

DR. NAKAGAWA: In an effort to determine the plasma concentration of testosterone, we tried to make some halo-alkyl derivatives of testosterone which are expected to be more sensitive to electron capture detector.

In this way, we found that there is a strong possibility that monoheptafluorobutyrate of testosterone which was reported by Drs. Clark and Wotiz (Steroids 2: 535, 1963) is the di-heptafluorobutyrate of the enol form of testosterone. This means that the compound which has relative retention time of about 0.6 to cholestane on XE-60 column is actually di-heptafluorobutyrate and the other compound which has relative retention time of about 3.1 to cholestane on the same column is mono-heptafluorobutyrate of testosterone. The molecular weight determination, elementary analysis and infrared spectra of these compounds supported this possibility.

The relative molar responses on peak area basis (peak area = peak height x peak width at half peak height), taking the responses of testosterone monochloroacetate as 1.00, were as follows:

|  | flame ionization | electron capture |
|---|---|---|
| monochloroacetate | 1.00 | 1.00 |
| monochlorodifluoroacetate | 0.74 | 8.00 |
| trifluoroacetate | 1.10 | 0.06 |
| pentafluoropropionate | 0.92 | 1.50 |
| heptafluorobutyrate | 1.35 | 6.25 |
| di-heptafluorobutyrate | 0.70 | 12.90 |
| perfluorooctanoate | 1.01 | 10.38 |

DR. WOTIZ: Dr. Nakagawa is quite correct. The substance Dr. Clark and I reported on was in fact the enolized testosterone diheptafluorobutyrate. The testosterone-17β-heptafluorobutyrate has a longer retention time on XE-60 and a significantly lower electron capture coefficient.

DR. EIK-NES: Drs. Friedlander and Rapp of my group have tested several steroid derivatives for their ability to capture electrons. Some brominated derivatives of the estrogens will indeed do so, but since these compounds are most difficult to prepare and to purify, I have little hope that this observation can be of any help in estrogen methodology. Dr. Rapp has found that acetates of some of the adrenocorticoids will capture electrons and low amounts of such steroids can be determined. It appears that this technique can be used for quantitation of some of the adrenocorticoids found in biological extracts provided that the steroid acetates are rigorously purified before GLC (Rapp, J. P. and Eik-Nes, K. B., in press, 1965).

DR. VAN DER MOLEN: I just want to say that it definitely is a drawback that steroid-chloroacetates have such long retention times on a specific column, like the XE-60 column. I noticed that there has been some confusion among people trying to use these derivatives on 6 and 12 ft columns with high concentrations of stationary phases. They obtained no results and consequently raised their column temperature up to or over, the maximum temperature permissible for the columns (and, by the way, far over the maximum temperature permissible for the available electron capture cells), and, I believe, they run into all the troubles associated with these less than ideal operating conditions. So I want to stress that we use a short (3 ft) column with a 1% stationary phase to obtain good peaks at reasonable times using reasonable temperatures (210-220 C) and I can say that under these conditions we also find for testosterone-chloroacetate a retention of 21.0 relative to that of cholestane.

DR. BROWNIE: I would like to go back, if I may, to the comparison of the values obtained by Dr. Goldfien and ourselves. Has he done the relatively simple experiment of taking the same sample of plasma, say 30 ml of testosterone and dividing it in half and adding about 5 m$\mu$g of testosterone to one-half and then analyzed both samples?

DR. GOLDFIEN: Yes, we have done this. In listening to the previous comments I am wondering whether we may on occasion lose some testosterone because we do not prime the columns, even though we precede and follow every sample by standards, somewhat in the range that we anticipate values. Most often we have mixed samples and there are relatively large amounts of testosterone going on the column in periods of 30 to 60 min.

DR. TAIT: I should like to go back to Dr. Brownie and his discussion of the discrepancy between his values and those of Dr. Goldfien for testosterone in the blood of women. If one measures androstenedione in female plasma, and as this is quite high it can be estimated with some certainty, and then estimates the conversion of androstenedione to blood testosterone by a rather direct method, as Dr. R. Horton has done, one can calculate a minimum value for blood testosterone as 0.02 $\mu$g per 100 ml.

DR. SOMMERVILLE: I should like to make a general comment and, if at all possible, to go into this matter in more detail in the progesterone session. We started in this field 2 1/2 years ago; we had the sort of difficulties with unknown peaks to which Dr. Wilson referred, and 2 years ago changed over to gas-liquid radiochromatography--the system being illustrated in Fig. 4. Until quite recently the proportional counter was used only for the measurement of $^{14}C$. Now, a stream of hydrogen is introduced as shown on the right of the diagram so that $^3H$ can also be determined without any evidence of a memory effect. This system has two main applications:

First, all quantitative methods, e. g., for the determination of steroids in body fluids or tissues, can be controlled by the addition of a $^{14}C$ or $^3H$ steroid standard and the results of all our assays are calculated on the basis of the integrated radioactivity of appropriate standards. Furthermore, with regard to specificity, it is valuable to have the simultaneous recording of the label derived from the combustion of the authentic steroid.

Secondly, the system is extremely useful in the study of steroid biosynthesis, in that relatively crude extracts can be analyzed for their content of radiometabolites, whereas the differential trace obtained by conventional GLC would be off-scale. Other methods for the analysis of such extracts and identification by re-crystallization to constant specific activity involve the addition of expected products whereas unexpected radiometabolites may be revealed by gas-liquid radiochromatography.

Finally, I should like to show two slides on the separation of urinary testosterone and epitestosterone. These results were obtained with a urinary method based upon that reported by Dr. R. V. Brooks (Steroids, 1964). This has been modified to achieve a higher degree of purity in the

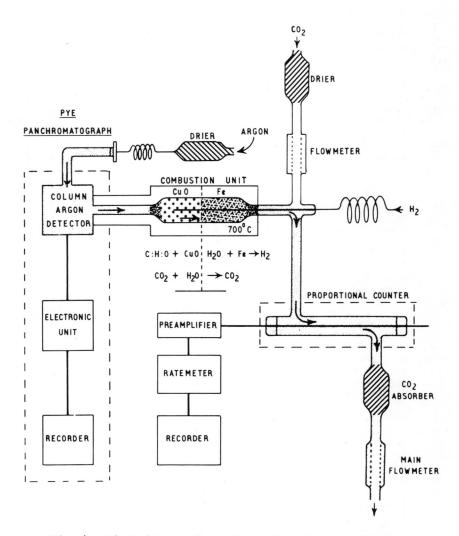

Fig. 4. Block diagram for radiogas-liquid chromatography.

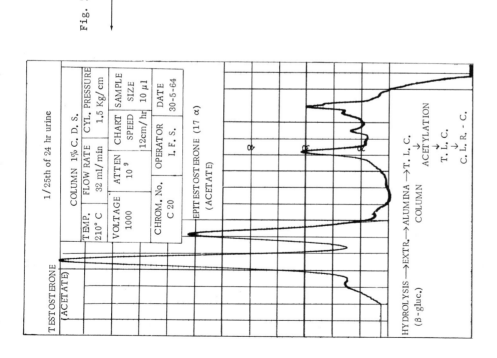

Fig. 5. Separation of urinary testosterone and epitestosterone as their acetates.

## URINARY TESTOSTERONE

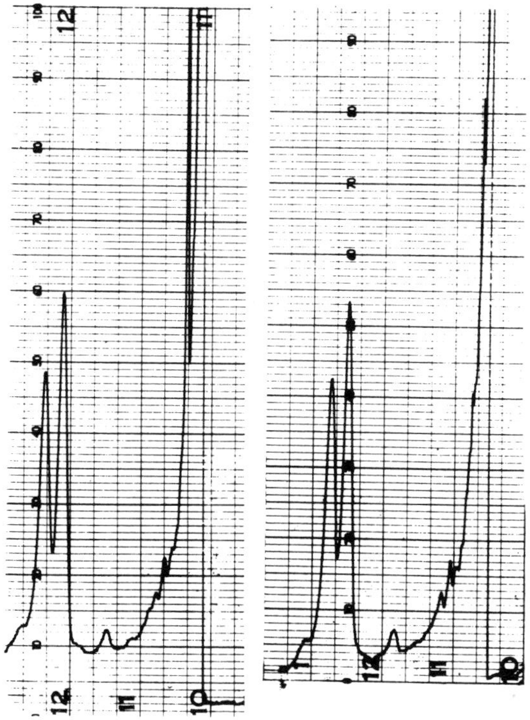

Fig. 6. Duplicate analyses of testosterone and epitestosterone acetate at higher column temperature than in Fig. 5.

final product and there is satisfactory separation of these two compounds on 1% cyclohexane-dimethanol succinate with convenient retention times. These slides were chosen to illustrate the separation and are from the urines of human subjects in whom there was an unusually high concentration of epitestosterone. You will see that the separation is better in Fig. 5 where the column temperature was lower. Figure 6 shows duplicate analyses on a urine sample and illustrates the reproducibility of the method.

# THE ESTIMATION OF UNCONJUGATED CORTISOL METABOLITES IN URINE BY GAS CHROMATOGRAPHY

E. Bailey, Rheumatism Research Unit, Nether Edge Hospital, Sheffield, England

## INTRODUCTION

In our rheumatism research unit we are interested in the excretion of unconjugated corticosteroid metabolites because they may provide some evidence of the extra-hepatic metabolism of corticosteroids. A method was developed for estimating cortisol metabolites in urine (1) which involved a preliminary separation of the extract by paper chromatography (Bush B5 system) and, after bismuthate oxidation, determination by GLC on a neopentyl glycol adipate column. This communication describes an improvement on the method and involves the following procedure.

1. Ethyl acetate extraction after the addition of sodium sulphate.

2. Separation of the extracts into three or four fractions by TLC.

3. Oxidation of each fraction with sodium bismuthate.

4. GLC on an XE-60 column.

5. Quantitation by reference to internal standards added to the urine before extraction.

## MATERIALS

Silica gel G (Merck, A. G. Darmstadt).

Ethyl acetate (B.D.H. Analar Grade) and ethylene dichloride (May and Baker) were each distilled twice.

Methanol and ethanol (Burrough's Analytical Grade) refluxed with 2:4 dinitrophenylhydrazine and conc HCl, filtered and distilled twice.

Sodium bismuthate (B.D.H. Analar Grade).

Hexamethyldisilazane (B.D.H.).

Trimethylchlorosilane (Hopkins and Williams).

$1,2-{}^3$H-cortisol (Radiochemical Centre, Amersham) repurified before use by TLC.

Scintillation fluid was prepared by adding 4 g of PPO and 100 mg POPOP to 1 liter toluene.

## METHODS

*Extraction.* As the excretion of unconjugated steroid metabolites is affected by the rate of urine flow (2) four hour morning specimens were used that had been excreted at 3-4 ml per min. The whole specimen was extracted unless ACTH or cortisol had been given. Two internal standards, prednisone and $1,2-{}^3$H-cortisol were added to the urine for the estimation of recoveries and for quantitation. The method of extraction was essentially that of Frantz, Katz and Jailer (3) for polar corticosteroids. To the urine was added 20% by weight of sodium sulphate and this was extracted twice with equal volumes of ethyl acetate. The extracts were washed twice with 1/20th vol 1 N NaOH containing 15% by weight of sodium sulphate and once with 1/20th vol of 0.5% acetic acid containing 15% by weight of sodium sulphate. The combined extract was finally dried over sodium sulphate and taken to dryness under reduced pressure. The extract was further purified by partition between light petroleum (100-120 C) and 70% aqueous methanol.

**TLC.** The urine extract dissolved in methanol was applied as a band 3 cm from the edge of a 20x20 cm glass plate coated with silica gel G (0.5 mm thickness). The following standards were applied on either side of the extract, cortisone (E), cortisol (F), tetrahydrocortisone (THE), tetrahydrocortisol (THF) and 20β-hydroxycortisol (20β(OH)F). The chromatogram was developed in a constant temperature room at 28 C in a fully saturated tank in the system chloroform:ethanol:water (174:26:2). The solvents were allowed to ascend 15 cm from the origin which took approximately 45 min. The plates were dried in air at room temperature and the standards sprayed with a 0.2% solution of rhodamine 6-G in ethanol. Under UV light (2537Å) the $\Delta^4$-3-ketosteroids were located as dark bands on a yellow background. Corticosteroids with a reducing side-chain were located by a second spray of 0.1% alkaline solution of blue tetrazolium. The urine extract was divided into four fractions which contained the areas located as follows:

fraction (1) from just above E to just above F;

fraction (2) from just above F to just above THF;

fraction (3) from just above THF to just below 20β(OH)F;

fraction (4) from just below 20β(OH)F down to and including the origin. Each fraction was removed with an extractor similar to that described by Matthews (4), eluted with 8 ml ethanol and taken to dryness.

**Oxidation.** The dried extracts from fractions 1, 2 and 3 were shaken in the dark for 1 hr with 1 ml 15% acetic acid and 25 mg sodium bismuthate. Two ml saturated $NaHCO_3$ were added and the reaction mixture extracted twice with 20 ml ethylene dichloride, washed with 4 ml N NaOH and 4 ml distilled water, dried over sodium sulphate and taken to dryness under reduced pressure. Fraction 4 was oxidized with periodic acid instead of sodium bismuthate in order to oxidize to 17-ketosteroids (17-KS) the corticosteroids with a glycerol side chain and not those with a dihydroxyacetone side chain. Each oxidized fraction was taken up into 1 ml ethanol and suitable aliquots used for GLC.

**Preparation of TMS ethers.** Dried aliquots representing half the oxidized fraction of 2 and of 3 were dissolved in 0.2 ml chloroform and 0.2 ml hexamethyldisilazane and a few drops trimethylchlorosilane added. The mixture was allowed to stand at room temperature overnight, or at 60 C for 1 hr, in a stoppered tube. The excess reagents were removed under nitrogen and the TMS ethers taken up into 0.5 ml hexane. Suitable aliquots were used for GLC.

**Isotope recoveries.** To estimate losses during the extraction, thin film chromatography and oxidation procedures, a known amount (trace quantity) of 1,2-$^3$H-cortisol was added to the urine and two aliquots of the final extract of fraction 2 were taken for counting in a Packard Tri-Carb scintillation spectrophotometer.

**GLC.** A Pye panchromatograph equipped with a $^{90}$Sr macro-argon detector was used. The injection port was modified to accommodate a flash heater so that a method of solid injection could be employed as previously described (1). Two U-shaped glass columns 5 ft x 4 mm i d were used. 100-120 mesh acid washed and siliconized celite was coated for one column with 1% XE-60 and the other with 2% SE-52. Coating was effected by the filtration technique and columns were packed under reduced pressure with the aid of a vibrator. The columns were conditioned for 48 hr at 250 C. The life of these columns was preserved by reducing the temperature to 175 C when not in use. Collection of peak material was carried out at room temperature in 10 cm x 1 mm i d glass capillary tubes connected to a heated outlet from the detector (5). Peak areas were determined as the product of peak height and width at half the height.

**Operating conditions.** (A) 1% XE-60 column temperature 240 C; detector temperature 250 C; flash heater temperature 280 C; argon pressure 27 lb psi giving a flow rate of 90 ml per min; detector voltage 1000 v; amplification $10^{-8}$ amps; chart speed 15 in per hr.

(B) 1% XE-60, conditions as in (A) except column temperature 195 C and amplification $3\times10^{-9}$ amp.

(C) 2% SE-52 conditions as in (A) except column temperature 210 C and argon pressure 10 psi giving a flow rate 36 ml per min.

Quantitation. Increasing amounts of the 17-KS standards (see Fig. 2), except for 6β-hydroxycortisone (6β(OH)E), were chromatographed and a mass/area response curve determined for each. These all gave almost identical curves which were linear over the range 0.5 to 2.5 μg. Having demonstrated a similar mass/response ratio in each case, quantitation of cortisol and its metabolites were made by reference to the known quantity of prednisone added to the urine. The necessary corrections for molecular weight differences being made. Quantitation of 6β(OH)F was carried out taking the mass/response ratio to be the same as that of the 17-KS of prednisone.

## RESULTS

In Table 1 are given the $R_f$ values for corticosteroids in the TLC system chloroform:ethanol:water (174:26:2). The separation of these steroids into four fractions is illustrated in Fig. 1. Fraction 1 contains E and prednisone (internal standard); fraction 2 contains F, 20β(OH)E, THE and 6β(OH)E; fraction 3 contains THF, 20β(OH)F and 6β(OH)F-dihydroxy F.

TABLE 1. $R_f$ values of corticosteroids in thin-layer silica gel; chloroform:ethanol:water (174:26:2)

| CORTICOSTEROIDS | $R_f$ |
|---|---|
| Cortisone | .61 |
| Prednisone | .57 |
| Cortisol | .42 |
| 20β-hydroxycortisone | .37 |
| Tetrahydrocortisone | .35 |
| 6β-hydroxycortisone | .34 |
| Tetrahydrocortisol | .24 |
| 6β-hydroxycortisol | .22 |
| 20β-hydroxycortisol | .21 |
| Cortolone | .18 |
| Cortol | .11 |

In Table 2 are given the retention times relative to the 17-KS of E, on both XE-60 and SE-52 phases of the steroid standards used. Also listed are the retention times of the TMS derivatives of six 17-KS relative to cholestane. The gas chromatographic separation, on 1% XE-60 phase, of a mixture of 17-KS standards is illustrated in Fig. 2.

Figures 3 to 5 show the gas chromatograms obtained from each of the three fractions after bismuthate oxidation. The conditions for these chromatograms and for the standards were identical (see under operating conditions (A). In fraction 1 (Fig. 3) peak 3 is derived from cortisone and peak 4 from prednisone (the internal standard). In fraction 2 (Fig. 4) peak 1 is from THE, peak 3 from 20β(OH)E, peak 5 from F and peak 7 from 60β(OH)E. In fraction 3 (Fig. 5) peak 2 is from THF, peak 5 from 20β(OH)F and peak 8 from 6β(OH)F. It will be noted that the retention time of the 17-KS of 6β-(OH)F is undesirably long for a routine procedure. To obviate this a preliminary study has been made of chromatographing this fraction as TMS ethers. Figure 6 shows such a chromatogram. It will be seen that the 17-KS of cortisol has not been affected by the reaction under the conditions described but the retention time for 6β(OH)F derivative has been reduced to 21 min. This procedure not only reduces the retention time but also results in the formation of a less polar derivative which is less suceptible to column adsorption. Gas chromatography of fraction 4 (periodic acid oxidation) showed peaks with retention times identical to those of 11-ketoetiocholanolone, 11-hydroxyetiocholanolone and the 17-KS of 6β(OH)F which were derived, presumably, from cortolones, cortols and 6β,20β-dihydroxy F respectively. Unconjugated urinary metabolites less polar than E on the TLC system were not studied.

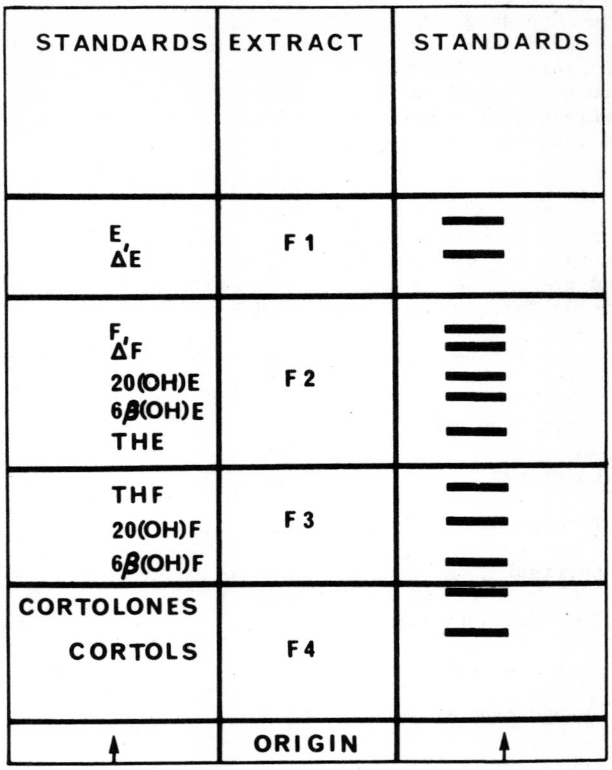

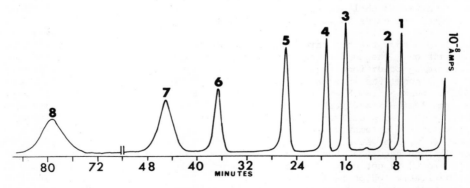

Fig. 1. Thin-layer chromatogram of standards and extract in the chloroform: ethanol:water (174:26:2) system showing the division of the urine extract into four fractions.

Fig. 2. Gas chromatograph of a mixture of 17-KS standards on 1% XE-60 column. The peaks are: 1. 11-ketoetiocholanolone. 2. 11-hydroxy-etiocholanolone. 3. 17-KS of E. 4. 17-KS of prednisone. 5. 17-KS of F. 6. 17-KS of prednisolone. 7. 17-KS of 6β(OH) E and 8. 17-KS of 6β(OH)F. Operating conditions (A).

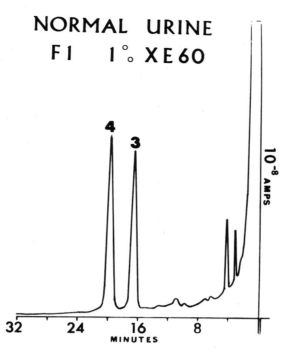

Fig. 3. Gas chromatogram of fraction F1 after bismuthate oxidation. (1/20th of a normal 4 hr urine). Peak 3: 17-KS of E derived from E and Peak 4: 17-KS of prednisone derived from prednisone (internal standard). Operating conditions (A).

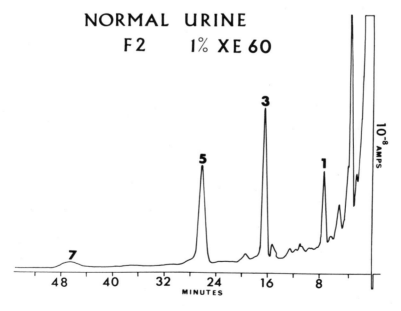

Fig. 4. Gas chromatogram of fraction F2 after bismuthate oxidation. (1/10th of a normal 4 hr urine). Peak 1: 11-ketoetiocholanolone derived from THE. Peak 3: 17-KS of E derived from 20β(OH)E. Peak 5: 17-KS of F derived from F and Peak 7: 17-KS of 6β(OH)E derived from 6β(OH)E. Operating conditions (A).

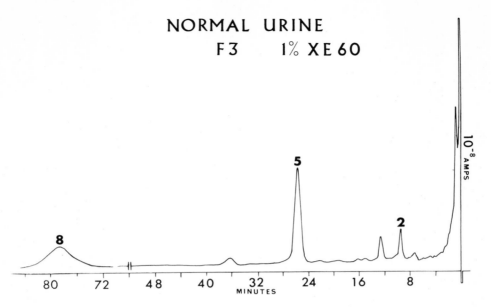

Fig. 5. Gas chromatogram of fraction F3 after bismuthate oxidation (1/20th of a normal 4 hr urine). Peak 2: 11-hydroxyetiocholanolone derived from THF. Peak 5: 17-KS of F derived from 20β(OH)F and 8: 17-KS of 6β(OH)F derived from 6β(OH)F. Operating conditions (A).

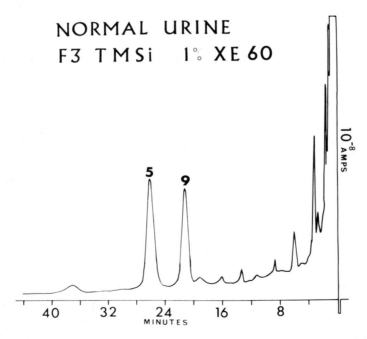

Fig. 6. Gas chromatogram of fraction F3 after bismuthate oxidation and TMS ether formation. (1/20th of a normal 4 hr urine). Peak 9: TMS derivative of the 17-KS of 6β(OH)F derived from 6β(OH)F and Peak 5: 17-KS of F derived from 20(OH)F.

TABLE 2. Relative retention times of steroids.

| STEROIDS | XE-60[a] | SE-52[c] |
|---|---|---|
| 11-Ketoetiocholanolone | .466 | .677 |
| 11-Hydroxyetiocholanolone | .603 | .925 |
| 17-KS of cortisone (andrenosterone) | 1 | 1 |
| 17-KS of prednisone | 1.18 | 1.06 |
| 17-KS of cortisol | 1.55 | 1.50 |
| 17-KS of prednisolone | 2.16 | 1.72 |
| 17-KS of 6β-hydroxycortisone | 2.73 | 1.42 |
| 17-KS of 6β-hydroxycortisol | 5.01 | 2.49 |
| 17-KS of 6β-hydroxycortisol-TMS* | 1.28 | 1.93 |

| | XE-60[b] |
|---|---|
| Cholestane | 1 |
| Androsterone-TMS | 1.01 |
| Etiocholanolone-TMS | 1.25 |
| Dehydroepiandrosterone-TMS | 1.47 |
| 11-Ketoetiocholanolone-TMS | 3.21 |
| 11-Hydroxyandrosterone-TMS | 3.59 |
| 11-Hydroxyetiocholanolone-TMS | 4.36 |

(a) Retention times relative to 17-KS of cortisone = 16.0 min 1% XE-60. 5 ft x 4 mm 240 C, 27 psi.

(b) Retention times relative to cholestane = 5.6 min 1% XE-60. 5 ft x 4 mm 195 C, 27 psi.

(c) Retention times relative to 17-KS of cortisone = 12.8 min 2% SE-52. 5 ft x 4 mm 210 C 10 psi.

* TMS ether.

<u>Specificity of method</u>. Urinary corticosteroids were located on the TLC by reference to appropriate standards rund on either side of the extract. GLC of the three bismuthate oxidized fractions gave in all cases peaks with identical retention times to those of authentic standards. The addition of these standards to the extracts produced in each case increases in peak size without change in peak symmetry. Additional evidence of identification was obtained by trapping peak material and re-running it on the non-polar SE-52 phase. The retention times found remained identical to those of the authentic standards (operating conditions (C)).

Further evidence of structure was derived from the formation of TMS ether derivatives. Thus from the oxidized fraction 2, the peak for the 17-KS of THE disappeared (operating conditions (A)) and was replaced when the column temperature was reduced to 195 C (operating conditions (B)) by a peak with a retention time identical to 11-ketoetiocholanolone-TMS. Similarly from the oxidized fraction 3 the peak for the 17-KS of THF disappeared and 2 peaks with the retention time of 11-hydroxyetiocholanolone-TMS and 11-hydroxyandrosterone-TMS appeared. This would suggest the presence of THF and allo-THF, the 17-KS of these isomers not being separated on the XE-60 phase.

Recoveries. Prednisone and 1,2-$^3$H-cortisol were added to 5 normal 4-hr urine specimens which were examined by the method described. Recoveries of prednisone were 82.2, 80.0, 78.1 and 76.4% (mean 97.8%). The recoveries of radioactivity to the stage of GLC were 80.1, 76.2, 80.2, 80.1, 82.9 (mean 79.9%).

Reproducibility. A number of normal 4 hr urine specimens were combined and divided into six equal aliquots of 450 ml. Each was studied by the method described. The results are shown in Table 3.

TABLE 3. Replicate determinations of corticosteroids in combined 4 hr urine specimens
($\mu$g per 450 ml aliquot)

| Aliquot | 1 | 2 | 3 | 4 | 5 | 6 | Mean | SD* |
|---|---|---|---|---|---|---|---|---|
| Tetrahydrocortisone | 5.6 | 4.6 | - | 6.0 | 5.8 | 5.3 | 5.5 | 0.6 |
| Tetrahydrocortisol | 9.8 | 9.3 | 11.2 | 8.0 | 8.6 | 10.1 | 9.5 | 1.24 |
| Cortisone | 37.3 | 35.8 | 34.0 | 34.7 | 32.1 | 33.8 | 34.6 | 2.3 |
| Cortisol | 25.0 | 20.1 | 19.8 | 22.5 | 22.7 | 21.2 | 21.9 | 1.7 |
| 20$\beta$-hydroxycortisone | 22.1 | 19.1 | 22.7 | 21.0 | 21.3 | 19.8 | 21.0 | 1.5 |
| 20$\beta$-hydroxycortisol | 51.3 | 49.8 | 57.2 | 52.2 | 56.3 | 54.3 | 53.3 | 3.6 |
| 6$\beta$-hydroxycortisol | 39.2 | 39.1 | 45.1 | 39.5 | 35.7 | 41.6 | 40.1 | 3.7 |

*Corrected for small numbers.

## DISCUSSION

Our purpose in developing GLC procedures for measuring cortisol, and some of its metabolites, was to improve on the speed, specificity, accuracy and sensitivity of the procedures we had previously in use, i.e., repeated paper chromatographic separations and the use of non-specific color reactions.

With the method described, it is possible to measure the seven corticosteroids referred to in two to three urine specimens each day. This represents a great saving of time on our previous methods but it should be stated that in our experience the accurate measurement of $\mu$g quantities of relatively polar steroids by GLC cannot be described as a simple procedure.

With regard to specificity, it cannot be said that substances will not be found in urine, in abnormal circumstances, that produce peaks in the vicinity of those being measured. So far, using mainly our previous technique (1), in some 100 or more specimens from normal controls, patients with rheumatoid arthritis and women in the third trimester of pregnancy, no serious interference from unknown substances has been detected.

No attempt has been made to determine how little of each steroid could be measured in the 4 hr morning specimens. The amount would be affected by the quantity of steroid present relative to other substances. In the case of specimens used in this study the minimum quantity of cortisol that could have been measured was approximately 2 $\mu$g per specimen.

The procedure is an improvement on the GLC method previously described (1) in the following respects. 1. The use of sodium sulphate improves the extraction of the polar steroids. 2. The use of TLC reduces the time taken for the preliminary fractionation, allows a more convenient fractionation for our purpose and results in higher recoveries. 3. The use of the XE-60 phase in place of neopentyl glycol adipate results in an improved separation of the steroids. It is also a more thermally stable phase.

Prior to the adoption of the methods described the possibility of separating corticosteroids with a glycerol side chain from those with a dihydroxyacetone side chain by periodic acid oxidation and the formation of carbomethoxy derivatives (6) was investigated. For our purpose the method proved to have more disadvantages than advantages.

## SUMMARY

A method is described using GLC for the measurement of free cortisol and some of its unconjugated metabolites in four hr morning urine specimens. Following extraction the corticosteroids were separated into four fractions by TLC. Oxidation of each fraction converts the corticosteroids to their ketosteroid derivatives which are then estimated by GLC on a 1% XE-60 column. The corticosteroids measured were cortisol, cortisone, 20β-hydroxycortisol, 20β-hydroxycortisone, tetrahydrocortisol, tetrahydrocortisone and 6β-hydroxycortisol. Evidence for the presence of 6β-hydroxycortisone, cortolones, cortols and 6β,20β-dihydroxycortisol is given. The specificity, reproducibility and accuracy of the method are discussed.

## ACKNOWLEDGEMENTS

Thanks are due to the Sheffield Regional Hospital Board for financial support and to Dr. H. F. West for help and encouragement.

## REFERENCES

1. Bailey, E., J Endocr 28: 131, 1964.

2. Murphy, D. and H. F. West, Acta Endocr Suppl 1965 (in press).

3. Frantz, A. G., F. H. Katz, and J. W. Jailer, J Clin Endocr 21: 1290, 1961.

4. Matthews, J. S., A. L. V. Pereda, and A. P. Aguilera, J Chromatogr 9: 331, 1962.

5. Brooks, S. C. and V. C. Godefroi, Anal Biochem 7: 135, 1964.

6. Kittinger, G. W., Steroids 3: 21, 1964.

# GAS CHROMATOGRAPHY OF CORTOLS AND CORTOLONES

R. S. Rosenfeld, Institute for Steroid Research, Montefiore Hospital, New York

In man, the epimeric cortols, cortol, β-cortol, cortolone, and β-cortolone, comprise about 20% of the metabolites of cortisol in the neutral steroid extracts of urine (1). This amount is about two-thirds of the quantity of hormone which is excreted as THE and THF, the metabolites most commonly measured. The absence of chromophoric groups in ring A and of α-ketolic function in the side chain have made the estimation of cortols and cortolones by the usual spectrophotometric methods difficult. Apart from reverse isotope dilution studies after administration of tracer amounts of labeled hormone, cortols and cortolones have been measured by oxidative cleavage of the side chain followed by the Zimmerman color reaction on the 11-oxy-17-ketosteroid produced (2, 3), and by the α-hydroxysteroid dehydrogenase procedure of Hurlock and Talalay (4). It is the purpose of this communication to report some gas chromatographic properties of TMS ether derivatives of epimeric cortols and cortolones and a limited number of analyses of non-ketonic extracts of urine for these compounds. The results indicate that GLC may be successfully used for the determination of urinary cortols and cortolones.

## EXPERIMENTAL

*Gas Chromatography.* Chromatography was carried out with a 1.8 m x 4 mm glass column packed with 100-140 mesh Gas Chrom P coated with QF-1 (3% by weight) which had been in intermittent use for about six months; the column temperature was 220 C during the studies with the pure compounds and was maintained at 232 C during analyses of urinary extracts. The four standard compounds were also chromatographed on a 3% SE-30 Gas Chrom P column at 235 C. The argon pressure was 30 psi and a β-ray detector with a radium source was used.

*Preparation of TMS derivatives.* Method A: From 4-25 mg of material was dissolved in 0.4 ml of pyridine. To this was added 40 μl of hexamethyldisilazane and 4 μl of trimethylchlorosilane and the mixture was allowed to stand at room temperature overnight. Solvent and reagents were removed under a stream of nitrogen at 60 C and the product was diluted to 1 ml with chloroform for chromatography. This is essentially the method reported by Chamberlain, Knights, and Thomas (5). Method B: The same quantity of material was dissolved in 0.4 ml of chloroform to which was added 100 μl of hexamethyldisilazane and 10 μl of trimethylchlorosilane. Drying under nitrogen and redilution to 1 ml was identical with A.

*Urine extracts.* Approximately 15 mg of non-ketonic fraction from enzyme hydrolyzed urine was applied along the origin line of a 20x20 cm plate coated with a 500 μ thick layer of silica gel G. At each side of the starting line was spotted a mixture of the cortols and cortolones as a marker. The plate was first run in ethyl acetate until the solvent front had traveled 10 cm from the origin. It was removed from this system, air dried, and then developed in ethyl acetate: methanol (70:30) until the solvent front had moved 10 cm from the origin. The plate was air dried and the area containing the cortols and cortolones was located by spraying the marker portion with phosphomolybdic acid. The appropriate area was removed from the plate, the silica gel was extracted several times with warm methanol and the extract was filtered through a celite pad. The methanol solution was concentrated and the residue was converted to TMS ether derivatives for GLC (Method A).

For obtaining larger quantities of the TMS ethers of pure compounds, the proportions of steroids to solvent and reagents were unchanged. Of the four TMS derivatives of the epimeric cortols and cortolones prepared by method A, one, cortol-penta TMS ether, has been isolated and characterized (6). The TMS ethers could be hydrolyzed to their parent compounds under mild acid conditions while in the case of buffered hydrolysis, partially hydrolyzed derivatives of cortol and β-cortol were obtained. The partially hydrolyzed derivative of β-cortol was isolated and characterized and proved to be β-cortol-mono TMS ether (6); thin-layer chromatographic mobilities indicated that in the case of cortol, the mono TMS ether had been formed from the penta TMS derivative.

## RESULTS AND DISCUSSION

Figure 1 shows the chromatographic behavior of the epimeric cortols and cortolones on the non-selective phase SE-30 after formation of TMS derivatives by method A or B. In both cases separation was obtained between cortols and cortolones with the cortolone derivative predictably having shorter retention times. Partial resolution was observed with the epimeric cortolones whose retention times were independent of the method of TMS formation. Both cortols emerged as a single peak, but derivatives prepared in pyridine had a longer retention time. These results show that cortol and β-cortol were not completely derivatized when TMS formation was carried out in chloroform solution but that in pyridine fully silylated compounds were formed; under these conditions cortol-penta TMS ether was isolated. Since the 11β-hydroxyl is usually unreactive in derivative formation it is probable that this function was not silylated when the pentol TMS ethers were prepared in chloroform. With the epimeric cortolones, where the 11-ketone is unaffected by the reagents, the same TMS derivatives were formed by either method.

When the mixtures of cortols and cortolones were derivatized by method A and B and chromatographed on QF-1, curves were obtained as shown in Fig. 2. The top curve shows good separation of all four substances where the compounds are fully silylated; with the selective phase QF-1 (5, 7), ketones have longer retention times than do the corresponding 11-trimethylsiloxy compounds. Chromatography of TMS ethers prepared with reagents in chloroform afforded the lower curve (Fig. 2), where the cortolone derivatives are identical with those in the top chromatogram but cortol and β-cortol TMS ethers possess free 11β-hydroxyl groups. When TMS ether mixtures prepared in chloroform were treated with trimethylchlorosilane-hexamethyldisilazane-pyridine, the products chromatographed identically with the upper curve.

It should be noted in Fig. 1 and 2 that chromatograms of the derivatives prepared in pyridine showed some small peaks about 7-8 min and 6 min respectively. Chromatography of the individual constituents demonstrated that a portion of the β-cortol and, to a lesser extent, β-cortolone underwent some alteration during preparation of TMS ethers; this did not occur when the compounds were derivatized in chloroform.

Figure 3 shows chromatographic patterns of the non-ketonic fraction from extracts of urine treated with β-glucuronidase (1). The fractions were reacted with the silylizing agents in pyridine. The top curve represents material from a patient with Cushing's syndrome while the middle curve is from a subject with normal adrenocortical function. They were compared with standards measured during the analysis (bottom curve). There is little doubt that the peaks so designated in the figure represent cortols and cortolones since characteristic retention times were also obtained on SE-30. Comparison of areas under the appropriate peaks with corresponding standards (Fig. 4) have given estimations of excretion rates which varied from 1.0 to 2.6 mg per day in six subjects with normal adrenocortical function and 10, 20, and 40 mg per day in three patients with adrenocortical hyperfunction. While the normal values were in reasonable agreement with results from enzymatic determinations (4), it is felt that more precise information might be obtained by TLC of the non-ketonic fraction or of the very polar material extracted from paper in the separation of $C_{21}$ triols; such experiments are underway and some preliminary results are shown in Fig. 5. In this case, the non-ketonic fraction was chromatographed on a silica gel plate prior to derivatization and GLC. Comparison with Fig. 3 shows that much interfering material has been removed by TLC as evidenced by the flatter base line. As it stands, the procedure is a rapid method for the detection and qualitative evaluation of these hydrocortisone metabolites.

## SUMMARY

Gas chromatographic properties of the trimethylsilyl ether derivatives of cortol (5β-pregnane-3α,11β,17,20α,21-pentol), β-cortol (5β-pregnane-3α,11β,17,20β,21-pentol), cortolone (3α,17,20α,21-tetrahydroxy-5β-pregnane-11-one), and β-cortolone (3α,17,20β,21-tetrahydroxy-5β-pregnane-11-one) have been reported. These compounds have been detected in the non-ketonic extracts of urine by gas chromatography.

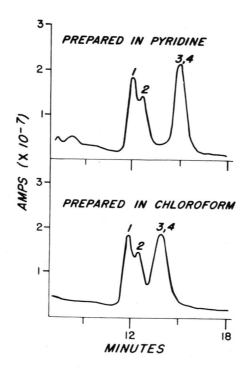

Fig. 1. Chromatography of TMS ether derivatives on SE-30; T = 235 C.
1, cortolone; 2, β-cortolone; 3, 4, mixture of epimeric cortols.
Reproduced by permission of Steroids.

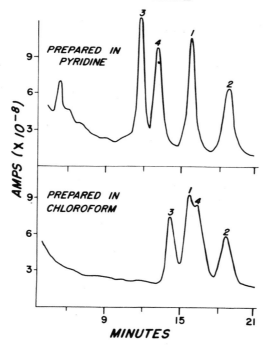

Fig. 2. Chromatography of TMS ether derivatives on QF-1; T= 220 C.
1, cortolone; 2, β-cortolone; 3, cortol; 4, β-cortol. Reproduced
by permission of Steroids.

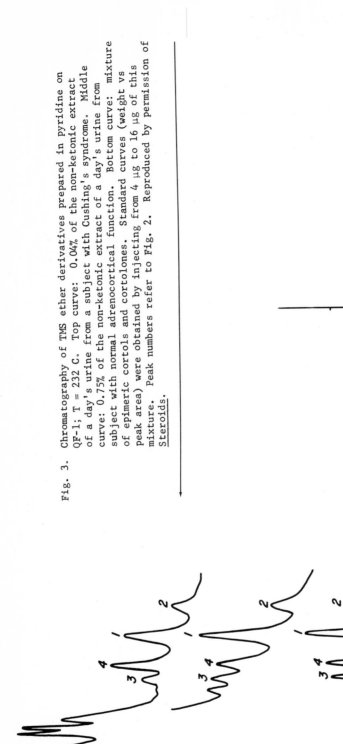

Fig. 3. Chromatography of TMS ether derivatives prepared in pyridine on QF-1; T = 232 C. Top curve: 0.04% of the non-ketonic extract of a day's urine from a subject with Cushing's syndrome. Middle curve: 0.75% of the non-ketonic extract of a day's urine from subject with normal adrenocortical function. Bottom curve: mixture of epimeric cortols and cortolones. Standard curves (weight vs peak area) were obtained by injecting from 4 μg to 16 μg of this mixture. Peak numbers refer to Fig. 2. Reproduced by permission of <u>Steroids</u>.

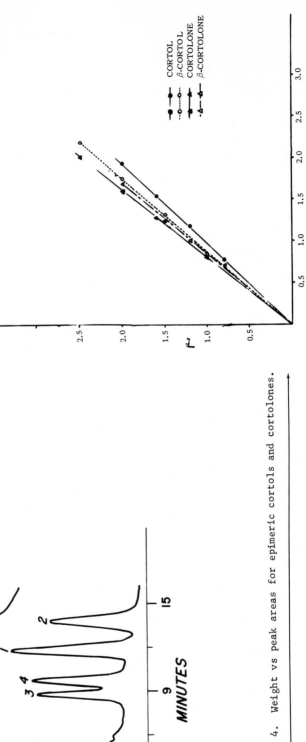

Fig. 4. Weight vs peak areas for epimeric cortols and cortolones.

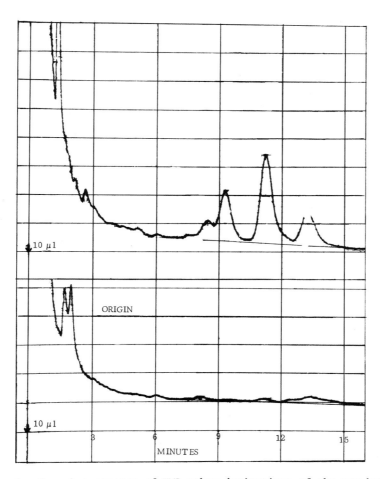

Fig. 5. Gas chromatogram of TMS ether derivatives of the non-ketonic extract after TLC.

## REFERENCES

1. Fukushima, D. K., H. L. Bradlow, L. Hellman, B. Zumoff, and T. F. Gallagher, J Biol Chem 235: 2246, 1960.

2. Michelakis, A. M., J Clin Endocr 22: 1071, 1962.

3. Appleby, J. I., G. Gibson, J. K. Norymberski, and R. D. Stubbs, Biochem J 60: 453, 1955.

4. Hurlock, B. and P. Talalay, J Biol Chem 227: 27, 1957.

5. Chamberlain, J., B. A. Knights, and G. H. Thomas, J Endocr 26: 367, 1963.

6. Rosenfeld, R. S., Steroids 4: 147, 1964.

7. VandenHeuvel, W. J. A., E. O. A. Haahti, and E. C. Horning, J Am Chem Soc 83: 1513, 1961.

# URINARY CORTICOIDS BY GAS CHROMATOGRAPHY

Milton G. Crane and John J. Harris, Department of Internal Medicine, Loma Linda University, Loma Linda, California

This investigation was supported by PHS Research Grant HE-4745, and Career Development Award No. K3-GM-7627 from the U. S. Public Health Service, Bethesda, Maryland.

## INTRODUCTION

The ideal tool in human steroid chemistry would be one that would separate a conglomerate of steroids and render a quantitative answer for each with a minimum of time and effort. We have not expected GLC to do this alone. Our aim has been to utilize preliminary solvent partition extraction; column, paper, or TLC; and GLC for isolation and quantitation of urinary steroids. Our approach has been to analyze the steroids as the free alcohols or ketones rather than as acetates or ethers. We recognize the advantage of derivatives, but we are also concerned about incomplete conversion of the steroids to esters and ethers. This paper is a short presentation of our experience with GLC of adrenal steroids.

## MATERIALS AND METHODS

A. <u>Apparatus</u>. These experiments were made with a Barber-Coleman Model 20 apparatus. The flash heater was remodeled to give a straight line through the flash heater application of the compound on the column, (Fig. 1). A straight length of stainless steel tubing 1/8 in o d by 6 in in length was mounted in a block of aluminum 2 x 2 x 6 in so that the injection could be made through the silicone rubber septum in one side arm of a Swagelok-T fitting, and <u>preheated</u> carrier gas, coming in at right angles from the side arm, carried the steroid through the 1/8 in tubing directly on the column.

Columns were all of coiled 1/4 in o d stainless steel tubing from 2 to 6 ft in length. The indicated acid washed silanized supports were coated with w/w ratios of commercially available SE-30 silicone rubber gum or XE-60 silicone nitrile polymer in the proportions indicated.

Either a radium or a $^{90}$Sr foil ionization detector was used; 1250 v were applied to the detector under all circumstances unless otherwise stated.

B. <u>Preparation of samples</u>. Purified preparations of the various steroid standards were obtained from U.S.P. Reference Standards, Mann Research Laboratories, or from the Upjohn Company. For the quantitation studies suitable quantities were weighed in 0.5 ml polyethylene stoppered glass tubes. Standard steroids were also used for isolation and quantitation of steroids on paper chromatograms.

Aliquots of urine for quantitation of cortisol, cortisone, and aldosterone were extracted with chloroform after pH 1.0 hydrolysis, washed with 0.1 N NaOH, and chromatographed on the first system shown on Table 1. The appropriate zones were eluted, transferred to the 0.5 ml glass tubes, and prepared for quantitation by GLC and by one further paper chromatographic system as previously described (1). 6β-hydroxycortisol was handled similarly except sodium sulfate was added to the urine and ethyl acetate was used for the extraction.

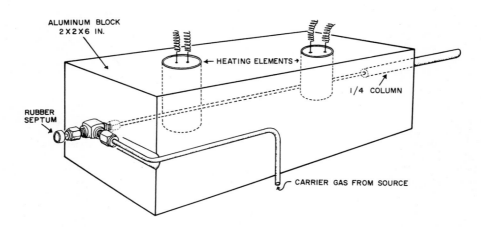

Fig. 1. Redesigned flash heater. Gas preheated in a length of tubing prior to entering the T-fitting. The entire unit encased in glass wool insulation.

TABLE 1. Paper chromatography systems

| System No. | Chemical composition | Ratio | Whatman Paper No. | Running time (hr) |
|---|---|---|---|---|
| 1 | Chloroform:formamide | Saturated | 1 | 4 |
| 2 | Benzene:Lt. Petrol :MeOH:$H_2O$ | 5:5:7:3 | 3 MM | 3 |
| 3 | Benzene:Et. Ac.:MeOH:$H_2O$ | 10:1:10:10 | 2 | 15 |
| 4 | Lt. Petrol:MeOH:$H_2O$ | 10:9:2 | 2 | 14 |
| 5 | Benzene:Tert. Butanol:$H_2O$ | 70:43:86 | 2 | 4 |

Aliquots of urine for quantitation of the tetrahydro metabolites were extracted with ethyl acetate:ethanol (3:1), incubated with β-glucuronidase for 15 hr at 47 C, extracted with ethyl acetate, and initially purified on a silica gel column. The urinary steroids were then chromatographed on one or more paper systems as indicated in Table 1. Appropriate zones were eluted from the papergram, transferred to the small polyethylene stoppered tubes, and prepared for quantitation by GLC.

The urinary 17-ketosteroids were prepared as follows: aliquots of 24 hr urine samples were extracted and subjected to β-glucuronidase hydrolysis in the same manner as that described above for the tetrahydro derivatives. After extraction of the hydrolyzed steroids the remaining 17-ketosteroids conjugates were subjected to solvolysis and extracted, and this extract was combined with that of the tetrahydro derivatives. After this extract was washed with 0.1 N NaOH and passed through a silica gel column for further purification, the residue containing the steroids was chromatographed in system No. 4 of Table 1. The appropriate zones on the paper were eluted and chromatographed on the GLC column.

The steroid standards and urinary extracts were dissolved for injection in acetone, chloroform or in the mixture of tetrahydrofuran and chloroform that would give the most rapid return to baseline of the solvent peak on a particular column. A minimum of three dilutions of the standards were made roughly 1X, 5X, and 10 X the lowest concentration. The urinary extracts were diluted as needed by trial to produce a peak for the desired steroid within the range of the three standard samples.

C. _Injection procedure._ In measuring the volume in the Hamilton microliter syringe for injection, we observed the following technique. A column of solution was drawn into the syringe without air bubbles; and, with the needle tip still in the solution, the plunger was adjusted to the desired volume under magnification. The solution in the needle was then withdrawn into the barrel of the syringe to prevent premature vaporization of the needle contents after penetration of the septum.

An injection of the steroid solution was made as follows. After complete insertion of the needle, the syringe was emptied as rapidly as possible, the electronic injection marker activated, an electric timer started, and the syringe removed all in rapid sequence. The amount of residual solution in the syringe and needle was next measured.

In the quantitation studies 2, 3, or 4 µl of the urine fractions and an identical volume of the three standard dilutions were injected from a Hamilton w/g 10 µl syringe into the flash heater. A minimum of two injections of each sample was made for statistical purposes and the average of the two was graphed to obtain the results in µg per ml.

## RESULTS AND DISCUSSION

A. _Injection studies._ We have had difficulty, as have other investigators, in obtaining reproducible peak heights with the µl syringe. Table 2 presents the comparison for the reproducibility of peak height and peak area (by planimetry) of several consecutive injections of seven different steroids. The column temperature was adjusted to make the retention times nearly the same for all steroids.

TABLE 2. Reproducibility of peak height - one μl per injection.

| Steroid | No. of tests | Column temp C | Average retention time (sec) | Average peak height (cm) | Average percent deviation | Average peak area (cm) | Average percent deviation | Average amp output X 10$^{-7}$ |
|---|---|---|---|---|---|---|---|---|
| Cortisol | 12 | 255 | 331.2 ± 2.2 | 16.1 ± 2.0 | 10.4 | 7.42 ± 0.7 | 8.0 | 19.9 |
| Cortisone | 7 | 243 | 326.9 ± 6.4 | 15.9 ± 2.0 | 10.9 | 8.42 ± 1.5 | 14.1 | 19.7 |
| DHEA | 8 | 227 | 337.8 ± 2.6 | 10.9 ± 1.05 | 7.2 | 4.03 ± 0.44 | 9.1 | 135.1 |
| Androsterone | 8 | 227 | 332.8 ± 3.4 | 8.65 ± 1.00 | 9.8 | 3.39 ± 0.37 | 9.0 | 107.2 |
| Etiocholanolone | 4 | 226 | 328.0 ± 3.6 | 9.20 ± 0.3 | 2.5 | 3.85 ± 0.02 | 4.3 | 114.0 |
| THF | 7 | 242 | 327.6 ± 3.4 | 15.4 ± 1.7 | 9.5 | 8.39 ± 1.3 | 12.3 | 19.1 |
| THS | 6 | 227 | 331.9 ± 6.0 | 19.6 ± 1.4 | 6.5 | 13.0 ± 1.0 | 7.4 | 24.3 |

10 μl Hamilton #701-N
5 μg of steroid
$^{90}$Sr detector - 260 C

5% SE-30 on 60-80 mesh silanized Gas Chrom P
1/4 in x 6 ft stainless steel column
Argon gas - 30 psi
133 ml per min

The average deviation for each steroid from the mean value ranged from 2.5 to 10.9%. The values for the peak areas showed essentially the same degree of error as did the peak height measurement. The variation in the retention time from one injection to the next (whether we measure the retention time from the electrical injection mark or from the onset of the solvent peak) was much less subject to error than was the peak height. The greatest standard deviation in retention time was ± 6.4 sec or ± 2.0%.

It was also noted, for example, that 5 µg of steroid injected in 1 µl of solvent resulted in a greater peak height than 5 µg of the same steroid in 5 µl. This finding, together with the inaccuracy in Table 2, suggested that the needle component, the portion contained in the needle after injection, was boiling out and contributing to a variable extent in the total quantity of solution delivered. (The needles of our Hamilton syringes contained approximately 0.9 µl.)

Figure 2 shows the results obtained when 1.5 to 7.5 µg of THF was injected and the quantity of steroid varied by (a) increasing the concentration of steroid or (b) increasing the volume of injection of a fixed concentration. The left hand portion of the figure is a plot of the peak height per µg of THF delivered <u>excluding</u> the needle component. By direct measurement we found that there was a residual of 0.2 to 0.4 µl of solution in the needle after injection. The right hand portion of the figure is a correction in the estimated micrograms delivered to <u>include</u> 0.7 µl of the needle contents (the needle boil-out portion).

Under certain circumstances the silicone rubber septum of the injection port may affect the column deleteriously. Wotiz has mentioned that the rubber septum may cause spurious peaks (2). Figure 3 presents a sequence of tracings before and after changing the silicone rubber septum. The attenuator setting, column temperature, gas pressure and flow rate, and volume of injection (1 µl acetone) were the same for all of the tests. You will notice that the last solvent peak before the rubber septum was changed and the first one through the new plug resulted in a solvent peak 3 min in duration. The subsequent solvent peaks for the next 48 hr required approximately 15-20 min to return to baseline. It required 72 hr with the column temperature kept at 240 C with a slow flow of argon between analyses for this defect to be corrected. In addition to needle penetration, squeezing or pressing the freshly changed, heated rubber septum will cause this amount or even greater changes in the column characteristics. In some instances the column may be ruined. If the column is disconnected from the flash heater while the septum is being changed or if the septum is not overheated, then we do not have this problem.

B. <u>Retention time of steroids</u>. Wotiz (2) has mentioned displacement of retention time of steroids in crude extracts from the values of pure standard steroid. We have observed a similar situation. Figure 4 and 5 illustrate the effect of mixtures of steroids on their retention times. The left hand portion of Fig. 4 is the THF peak. The peak for THF was sketched in (dotted line) to show the relative retention times of the two on separate injection. The right hand portion of the figure shows the recording of a mixture of THS, THE, and THF. The retention time of THE was shortened and of THF was prolonged when they were injected as a mixture resulting, in this instance, a 20 sec increase in the difference between the two.

DHEA and androsterone behave differently as a mixture as shown in Fig. 5. The left hand portion of the figure shows a peak obtained with androsterone alone with that obtained for DHEA sketched in (dotted line) to show the relative position of the two on separate injections. The right hand portion of the figure shows the chromatogram of a mixture of androsterone and DHEA. As a mixture with androsterone, DHEA has the same retention time as androsterone alone resulting in a single peak.

C. <u>Problems of apparatus</u>. (1) <u>Nonlinearity of apparatus</u>. Kirschner and Lipsett (8) have mentioned that for 17-ketosteroids the graph of peak height versus mass was linear, but the slope became slightly greater at about the 3 µg level. Figure 6 is a plot of amperes output per µg of androsterone, dehydroepiandrosterone, and etiocholanolone. The results for all three are an S-shaped curve with only the results from about 7 to 35 µg linear.

Quantities of steroid greater than approximately 80 µg caused "folding" of the recording at 1250 v. Quantities greater than 80 µg were analyzed at 660 v (Fig. 6) and 1000 v (Fig. 7) and values prorated to 1250 v by comparison of the peak height obtained from measurement of suitable quantities at the two voltages.

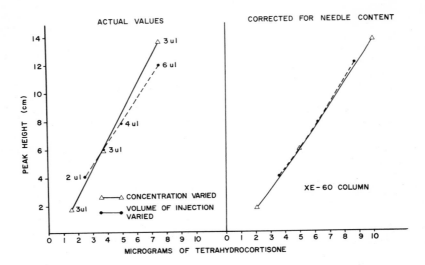

Fig. 2. Graph of peak height per μg of tetrahydrocortisone in two ways. The left hand portion omitting the needle contents and the right hand portion including 75% of needle volume, - the boil-out portion.

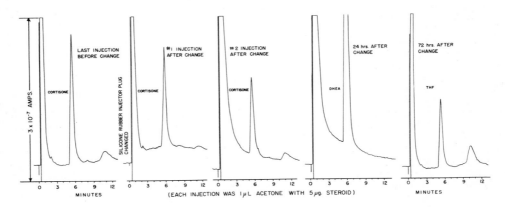

Fig. 3. Effect of changing rubber septum on column characteristics. Column conditions: 5% SE-30 on 60-70 mesh silanized Gas Chrom P in 1/4 in x 6 ft tubing; flash heater 350 C, detector 265 C, column 242 C; argon 133 ml per min at 30 psi.

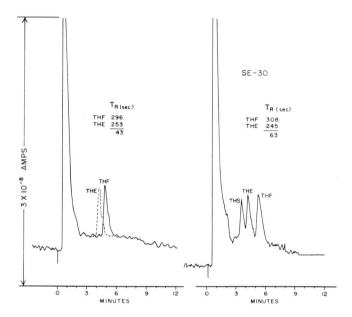

Fig. 4. Shift in retention time of tetrahydrocortisone and tetrahydrocortisol injected separately and as mixtures. Column conditions: 2.5% SE-30 on silanized Chromosorb W in 1/4 in x 5 ft stainless steel tubing; flash heater 315 C, detector 270 C; argon 20 ml per min plus 85 ml per min scavenger at 14 psi.

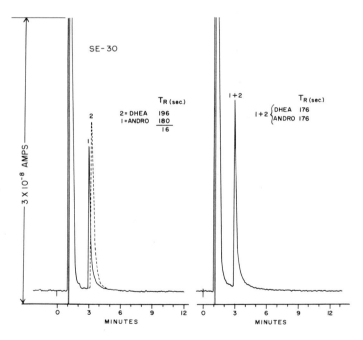

Fig. 5. Shift in retention time of dehydroepiandrosterone. Column conditions: 0.5% SE-30 on Gas Chrom P in 1/4 in x 6 ft stainless steel tubing; flash heater 325 C, detector 252 C, column 230 C; argon 30 ml per min plus 90 ml per min scavenger at 20 psi.

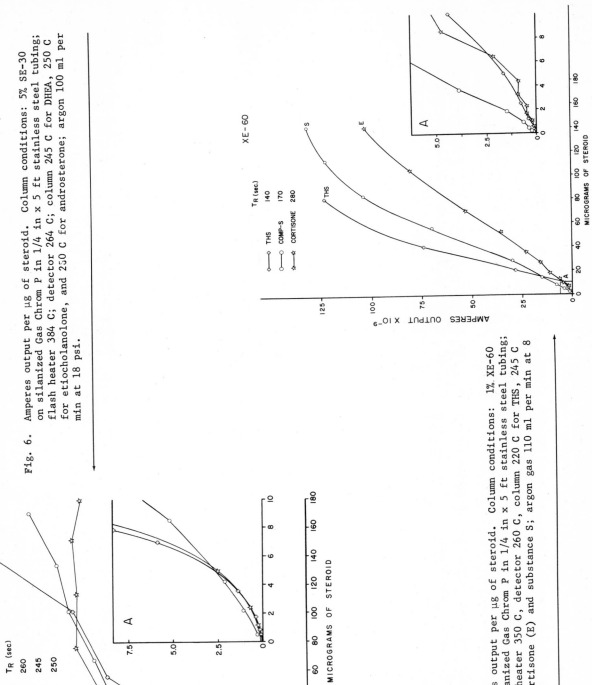

Fig. 6. Amperes output per μg of steroid. Column conditions: 5% SE-30 on silanized Gas Chrom P in 1/4 in x 5 ft stainless steel tubing; flash heater 384 C; detector 264 C; column 245 C for DHEA, 250 C for etiocholanolone, and 250 C for androsterone; argon 100 ml per min at 18 psi.

Fig. 7. Amperes output per μg of steroid. Column conditions: 1% XE-60 on silanized Gas Chrom P in 1/4 in x 5 ft stainless steel tubing; flash heater 350 C, detector 260 C, column 220 C for THS, 245 C for cortisone (E) and substance S; argon gas 110 ml per min at 8 psi.

Figure 7 is a similar graph of values for tetrahydro-11-desoxycortisol, 11-desoxycortisol, and cortisol. The results are quite similar to thos of Fig. 6 although less dramatic. The plot of mass vs. peak height for 11-desoxycortisol gave essentially a straight line from 1.0 to 80 µg of steroid.

Not only were the peak-height response curves non-linear in portions, but there was significant variation from day to day in the slope of the curves and in the sensitivity of the GLC system.

(2) <u>Detector responsiveness</u>. It has been observed by other investigators (4) that the argon ionization detector gave a greater response to some steroids than others. Table 2 introduces our results in this connection. Peak height per µg was used for comparisons among steroids. In order to make the peak heights comparable, column temperature was adjusted so that the retention time was essentially the same for all steroids. The average amperes output for the 17-ketosteroids was 5.5 to 7.0 times that of cortisol, cortisone, and THF. Also the peak height of THS was 1.27 times the peak height of THF. The same general pattern of response is also evident from the data of Fig. 6 and 7 although the retention times and columns differed. These data are consistent with the observation (9) that the steroids with higher oxygen content gave lower ionization detector response. We recognize that there are other factors in addition to oxygen content such as changes in the side chain of the steroid that may affect the response of the detector.

(3) <u>Column specificity</u>. For reference purposes Table 3 presents the retention times of a group of steroids on a 22 in XE-60 column and a 60 in SE-30 column with the conditions kept constant on each column for all the steroids. It is interesting to note the results for some of the steroids and their corresponding 17-KS. THS and etiocholanolone; THE, cortolone, and 3α-hydroxy-etiocholane-11, 17-dione; THF and 3α,11β-dihydroxyetiocholane have essentially identical retention times. This supports the observations by VandenHeuvel and Horning (10) that adrenal hormones of the cortisone group with a 17-hydroxy radical undergo quantitative pyrolysis with loss of the side chain to become the corresponding 17-KS. It is noted that the retention time of THDOC is much longer than that of etiocholanolone. Table 3 also presents the usual specificity of separation that we have obtained on the XE-60 and SE-30 columns in our apparatus.

D. <u>Quantitation of specific steroids</u>. The next few figures are representative recordings of various steroids from 24 hr urine samples for quantitation purposes. The extracts were chromatographed on one or more paper chromatograms as indicated.

(1) <u>Cortisol</u>. Figure 8 shows a recording of standard cortisol on the left and the extract of patient L.M. with Cushing's syndrome after chromatography on Zaffaroni chloroform:formamide system. The injection represents 1/450 of a 24 hr excretion.

Table 4 presents the data for ten measurements of urinary cortisol by GLC in comparison with the values by quantitation after one further paper chromatogram and blue tetrazolium reaction (6). There was no consistent difference between the results. (By direct measurement of known amounts of steroids by blue tetrazolium on the paper chromatogram we have obtained a standard deviation of $\pm$ 9.0% of the correct value.)

(2) <u>6β-Hydroxycortisol</u>. Figure 9 is a recording of the 6β-hydroxycortisol zone from the urine of a patient with Cushing's syndrome due to an adrenalcortical carcinoma. This figure illustrates the suggested technique for the initial injection into the gas chromatogram for quantitation. The curve on the left represents 1/400 of the 24 hr volume. The one on the right is after further sample dilution based upon the results of the initial injection. It is helpful to err on the side of having the steroid too concentrated on the initial injection rather than too dilute. An approximation from the first injection will almost always indicate the proper dilution to bring the sample peak into the range of the standard concentrations. Notice that the attenuator setting was stepped down in sensitivity to obtain the tip of the peak and then returned to the original setting. A measurement of the base of the peak at $3 \times 10^{-7}$ setting must also be obtained for peak height measurement.

(3) <u>Tetrahydrocortisone and tetrahydrocortisol</u>. THE and THF may be quantitated either together as shown in Fig. 10 or separately as shown in Fig. 11 and 12. The left hand portion of these figures presents the standard peaks and the right hand presents the peaks of the urinary extracts after two paper chromatograms. Occasionally, as shown in Fig. 10, the desired steroid appears on the downslope of the solvent peak or adjacent to another peak which decreases the accuracy of quantitation.

TABLE 3. Retention times of standard steroids.

| Steroid | Retention time (sec) | |
|---|---|---|
| | XE-60 | SE-30 |
| Cholestane | 90 | 310 |
| Pregnanetriol | 150 | |
| | 280 | |
| | 450 | |
| Androsterone | 243 | 162 |
| THS | 249 | 147 |
| Etiocholanolone | 250 | 161 |
| DHEA | 260 | 163 |
| Pregnanediol | 290 | |
| Androstenolone | 300 | |
| Testosterone | 460 | |
| Estradiol | 482 | |
| THE | 495 | 175 |
| 17β-estradiol | 507 | |
| | 510 | |
| Cortolone | 515 | 237 |
| Estrone | 530 | 197 |
| DOC | 535 | |
| Compound S | 545 | 205 |
| THF | 635 | 217 |
| | 640 | |
| Progesterone | 730 | |
| THDOC | 735 | |
| Cortisone | 1190 | 231 |
| Cortisol | 1580 | 318 |
| 6β-OH-F | 3200 | 400 |

| | | |
|---|---|---|
| Stationary phase | | |
| Column - 1/4 in stainless steel - length | 22 in | 60 in |
| Support silanized | Gas Chrom-P | Chrom-W |
| Mesh | 80-100 | 80-100 |
| Per cent stationary phase | 7.5 | 5.0 |
| Flash heater C | 370 | 370 |
| Column temperature C | 240 | 258 |
| Argon psi | 4.0 | 14 |
| Column flow ml per min | 94 | 20 |
| Scavenger flow ml per min | 0 | 90 |

TABLE 4. Quantitation of urinary cortisol.

| Patient Sample | Micrograms cortisol per 24 hours | | |
|---|---|---|---|
| | By paper chromatography | By gas chromatography | Percent difference |
| F.S. 1 | 2000 | 2465 | + 23.2 |
| F.S. 2 | 1600 | 1570 | - 1.9 |
| B.W. 1 | 1120 | 1100 | - 1.8 |
| B.W. 2 | 640 | 740 | + 15.6 |
| B.W. 3 | 800 | 870 | + 8.8 |
| B.W. 4 | 480 | 510 | + 6.3 |
| A.B. 1 | 800 | 840 | + 5.0 |
| A.B. 2 | 1280 | 1150 | - 10.3 |
| C.S. 1 | 144 | 123 | - 14.6 |
| C.S. 2 | 120 | 88 | - 26.7 |

Average + 0.36

S D ± 14.8

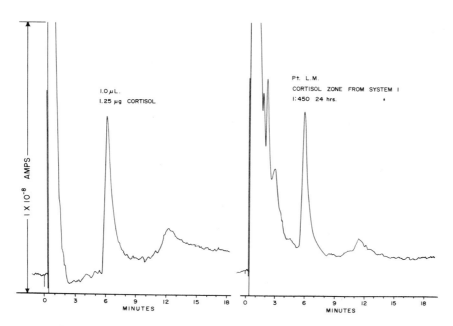

Fig. 8. Quantitation of urinary excretion of cortisol of patient L.M. with Cushing's syndrome. Column conditions: 5% SE-30 on 60-70 mesh silanized Gas Chrom P in 1/4 in x 6 ft stainless steel tubing; flash heater 350 C, detector 263 C, column 253 C; argon 133 ml per min at 30 psi.

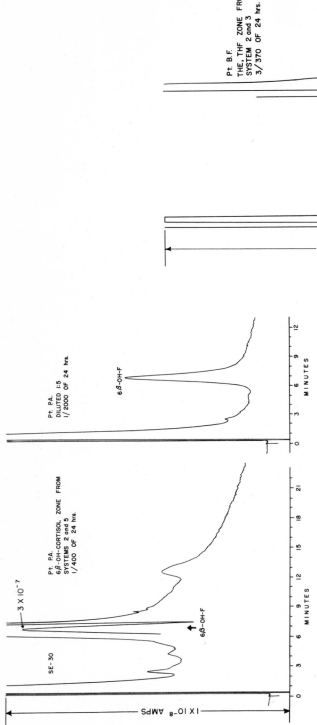

Fig. 9. Quantitation of urinary 6β-hydroxycortisol of patient P.A. with Cushing's syndrome. Column conditions: 5% SE-30 on silanized 90-100 mesh Chromosorb W in 1/4 in x 6 ft stainless steel tubing; flash heater 350 C, detector 265 C, column 258 C; argon 135 ml per min at 30 psi.

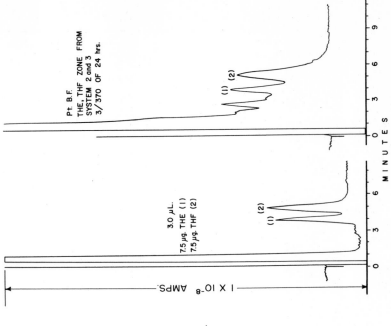

Fig. 10. Quantitation of urinary tetrahydrocortisol and tetrahydrocortisone as a mixture from patient B.F. with Cushing's syndrome. Column conditions: 1% XE-60 on Gas Chrom P in 1/4 in x 30 in stainless steel column, flash heater 365 C; argon gas 110 ml per min at 8 psi.

Fig. 11. Quantitation of urinary tetrahydrocortisone on normal subject G.J. Column conditions: 1% XE-60 on silanized Gas Chrom P in 1/4 in x 24 in stainless steel tubing; flash heater 350 C, detector 265 C, column 245 C; argon gas 100 ml per min at 4.5 psi.

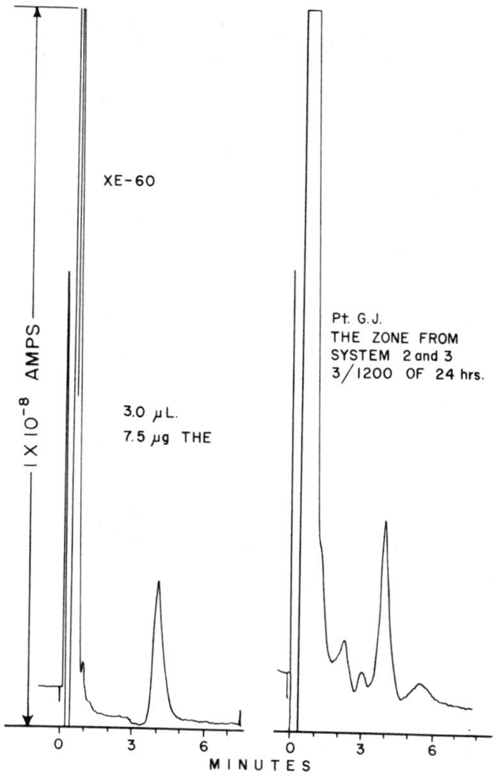

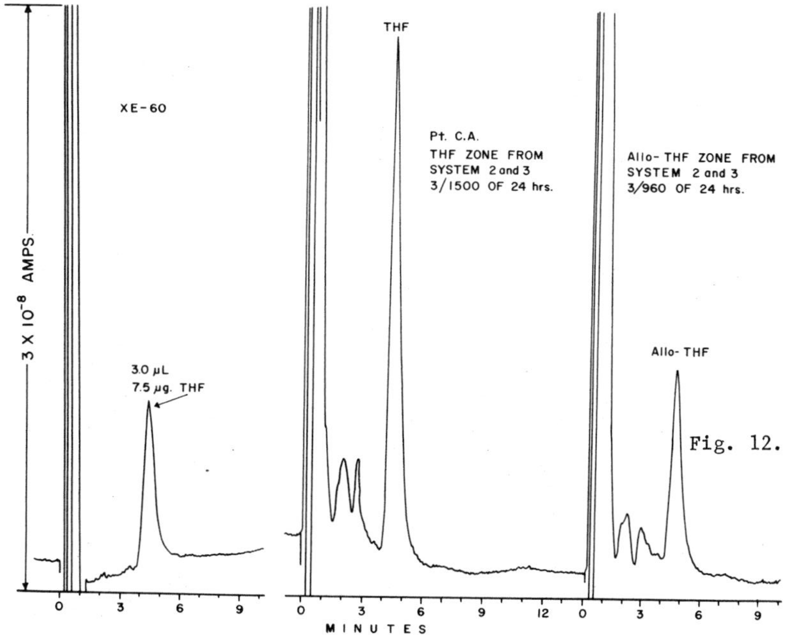

Fig. 12. Quantitation of urinary tetrahydrocortisol and allo-tetrahydrocortisol of patient C.A. with Cushing's syndrome. Column conditions: 1% XE-60 on silanized Gas Chrom P in 1/4 in x 22 in stainless steel tubing; flash heater 365 C, detector 270 C, column 250 C; argon gas 100 ml per min at 4.5 psi.

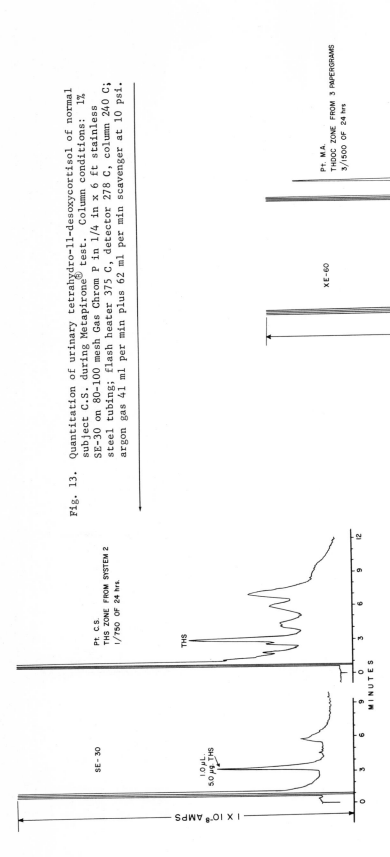

Fig. 13. Quantitation of urinary tetrahydro-11-desoxycortisol of normal subject C.S. during Metapirone® test. Column conditions: 1% SE-30 on 80-100 mesh Gas Chrom P in 1/4 in x 6 ft stainless steel tubing; flash heater 375 C, detector 278 C, column 240 C; argon gas 41 ml per min plus 62 ml per min scavenger at 10 psi.

Fig. 14. Quantitation of urinary tetrahydrodesoxycorticosterone of normal subject M.A. during Metapirone® test. Column conditions: 1% XE-60 on silanized Gas Chrom P in 1/4 in x 22 in stainless steel tubing; flash heater 360 C, detector 265 C, column 245 C; argon gas 180 ml per min at 7 psi.

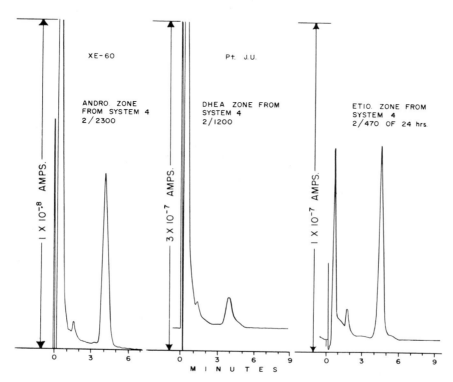

Fig. 15. Quantitation of urinary androsterone, DHEA, and etiocholanolone of normal male adult after one paper chromatogram. Column conditions: 1% XE-60 on silanized Gas Chrom P in 1/4 in x 22 in stainless steel tubing; flash heater 370 C, detector 265 C, column 240 C; argon gas 100 ml per min at 3.5 psi.

(4) _Allotetrahydrocortisol_. Since alloTHF had the same retention time as THF on our XE-60 column, it was necessary to separate the two steroids by paper chromatography and inject them separately. Figure 12 presents the recordings of standard THF and the fractions of the urinary extracts of THF and alloTHF from a patient with Cushing's syndrome after chromatography on two paper systems.

(5) _Tetrahydro substance S_. Figure 13 presents the recordings of standard THS (on the left) and a fraction of a 24 hr urine sample of a normal subject receiving Metopirone®. The patient's sample had been chromatographed on only one paper chromatogram and contains THS and several unidentified peaks.

(6) _Tetrahydrodesoxycorticosterone_. Figure 14 presents the recording of standard THDOC (on the left) and a fraction of a 24 hr urine of a normal subject receiving Metapirone®. The patient's sample had been chromatographed on three paper chromatograms for isolation of the THDOC but still contained several unidentified peaks. (We are indebted to Dr. David Fukushima for the standard THDOC.)

(7) _17-Ketosteroids_. Figure 15 presents the recording of a fraction of androsterone, DHEA, and etiocholanolone after their separation on one paper chromatogram. There is a noticeable lack of interfering peaks in all three steroid samples.

## CONCLUSION

GLC has the advantage of being rapid and precise. Before it will reach its greatest application in steroid analysis improvements are necessary. A more accurate injection system must be devised with the convenience of the microliter syringe. The problem of contamination of the ionization detector with liquid phase must be solved. We have spent considerable time and effort in cleaning the detector. We are also anticipating a time when standardized, tested, column packings will be available more economically.

## ACKNOWLEDGEMENT

The authors acknowledge with pleasure the assistance of David and Clive Possinger, Jr.

## REFERENCES

1. Harris, J. J. and M. G. Crane, _Metabolism_ 13: 45, 1964.

2. Wotiz, H. H., _Recent Progr in Hormone Res_ 19: 92, 1964.

3. Kirschner, M. A. and M. B. Lipsett, _J Clin Endocr_ 23: 255, 1963.

4. Sweeley, C. C. and T.-C. LoChang, _Anal Chem_ 33: 1860, 1961.

5. VandenHeuvel, W. J. A. and E. C. Horning, _Biochem Biophys Res Comm_ 3: 356, 1960.

# A SPECIFIC METHOD FOR THE GAS CHROMATOGRAPHIC DETERMINATION OF ALDOSTERONE AND ADRENOCORTICAL STERIODS: APPLICATION TO ADRENAL VENOUS BLOOD OF VERTEBRATES.

H. Gottfried, Department of Zoology, The University, Sheffield, England.

## INTRODUCTION

This report discusses the preparation of corticosteroid derivatives to facilitate determination of these naturally occurring compounds by GLC. The thermal instability of corticosteroids, and their derivatives which have been studied so far, made it seem unlikely that ester derivatives (acetate, formates, 3,5-dinitrobenzoates) would survive GLC. In view of the results obtained for steroids of the 21,17α-dihydroxy-20-oxo series, by masking the active hydroxy groups by formation of the bis-methylene dioxy derivative it seemed plausible that ether formation at hydroxylated centers might give products suitable for GLC analysis. The TMS ether derivatives of steroids of all other classes have been successfully determined by GLC methods; consequently corticosteroid derivatives of this type were prepared and their GLC properties studied. The derivatives that were investigated satisfied three basic criteria essential for successful GLC (1) easily prepared with a minimum of chemical manipulation; (2) reasonable stability during thin-layer purification, and (3) stability in subsequent GLC.

## MATERIALS

All solvents were of Analar Grade, and were re-distilled before use. Corticosteroids used in these studies were gifts of Ciba Ltd., Basel, Switzerland (courtesy of Dr. R. Neher). Thin-layer chromatographic apparatus and silica gel G (Merck, Darmstadt, Germany), were obtained from the Shandon Scientific Co., London.

## METHODS

Thin-layer chromatographic purifications were carried out on silica gel G plates, spread to a thickness of 250 μ, on 20x20 cm and 10x20 cm glass plates of uniform thickness. A "Unoplan" applicator was used for preparation of chromatoplates. Two solvent systems were employed: light pretroleum:benzene:ethyl acetate (1:1:4) and chloroform:methanol (94:6).

*Preparation of TMS ethers (mg quantities).* The corticosteroid (5-25 mg) was dissolved in one of three solvents: pyridine (Method A), a 1:1 mixture of chloroform-pyridine (Method B) or chloroform (Method C). After solution was achieved, hexamethyldisilazane (500 μl) and trimethylchlorosilane (50 μl) were added in rapid succession, the reaction mixture was tightly stoppered to prevent the entrance of moisture and allowed to stand at room temperature for 12 hr. Following in vacuo evaporation of the volatile reactants, the residue was extracted with ethyl acetate (5 x 1 ml), and the combined organic extracts were concentrated in vacuo and applied as streaks to the origins of one or two chromatoplates, depending on the amount of material used.

*Preparation of μg amounts of corticosteroid TMS derivatives.* The methods employed were entirely similar to those described for milligram quantities; the only variation concerned the amounts of solvents and reactants employed. The corticosteroid (10-100 μg) in solvent (500 μl) - Methods A, B, or C - was treated with hexamethyldisilazane (200 μl) and trimethylchlorosilane (5 drops); the reaction then allowed to proceed, and was worked up as described previously.

*Preparation of 20β-dihydrocorticosteroids.* Five mg of the steroid in 1 ml methanol was treated with a freshly prepared solution of sodium borohydride (15 mg in 1 ml methanol) at 0 C for 1.5 hr. Reaction was stopped by the addition of 50 μl of glacial acetic acid, the mixture evaporated in vacuo, extracted five times with 0.5 ml chloroform/ethyl acetate (2:1 v/v) and the combined extracts concentrated in vacuo. The crude 20β-dihydrocorticosteroid, was subsequently purified by TLC in the system chloroform:methanol (94:6).

*Hydrolysis of corticosteroid TMS derivatives.* The TMS derivatives of cortisol prepared by methods A, B, and C were hydrolyzed by treatment with 80% ethanol/concentrated hydrochloric acid/pyridine (250 μl:50 μl:500 μl) at room temperature for 2 hr; the reaction products were isolated by extraction with ethyl acetate (7).

Aldosterone TMS derivatives were hydrolyzed by dissolving the derivative in 70% methanol (500 μl), and allowing the mixture to stand at room temperature for 12 hr. The products were isolated by evaporating the reaction mixture to dryness *in vacuo*, and dissolving the residue in methanol. Identification was based on subsequent TLC of the methanolic extract.

Corticosteroid TMS derivatives were detected on chromatoplates by viewing under UV light (254 mμ source). Elution of these derivatives was accomplished with ethyl acetate: their hydrolysis products were also detected by viewing chromatoplates under UV light, and then confirming the position of UV absorbing products by spraying with the sulphuric acid-ethanol (1:1) reagent, and observing the UV fluorescence produced after heating.

*Gas chromatography of corticosteroid TMS derivatives.* An F&M 400 Biomedical Gas Chromatograph was employed for GLC investigations. The stationary phase consisted of SE-30 (3.8%) deposited on siliconized support material Diatoport S (80-100 mesh), contained in a 4 ft U tube. A hydrogen flame ionization detector was used, and results were recorded graphically on a 1mV Minneapolis Honeywell Recorder at a chart speed of 12 in per hr. The eluant gas was nitrogen at an inlet pressure of 30 psi which by means of rotameter control maintained a flow rate of 45 ml per min. Throughout these investigations the column was maintained at a temperature of 275 C, the detector and flash heater unit at 315 C respectively.

Corticosteroid TMS derivatives were dissolved in ethyl acetate for GLC in order to avoid the onset of hydrolytic decomposition. Samples were injected by means of a Hamilton Microsyringe, and the sample volumes varied from 0.5-5.0 μl.

Areas under peaks were assessed by the product of peak height and width at half peak height values. Column efficiencies were 2840 and 2688 theoretical plates for the TMS derivatives of 20β-dihydrocorticosterone and aldosterone derivative 1, respectively.

## RESULTS AND DISCUSSION

Initial studies on the preparation of aldosterone TMS derivatives by Method A, followed by thin-layer purification, indicated that the major reaction product was a non-polar ($R_f$ 0.93 in both systems employed) UV absorbing material; a minor UV absorbing component ($R_f$ 0.53) was also observed. The yields of the major and minor aldosterone TMS derivatives in this reaction were 50% and 8% respectively. Subsequent GLC examination of the major product indicated a main peak $R_t$ (cholestane) 2.33 and a minor peak $R_t$ 3.08; the minor product appeared as two peaks $R_t$ 1.86 (aldosterone-γ-lactone) and 2.28 respectively. This evidence suggested that the major TMS product of aldosterone might have undergone hydrolysis before the onset of GLC. The main product was subsequently identified as aldosterone.

TLC of the major aldosterone product in the light petroleum:benzene:ethyl acetate system showed three UV absorbing components, ($R_f$ 0.05, 0.46, 0.57), and no trace of the original non-polar material subjected to GLC. The polar compound ($R_f$ 0.05) was shown to be aldosterone. Hydrolysis of the other materials with 70% methanol, followed by TLC gave aldosterone as the sole product. A similar result was obtained when the minor aldosterone TMS derivative was hydrolyzed. These data were interpreted as indicating firstly, that no basic alteration had occurred in the aldosterone nucleus as a result of TMS formation, and secondly that hydrolytic decomposition of the major TMS derivative occurred at some stage before or after TLC purification, into aldosterone via the minor TMS product. In order to elucidate these results, a further series of preparations were undertaken.

In three separate experiments aldosterone (100 μg) was derivatized as described previously. All the reaction mixtures were worked up after 12 hr, and chromatographed immediately in the chloroform:methanol system. Viewing the plates under UV light indicated conversion to a single absorbing component whose $R_f$ value was identical to that of the major product obtained previously; the yields appeared near-quantitative. The material from one chromatoplate was eluted immediately, and subjected to GLC; a single peak was observed $R_t$ 2.33. A peak area/mass response graph was constructed and appears in Fig. 1.

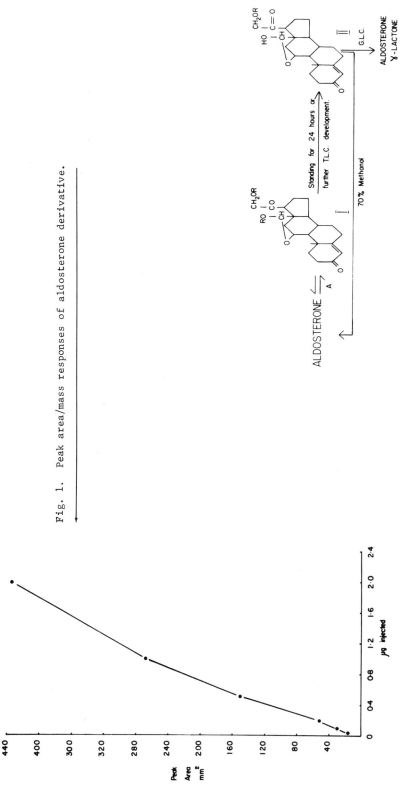

Fig. 1. Peak area/mass responses of aldosterone derivative.

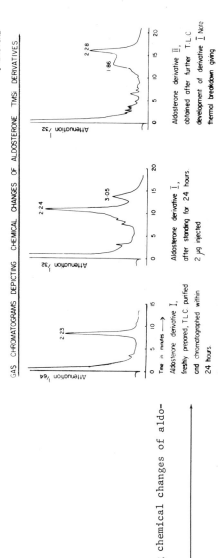

Fig. 2. Gas chromatograms depicting chemical changes of aldosterone, TMS derivatives.

Following GLC, the residual material was hydrolyzed with 70% methanol for 12 hr and yielded aldosterone only. The material on the other two chromatoplates was treated in the following manner: one chromatoplate was allowed to stand at room temperature for a further 12 hr after which the UV absorbing material was eluted with ethyl acetate, and then made up in an appropriate volume of this solvent for GLC analysis; the UV absorbing material on the third chromatoplate was eluted immediately after completion of chromatography, evaporated to dryness, allowed to stand in a stoppered flask for 12 hr and then dissolved in ethyl acetate for GLC studies. The results of GLC analysis indicated that the material left on the chromatoplate for 12 hr after TMS formation was decomposing; this was based on the fact that while the main peak due to the derivative itself was still discernible there were other peaks evident on this chromatogram which did not appear on that of the freshly prepared derivative described earlier. Further investigation of this partially decomposed aldosterone derivative by TLC indicated formation of the UV absorbing component $R_f$ 0.57, whose unsuitability for GLC has already been discussed. The TMS derivative which had been stored in the dry state for 12 hr after TLC purification was also analyzed by GLC. The results clearly indicated complete decomposition to the minor product isolated from the large-scale preparation; relative retention times of the two peaks were 1.86 (aldosterone-γ-lactone) and 2.28 respectively. TLC of the residual material confirmed the decomposition, since its $R_f$ value was 0.57; hydrolysis yielded aldosterone as the sole product. The results of these interconversions and the gas chromatograms associated with them appear in Fig. 2.

In summary, therefore, the following points are emphasized: 1. The reaction leading to aldosterone derivative 1 (Fig. 2) must be completed within 12 hr, and the product purified by TLC immediately. 2. Provided the elution is accomplished within 12 hr of TMS formation, subsequent GLC yields a single peak due only to the derivative itself. 3. The peak area/mass responses of aldosterone derivative 1 are very high, and in practice permits the detection of 0.025 μg, these results being based on preparations ranging from 10-100 μg, which gave near-quantitative yields. 4. Structural designation to aldosterone derivative 1 as the di-18,21-TMS-ether derivative, and to derivative 2 as the mono-21-TMS-ether, are based not only on their respective TLC and GLC properties, but that both compounds yield only aldosterone on hydrolysis. These results also appear in Table 1. 5. The di-TMS-ether of aldosterone cannot be eluted at column temperatures under 275 C necessitating high injection port and detector temperatures, and indicating the high thermal stability of this derivative. In view of these findings this method seemed applicable to the determination and identification of naturally occurring submicrogram amounts of aldosterone.

The TMS derivatives formed by cortisol (10-25 mg), when reacted by Methods A, B, and C were investigated. These large-scale preparations, employing Methods A and B, each afforded three cortisol TMS derivatives whose TLC and GLC properties were identical. All these reaction mixtures had been allowed to stand for 24 hr before they were extracted and purified by TLC. Their GLC properties are recorded in Table 1. From these data two rather interesting points emerged: firstly, all these UV absorbing compounds gave two peaks on GLC analysis, and the peak with the shorter retention time appeared common to all of these products; secondly, the polarity of the major products isolated from the reactions was higher than expected. Since two peaks were obtained on GLC of these compounds it seemed probable that thermal decomposition was occurring; the peak common to these materials $R_t$ 1.03 was identified as 11β-hydroxyandrostenedione-TMS ether by an independent preparation (Method A - see Table 1). The fact that the polarity of the major products prepared by Methods A and B was higher than expected (TLC-$R_f$ and GLC-$R_t$ values respectively given in Table 1), and the isolation of minor amounts of lower polarity materials, tended to suggest that the major products were not formed directly from cortisol, but resulted from the more highly substituted products. Therefore, the possibility existed that the major cortisol-TMS derivatives isolated from these reactions, were formed as a result of <u>in situ</u> decomposition of fully etherified, and consequently less polar derivavie. Reference to Table 1 indicates that the increase in retention times of the derivatives parallels their increment in chromatographic mobility; this confirms the increase in molecular weight expected with increasing substitution. The position of one etherified group in the cortisol molecule at position 11 was confirmed in all these derivatives by the identification of the thermal breakdown product, 11β-hydroxyandrostenedione TMS ether. However, the possibility existed that disruption of the cortisol nucleus could have occurred during derivative formation. The integrity of the cortisol nucleus was confirmed in all these derivatives by hydrolysis, which afforded the parent corticosteroid in good yield.

The information gained from the large-scale preparations served to suggest that the most stable product formed (major product 5 and 8 - Table 1), would also be suitable for GLC analysis. Investigations were then undertaken to obtain quantitative yields of this compound from cortisol, and to gain more information about the influence of reaction time, and the nature of the solvent used, upon the products formed.

TABLE 1. The gas chromatographic properties of corticosteroid trimethylsilyl-oxy ethers.

| Steroid | Reaction type | Properties on silica gel G in the system light/petroleum/benzene/ethyl acetate (1:1:4) | | Retention time relative to cholestane (1.00) | Hydrolysis products |
|---|---|---|---|---|---|
| | | No. of UV absorbing components | $R_f$ value | | |
| Aldosterone | A | 2 | 0.93 (major product) (1) | 2.33 | 70% methanol for 12 hr - aldosterone |
| | | | 0.53 (2) | 1.86, 2.28 | |
| Cortisol | A | 3 | 0.94 (3) | 1.04, 2.38 | Conc. hydrochloric acid/pyridine/80% ethanol at room temperature for 2 hr - Cortisol |
| | | | 0.84 (4) | 1.05, 1.97 | |
| | | | 0.56 (major product) (5) | 1.02, 1.90 | |
| | B | 3 | 0.95 (6) | 1.01, 2.33 | |
| | | | 0.82 (7) | 1.04, 2.06 | |
| | | | 0.63 (major product) (8) | 1.03, 1.09 | |
| | C | 2 | 0.95 (9) | 0.79, 0.89, 1.74 | |
| | | | 0.56 (major product) (10) | 0.88, 1.73 | |
| | Aliquots of major products from A, B and C mixed | | | | |
| 11β-hydroxy-androst-4-ene-3,17-dione | A | - | - | minor peak 0.91<br>major peak 1.03 | |

Reaction types: A: Hexamethyldisilazane/trimethylchlorosilane/pyridine; B: Hexamethyldisilazane/trimethylchlorosilane/chloroform/pyridine; C: Hexamethyldisilazane/trimethylchlorosilane/chloroform. Structural designation: (1) 21-trimethylsilyl-oxy-pregn-4-en-18-ol-11,18-hemiacetal-trymethylsilyl-oxy-3,20-dione. (2) - . (3) 3ξ,17α,11β-tri-(trimethylsilyl-oxy)-pregn-4-ene-3,20-dione. (4) 17α,11β-di-(trimethylsilyl-oxy)-21-hydroxy-pregn-4-ene-3,20-dione. (5) 21,11β-di-(trimethylsilyl-oxy)-17α-hydroxy-pregn-4-ene-3,20-dione. (6) -(8) as for reaction products from type A. (9) - . (10) 21-mono-(trimethylsilyl(oxy)-17α-11β-dihydroxy-pregn-4-ene-3,20-dione.

Cortisol (100 µg amounts) were derivatized by Methods A, B, and C; the products were purified in the manner described. Methods A and B yielded two UV absorbing products; derivative 1 ($R_f$ 0.95) and 90-95% yield and derivative 1a ($R_f$ 0.84) in trace amounts. GLC investigation of derivatives 1 and 1a, indicated that they were, respectively, identical with compounds 3 and 4 (Table 1); this was confirmed by development of mixed chromatograms. Hydrolysis of these derivatives with ethanol/pyridine:hydrochloric acid yielded cortisol. When derivative 1 was allowed to stand in the dry state for 12 hr, or hydrolyzed with 70% methanol, it was quantitatively converted to a UV absorbing material ($R_f$ 0.56), derivative 2; the latter compound was found to be identical to compound 5 (Table 1) on the basis of individual and mixed gas chromatograms. Derivative 2 was converted to cortisol by the procedure described for its parent compound.

The cortisol TMS products obtained by Method C differed from those prepared by Methods A and B. The major UV absorbing material obtained from this reaction, derivative 3, gave two peaks on GLC $R_t$ 0.88 (identified as 11β-hydroxyandrostenedione) and $R_t$ 1.73 due to the undissociated derivative. Since derivative 3 dissociated partially into 11β-hydroxyandrostenedione during GLC, it seemed likely that formation of cortisol-TMS ethers in chloroform did not lead to etherification of the 11β-hydroxy group. This difference between derivatives 2 and 3 was confirmed by mixed GLC, and the results are recorded in Table 1.

The interconversions between the various cortisol-TMS derivatives, and the structural changes believed to be associated with them appear in Fig. 3. The chromatograms obtained from derivatives 1a, 2 and 3, which are respectively identical to compounds 4, 5 and 10 whose structural designations appear in Table 1, are shown in Fig. 4. Structure assignments are made on the basis of hydrolytic degradation, TLC and GLC properties.

The course taken by the reactions leading to the cortisol TMS derivatives that have been described, indicate the influence of both reaction time and solvent upon the final products. If the reactions are carried out in pyridine, or 1:1 pyridine/chloroform during a 12 hr period, an almost quantitative yield of the 11β,17α,21-triTMS derivative is obtained; although this compound is unsuitable for GLC analysis due to its low volatility, it can be quantitatively converted to the 11β,21-dietherified derivative by mild hydrolysis. The latter compound (derivative 2) appeared suitable for the gas chromatographic determination of cortisol, despite the fact that it underwent partial thermal dissociation into 11β-hydroxyandrostenedione TMS ether. As already indicated derivative 2 gives two very well defined peaks, dissociation product and the intact derivative; it was possible in practice to detect 0.025 µg of this cortisol derivative. For the purposes of quantitation it was necessary to determine the extent of the thermal dissociation, and if this proved to be reasonably constant then it would necessitate only the quantitation of the 11β-hydroxyandrostenedione-TMS peak to determine the total amount injected. When 11β-hydroxyandrostenedione was converted to its TMS derivative by Method A, using 100 µg of starting material, subsequent GLC analysis indicated 80% formation of the derivative and 20% unchanged steroid. A peak area/mass response graph was constructed on this basis and appears in Fig. 5. The dissociation of derivative 2 was determined and found to average 5% for repeated injections of 3 and 4 µg amounts of derivative 2. This derivative can be stored under reasonably anhydrous conditions for four months without undergoing any chemical deterioration as determined by GLC and TLC.

The influence of the solvent upon the type of substitution which occurs in the cortisol molecule during TMS formation can be summarized as follows: the reaction in pure chloroform leads mainly to C-21 substitution; a fully substituted derivative (11β,17α,21) results from reaction in pure pyridine or 1:1 pyridine/chloroform. No TMS formation occurred when 11β-hydroxy-androstenedione was reacted in chloroform, whereas in pyridine an 80% yield of derivative was realized. These results agree very closely with those of another worker, who also reported that no 11β-hydroxy-etherification occurred when the isomeric cortols were reacted in chloroform, but substitution at the 11β-hydroxy group took place in the presence of pyridine (7).

Other corticosteroid TMS derivatives which were studied during the course of these investigations, included those of cortisone, 11-deoxycorticosterone, 11-dehydrocorticosterone and corticosterone. In all cases the resultant products formed in low yield were shown to consist of two or three component mixtures by TLC. GLC studies of these materials again indicated their low volatility, and like their aldosterone and cortisol counterparts, they could not be eluted at column temperatures below 275 C. However, the direct preparation of TMS derivatives from these steroids was abandoned, since the gas chromatograms obtained showed clear evidence of complete thermal breakdown; materials which had been repeatedly purified by TLC and shown to be homogeneous, decomposed during GLC to give three main peaks. Obviously it was impossible to assign any peak as being that due to the derivative.

Fig. 3. Chemical modifications of cortisol TMS derivatives.

Fig. 4. Gas chromatograms of cortisol TMS derivatives.

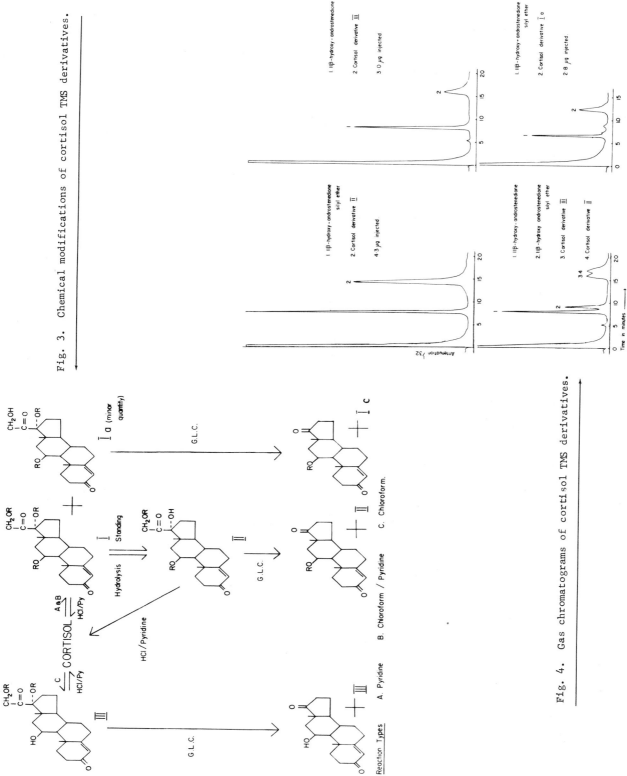

The thermal stability of corticosteroid TMS derivatives seemed to be increased when at least one oxo group in the parent nucleus was reduced to a hydroxy function prior to TMS formation. This evidence was obtained by studying a model compound, 11β,20β,21-trihydroxy-4-pregnene-3-one, prepared by sodium borohydride reduction of 11-deoxycorticosterone. TMS ether formation and GLC of thin-layer purified material, indicated thermal stability and homogeneity, since a single sharp peak was obtained. Therefore, the GLC properties of the 20β-dihydro-TMS derivatives of other corticosteroids were studied.

The 20β-dihydro derivatives of corticosterone, 11-dehydrocorticosterone, 11-deoxycorticosterone and cortisone were prepared by reduction with sodium borohydride by the method already described. The reduction products so obtained were purified by TLC and afforded homogeneous UV absorbing material in each case; the yields in several preparations from each steroid ranged from 75-85%. The reduced steroids (1 mg amounts) were converted to their TMS derivatives by Method C; this avoided etherification at the 11β-hydroxy group in the case of corticosterone. After derivative formation was complete (12 hr), the reaction mixtures were extracted by the method indicated earlier, and the TMS derivatives purified by TLC. In all cases single, non-polar UV absorbing materials were detected on the chromatoplates. GLC analysis of 20β,21-dihydroxy-4-pregnene-3-one, 20β,21-dihydroxy-4-pregnene-3,20-dione, and 11β,20β,21-trihydroxy-4-pregnene-3-one TMS derivatives gave single, sharply defined peaks, whose respective retention times relative to 5α-cholestane were 2.03, 2.17 and 3.10. The increase in retention time from the TMS derivative of 20β,21-dihydroxy-4-pregnene-3-one to that of 11β,20β,21-trihydroxy-4-pregnene-3-one reflects the expected increment due to differences in their respective molecular weights. The peak area/mass responses of these three corticosteroid derivatives was very high and permitted a sensitivity limit of 0.025 μg. Their high thermal stability and homogeneity is manifest in the results obtained. Details of the GLC properties and structural assignments appear in Table 2. Mixtures of these three derivatives were separable and chromatograms of separations achieved are shown in Fig. 6.

The importance of thin-layer purification of the foregoing TMS derivatives prior to submitting them to GLC is depicted in Fig. 7. Although the main peak due to the derivative itself can be discerned, there might be errors in quantification due to impurities in the reaction mixture whose retention times might coincide with that of the derivative, thereby causing peak elevation.

Although the methods described in this report are clearly subject to criticism and offer room for improvement, there are certain distinct advantages. The derivatives described are chemically accessible, and they can readily be purified by TLC. Furthermore, GLC affords, in all cases but cortisol, a single well-defined peak with high mass/area responses that permits detection at the 0.025 μg level. Derivatives obtained from aldosterone and cortisol can be hydrolyzed back to the parent corticosteroid with facility, thereby making them available for other determinations. A similar argument can be put forward for those corticosteroid derivatives which are reduced at C-20. Finally, the high retention times of all the corticosteroid TMS derivatives described greatly reduces the possibility of interference from traces of lipid material, which tend to contaminate them, despite rigorous purification of fractions from natural sources.

The disadvantages of the methods described must also not be overlooked. Corticosteroid TMS derivatives require very high column temperatures for elution, and as has been emphasized earlier, they are subject to hydrolytic cleavage by atmospheric moisture. This necessitates rapid handling of these materials between derivative formation, thin-layer purification and subsequent analysis by GLC.

These materials have been used to study adrenal vein cortisol in the hamster. After extraction and purification by TLC, the entire fraction was derivatized according to the methods given, and was purified by TLC to furnish the adrenal blood cortisol derivative 1. This derivative was selectively hydrolyzed with 70% methanol to adrenal blood cortisol derivative 2 and purified by TLC together with authentic cortisol derivative 2. Elution of authentic cortisol derivative 2 and the zone of the adrenal effluent derivative immediately level with it, afforded fractions which were, respectively, compared by GLC and hydrolytic cleavage.

GLC of authentic cortisol derivative 2 (range of three determinations) indicated two peaks $R_t$ 1.01 (11β-hydroxyandrostenedione-TMS ether) and $R_t$ 1.94 (undissociated derivative). The hamster cortisol derivative 2 (two determinations 1/3 aliquots) gave peaks $R_t$ 0.93 (11β-hydroxyandrostenedione-TMS ether), $R_t$ 1.96 (undissociated derivative), $R_t$ 2.45, and $R_t$ 2.85 (unknown impurities). These chromatograms are depicted in Fig. 8. The presence of extraneous material in the hamster cortisol fraction was not unexpected, since it has previously been pointed out that despite rigorous TLC purification of derivatives, such materials do occasionally occur in fractions from natural sources. In this instance, contaminating substances were adequately resolved from the cortisol derivative by GLC; the peaks due to the corticosteroid derivative were not masked by the presence of impurities, since the retention time of the latter were much higher than those of the cortisol derivative.

TABLE 2. The gas chromatographic properties of the trimethylsilyl-oxy derivatives of 20β-dihydrocorticosteroids.

| Steroid | No. of UV absorbing products after TLC purification | | Retention time relative to cholestane (1.00) | Structural designation |
|---|---|---|---|---|
| | NaBH$_4$ | TMS ethers | | |
| 11-Deoxycorticosterone | 1 | 1 | 2.03 | 21,20β-di-(trimethylsilyl-oxy)-pregn-4-en-3-one |
| Corticosterone | 1 | 1 | 3.10 | 21,20β-di-(trimethylsilyl-oxy)-11β-hydroxy-pregn-4-en-3-one |
| 11-Dehydrocorticosterone | 1 | 1 | 2.17 | 21,20β-di-(trimethylsilyl-oxy)-pregn-4-ene-3,11-dione |
| Cortisone | 1 | 1 | 2.93 | 21,20β-di-(trimethylsilyl-oxy)-17α-hydroxy-pregn-4-ene-3,11-dione |
| | | | 3.55 | 21,20β,17α-tri-(trimethylsilyl-oxy)-pregn-4-ene-3,11-dione |

| Amount injected µg | Corrected amount µg | Peak area at 1/32 |
|---|---|---|
| 1·250 | 1·00 | 1,112 |
| 1·000 | 0·80 | 766 |
| 0·750 | 0·60 | 546 |
| 0·250 | 0·20 | 259 |
| 0·125 | 0·10 | 177 |
| 0·100 | 0·08 | 139·8 |
| 0·075 | 0·06 | 112 |
| 0·050 | 0·04 | 67 |
| 0·025 | 0·02 | 36·8 |

Peak, $R_T$ 0·89, is equivalent to 0·180 µg unchanged 11β-hydroxy-androstene dione (18%). Mean for several determinations is (20%).

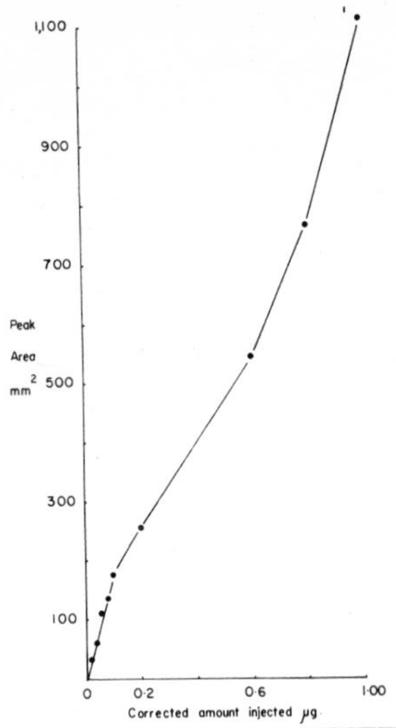

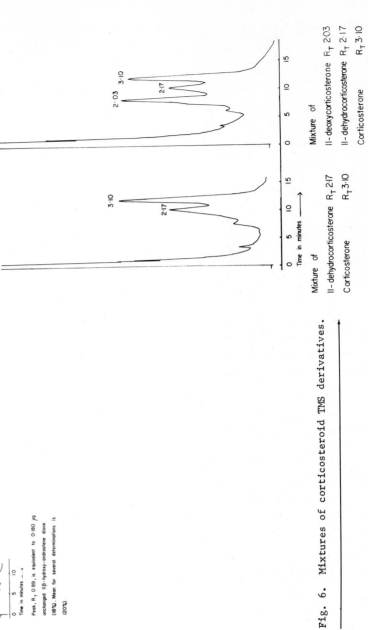

Fig. 5. Responses of 11β-hydroxyandrostenedione silyl ether.

Mixture of
11-dehydrocorticosterone $R_T$ 2·17
Corticosterone $R_T$ 3·10

4 µg of each steroid.

Mixture of
11-deoxycorticosterone $R_T$ 2·03
11-dehydrocorticosterone $R_T$ 2·17
Corticosterone $R_T$ 3·10

2 µg of each steroid.

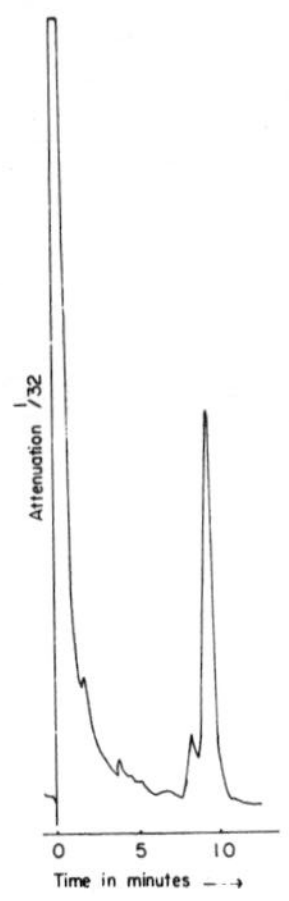

Fig. 6. Mixtures of corticosteroid TMS derivatives.

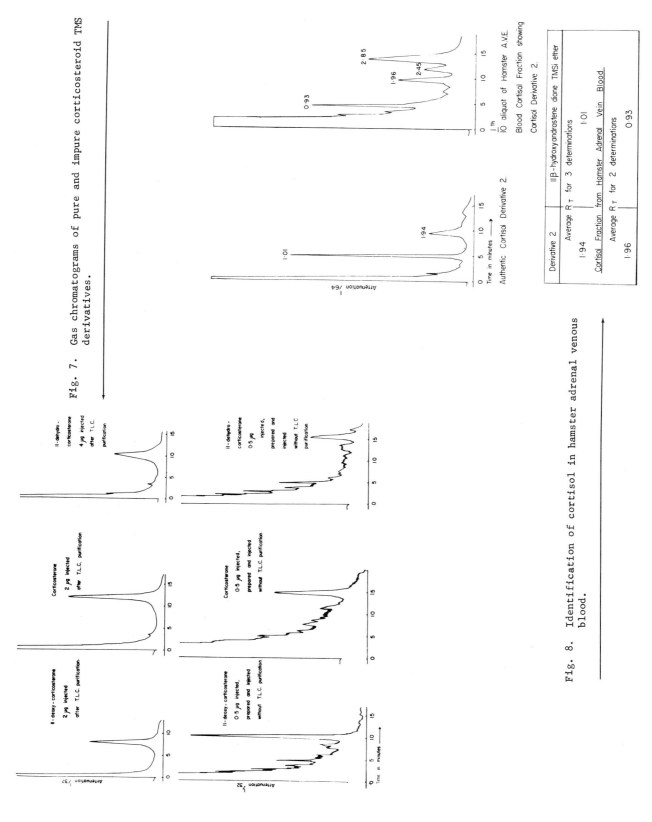

Fig. 7. Gas chromatograms of pure and impure corticosteroid TMS derivatives.

Fig. 8. Identification of cortisol in hamster adrenal venous blood.

## ACKNOWLEDGEMENTS

The author is indebted to Professor I. Chester Jones for his valued criticism, and also thanks Dr. P. A. Janssens for collecting the hamster adrenal vein blood.

## REFERENCES

1. Wotiz, H. H., I. Naukkarinen, and H. E. Carr, Biochim Biophys Acta 53: 449, 1961.

2. Kliman, B. and D. W. Foster, Anal Biochem 3: 403, 1962.

3. Merits, I., J Lipid Res 3: 126, 1962.

4. Bailey, E., J Endocr 28: 111, 1963.

5. Kirschner, M. A. and H. M. Fales, Anal Chem 34: 1548, 1962.

6. Brooks, C. J. W., Biochem J 92: 8P, 1964.

7. Rosenfeld, R. S., Steroids 4: 147, 1964.

8. Schindler, W. J. and K. M. Knigse, Endocr 65: 739, 1959.

9. Denyes, A. and R. H. Horwood, Can J Biochem Physiol 38: 1479, 1960.

# ANALYSIS OF ALDOSTERONE IN URINE BY DOUBLE ISOTOPE DILUTION AND GAS-LIQUID CHROMATOGRAPHY

Bernard Kliman, The Endocrine Unit, Department of Medicine, Massachusetts General Hospital, Boston, Mass.

Supported by U.S.P.H.S. Grant AM-06731 from the National Institute of Arthritis and Metabolic Diseases, Bethesda, Md.

## INTRODUCTION

The instability of corticosteroids, including aldosterone, during GLC was recognized by VandenHeuvel and Horning (1). Protection of aldosterone was provided by prior derivative formation as the gamma lactone (2, 3) and as the 18,21-diacetate (3, 4). Elimination of the 18 acetate group was found by Kliman and Foster (3) using 2%, 6 ft SE-30 columns, but not by Wotiz, Naukarinen and Carr who used 3%, 2-4 ft SE-30 columns (4). The application of GLC to the assay of biological samples containing aldosterone was reported by Kittinger (5), using gamma lactone formation after extraction of incubation fluid from rat adrenal glands. Because of prolonged retention time and low sensitivity of detection, 0.05-0.10 μg of aldosterone derivatives were required for analysis. The low concentration of aldosterone in biological fluids further limited the application of this new technique. The purpose of this investigation was to develop preparative and analytical methods allowing analysis of submicrogram amounts of aldosterone in human urine. Results using GLC were compared to the standard, sensitive, double isotope derivative method of Kliman and Peterson (6). The sensitivity of detection and identification of fractions eluting from the chromatograph were enhanced by simultaneous mass detection and collection of fractions for analysis of carbon-14 and tritium, by a modification of the method of Karmen et al. (7). The accuracy, precision, sensitivity, and utility of several methods of analysis were compared. The method of choice allows analysis in the millimicrogram range when radioactivity measurement is substituted as a detection system for GLC of aldosterone diacetate.

## MATERIALS

Solvents for paper and TLC systems were reagent grade. Chloroform was spectrophotometric quality. Pyridine was redistilled after refluxing over barium oxide. Acetic anhydride was redistilled. Crystalline d-aldosterone was purchased from Ciba. 4-$^{14}$C-aldosterone was obtained by courtesy of the Endocrine Study Section, National Institutes of Health. Tritium labeled acetic anhydride, 100 mc per mM, 20% in benzene, freshly redistilled, was purchased from New England Nuclear Corp., Boston, Mass. Whatman No. 1 paper for chromatography and silica gel G were obtained from Fisher Scientific Co., Boston, Mass. Rhodamine-6-G was obtained from National Aniline Corp., New York City, N. Y. Paraterphenyl, scintillation grade, was purchased from Packard Instrument Co., LaGrange, Ill. Liquifluor®, a stock solution phosphor, was obtained from Pilot Chemical Co., Watertown, Mass., and diluted with toluene. The scintillator solution contained 4 g PPO and 50 mg POPOP per l of toluene.

## METHODS

TLC was done with silica gel plates, 20 cm x 20 cm, spread 0.4 mm thick with a mixture of 40 gm silica gel G and 80 ml distilled water. For two dimensional chromatography, 8 ml of 0.01% aqueous solution of Rhodamine-6-G was diluted to 80 ml. Solvent systems were (I) methanol:chloroform (5:95) and (II) hexane:ethyl acetate (1:3). The plates were dried in air. Residues were applied by solution in 4, 2, 1 drops of ethanol:methylene chloride (1:1) and dried under a gentle air stream. The glass tanks were lined with Whatman No. 3 paper, and fresh solvent was added to a depth of 1 cm. The solvent was allowed to ascend to the upper end of the plate, and this required 40-50 min at 25 C. Steroid zones were removed by suction into the upper end of a disposable pasteur pipette containing a small wad of tissue paper as a filter. The pipette was placed upright in a 12 ml centrifuge tube and filled twice with acetone to elute the steroid into the tube.

<u>Paper chromatography</u>. The chromatography tanks were 12 x 24 in and walls were lined with Whatman No. 3 paper. A large petri dish at the bottom contained mobile phase, and the tank contained stationary phase to a depth of 4 cm. Whatman No. 1 paper was cut to 18 x 50 cm. After application of samples the papers were equilibrated in the tanks for 45 min before adding 50 ml mobile phase to each tray. Systems used were:

<u>Bush C</u>: Toluene:ethyl acetate/methanol:water (90:10/50:50).

<u>K-P-1</u>: Cyclohexane:benzene/methanol:water (100:40/100:20).

<u>K-P-2</u>: Cyclohexane:dioxane/methanol:water (100:80/100:40).

<u>K-P-3</u>: Cyclohexane:benzene/methanol:water (100:50/100:25).

<u>K-P-4</u>: Cyclohexane:dioxane/methanol:water (100:60/100:25).

<u>Acetylation</u>. The concentrated urine extract or first chromatogram eluate was dried in a conical, glass-stoppered tube of 6.5 ml capacity (Kontes Glass, Vineland, N. J.). For non-radioactive acetylation, 0.05 ml of pyridine and 0.05 ml acetic anhydride were added. For radioactive acetylation, 0.02 ml pyridine and 0.02 ml of a 20% solution of radioactive acetic anhydride in benzene were added. The stoppers were sealed with several drops of pyridine and covered firmly with parafilm. The tubes were incubated at 37 C for 18 hr or more. One ml of 20% ethanol in water was added to terminate the acetylation and the steroid was extracted into methylene chloride, 5 ml, washed once with 1 ml water and dried under a stream of filtered compressed air in a 37 C water bath. The specific activity of the reagents was determined as described previously (6).

<u>Oxidation procedure</u>. A stock solution of 12.5 ml distilled water in 1000 ml redistilled glacial acetic acid (Merck and Co.) was prepared. Fresh reagent was made each week by dissolving 50 mg chromium trioxide in 10 ml of stock solution, and was stored at 4 C. The reagent was light orange in color and was discarded if the color darkened; 0.2 ml of reagent was added to dry steroid residue in a 16 ml stoppered glass tube. After 5 min, 2 ml of 20% ethanol were added, the steroid product was extracted with 10 ml methylene chloride, washed once with 2 ml water, and dried under an air stream at 37 C.

<u>Gas chromatography</u>. The Chromalab Model A-110 gas chromatograph (Glowall Corp., Willow Grove, Pa.) was equipped with a radium detection cell, and 6 ft x 1/2 in coiled glass columns. Connections were made with 16 gauge teflon tubing, stainless steel hypodermic tubing, and silicone rubber stoppers. Columns were packed by suction and contained 2% SE-30 on 30-60 mesh acid washed chromosorb-W. The detection cell was cleaned at 6-8 week intervals by flushing with acetone, and the anode was polished with fine emory cloth. Injection stoppers were replaced after 10-15 punctures. Hamilton 10 $\mu$l syringes were used. Residues were dissolved in 10-20 $\mu$l of chloroform, and were concentrated to the end of the tube by a 30 second rotation in a clinical centrifuge. The syringe was held in the injection cap until the solvent trace appeared on the recorder. Usual conditions of analysis were: flash heater 280 C, column 245 C, cell 260 C. Argon carrier gas 20 psi with 50 ml per min flow rate. Chart speed was 20 in per hr. Peak areas were measured by one or more methods: (a) triangulation, (b) weight of the peak area cut out from the chart, (c) planimetry, using a K & E compensating polar planimeter. The triangulation method was adequate for peaks with area of 25 mm$^2$ or greater.

<u>Collection of gas chromatograph effluent</u>. The fraction collection device was a 50 sample, manual, model 830 Gas Fraction Collector (Packard Instrument Co., LaGrange, Ill.). The heated outlet tube was connected inside the detector oven to the detection cell outlet, and maintained at 250 C. The glass collection tubes were filled by suction with uncoated paraterphenyl crystals, scintillation grade, obtained from Packard Instrument Co. The fraction collector was indexed by electrical signal cord when the operator observed the appearance of the steroid peak on the chart recorder. The entire peak was collected and the next analysis was started after at least a 5 min interval. Surface melting of the crystals confirmed temperature above 212 C in the collection cartridge. The crystals were expelled into phosphor solution for liquid scintillation counting for carbon-14 and tritium.

Radioactivity counting and calculations. Carbon-14 and tritium were measured by simultaneous liquid scintillation counting in a Model 314-EX-Tri-Carb (Packard Instrument Co., LaGrange, Ill.). Conditions used were: gain 5%, discriminator 100-500 for the carbon-14 channel; gain 100%, discriminator 100-750 for the tritium channel, with 1190 v and wide window for both channels. Carbon-14 efficiency was 47.6% and tritium efficiency was 23.7%. Less than 0.1% of tritium counts appeared in the carbon-14 channel, and 36% of the carbon-14 counts appeared in the tritium channel. Each sample was counted for a minimum of 30 min. The data was processed with a PDP-14 digital computer (Digital Equipment Corp., Maynard, Mass.). Standard isotope dilution formulae were used (9), and were incorporated in a program developed with the Massachusetts General Computer Group to yield results which were expressed as µg per sample and µg per 24 hr.

Assay procedure: Single isotope dilution. 250 ml of urine were placed in a 1000 ml glass stoppered graduate; 4 ml of concentrated HCl were added and the pH was adjusted to 1.0 with additional acid. After 24 hr at room temperature, 5000 cpm (0.06 µg) of $4-^{14}C$-aldosterone, 40 mc per mM were added and the urine was extracted with 800 ml of methylene chloride. The urine layer was discarded and the solvent was shaken with 200 ml of 0.05 N NaOH, two or more times, until washings were clear. A final wash with 200 ml of 0.1 N HAc was done, and the solvent was dried in vacuo at 40 C. The residue was transferred to a tube with three washes of 2 ml ethanol and methylene chloride 1:1 and dried under an air stream at 37 C. The crude extract was chromatographed from a 10 cm streak applied to the center of the paper with aldosterone standard markers, 10 µg each, on opposite sides well separated from the urine extract. The Bush C system was allowed to run for 3-4 hr. The aldosterone zone was cut out, and eluted in a 10 ml centrifuge tube with 5-8 ml ethanol for at least one hr. The solution was again dried and acetylated with unlabeled acetic anhydride, for 18 hr at 37 C. Ethanol, 0.5 ml, was added and the solution was dried under an air stream. Further chromatography was done with paper system KP1, 16 hr, and with TLC systems I and II. Aliquots of 4-8 µl from the final eluate residue in 20 µl chloroform were taken for isotope counting and GLC.

Secretion rate analysis by single isotope dilution was performed in the same fashion except that 5 µc of aldosterone $1,2-^3H$, 100 µc per µg, (New England Nuclear Corp., Boston, Mass.) were prepared in sterile solution and injected intravenously into patients under study. No tracer was therefore added to the urine. Aldosterone diacetate, 20 µg was added to each sample after acetylation.

Assay procedure: Double isotope derivative dilution. 1. For secretion rate studies, 250 ml of urine was processed as described above, except that acetylation was performed with $^{14}C$-acetic anhydride, 1 mc per mm. Unlabeled aldosterone diacetate 20 µg was added to each sample after acetylation. No aliquots were required for isotope counting prior to GLC. The entire peak fraction was collected for double isotope analysis.

The standard secretion rate method was similar to the above for Bush C and KP1 chromatography. This was followed by oxidation and chromatography in paper systems KP-2 and KP-3, 16 hr at room temperature.

2. For assay of urine content of acid-released aldosterone, 15 ml of urine were placed in a 50 ml extraction tube and acidified to pH 1.0 with concentrated HCl. After 24 hr at room temperature, 1000 cpm (12 mµg) $4-^{14}C$-aldosterone in 25 µl ethanol were added and the extraction was done with 35 ml methylene chloride followed by 10 ml washes with NaOH and HAc similar to the procedure for 250 ml samples. Acetylation was performed with 100 mc per mm $^3H$-acetic anhydride. Aldosterone diacetate, 20 µg was used as an UV visible standard. Bush C chromatography was not required. Samples for GLC were processed in paper system KP-1 (16 hr) and TLC systems I and II.

Samples for standard method assay were processed in paper system KP-1 and KP-4, 16 hr each, oxidized, and run in system KP-3 for 16 hr. This procedure was found to be adequate using acetic anhydride which had been redistilled by the manufacturer within the proceding 8 weeks. If fresh reagent is not available, an additional chromatography in system KP-2 is advised.

## RESULTS AND CONCLUSIONS

1. <u>Direct Analysis of Aldosterone in Urine</u>. Previous studies indicated a linear response of the beta ionization detector to aldosterone diacetate standard of 0.05 μg or greater (3). This result was consistent, and a peak suitable for triangulation or planimetry is shown in Fig. 1. When a 6 ft column, 2% SE-30 on chromosorb W was used, the retention time at 245 C was 5 min. Columns of 1600 theoretical plates for testosterone acetate usually demonstrated a value of 600 theoretical plates for aldosterone diacetate. Significant loss of injected steroid was noted unless priming doses of 2 μg were injected onto the column on the day of use, with 0.2 μg standards repeated between subsequent analyses of biological samples. The analysis of urine extract containing aldosterone is shown in Fig. 2.

The direct analysis of aldosterone was limited by several factors. The amount of steroid obtained after acid treatment of urine is 5-20 μg per 24 hr, at an average concentration of 2.5-10 μg per l. A 250 ml urine aliquot, containing 0.6-2.5 μg, yielded over 10 mg of oily residue which required at least 3 stages of paper or TLC to obtain 0.2-0.8 μg of aldosterone in a form suitable for GLC. Purification loss was too variable and excessive to allow direct analysis without the use of a radioactive tracer. Attempts to reduce the preparative steps to a single paper chromatography and acetylation yielded a product which analyzed more than 4-fold in excess of the known concentration of aldosterone. Further studies were therefore limited to the use of isotope dilution with addition of high specific activity $4-^{14}C-$ or $1,2-^{3}H-$aldosterone to the acidified urine prior to the initial extraction. Direct analysis is henceforth considered as an isotope dilution procedure in which recovery of the tracer is determined by radioassay of an aliquot of the solution injected into the gas chromatograph.

2. <u>Isotope Fractionation and Identity of the Aldosterone Diacetate Product</u>. The suitability of radioactive aldosterone diacetate derivatives for GLC was studied by collection of a purified standard. Aldosterone-$4-^{14}C$ was acetylated with $^{3}H$-acetic anhydride after admixture with 20 μg of carrier aldosterone. The double labeled steroid was chromatographed to constant isotope ratio and an aliquot was injected into the gas chromatograph. Serial fractions were collected on terphenyl crystals and pooled to provide comparable amounts of radioactivity in adjacent fractions. Addition of Carrier aldosterone diacetate to such samples did not affect the isotope ratio. The 49.7% decrease in relative tritium content confirmed the previous report (3) of quantitative loss of one acetate group (Table 1). Tritium enrichment of the initial and final eluted fractions confirmed presence of isotope fractionation similar to that observed for $C_{19}$-steroid acetates by Kirschner and Lipsett (8).

The collection of $4-^{14}C$ radioactivity from seven separate injections in which the entire peak was collected was 64 ± 10% (mean ± SD) with a range of 44-76%. Under similar conditions, 80-100% of testosterone acetate or androsterone diacetate onto the column used for these experiments yielded no significant radioactivity. The collection of steroid was sufficient for analysis of double isotopes, by isotope ratio, but not for single isotope assay except as an identification procedure.

3. <u>Single and Double Isotope Dilution, Using 250 ml Urine</u>. The measurement of aldosterone in larger urine aliquots is not useful when urine level only is required since other analytical methods are capable of similar measurements. The procedure of greatest interest in clinical investigation is the measurement of <u>in vivo</u> secretion rate. Tritium labeled aldosterone is injected intravenously and the 24 hr urine is analyzed for specific radioactivity of an aldosterone metabolite (9). This is an isotope dilution procedure, and similar results may be obtained for urine levels of aldosterone by addition of the tracer to urine. Urine was processed by a standard method (10) in which double isotope labeling was done with $^{14}C$-acetic anhydride followed by multiple paper chromatography. Results were compared to single isotope measurement and GLC (Table 2, Section A). The urines were chosen to include the subnormal range (secretion rate less than 50 μg per 24 hr) and normal range (50-250 μg per 24 hr). Metabolite concentration is usually 5-10% of the secretion rate value. Despite preparation with two paper and one thin-layer chromatogram, the gas chromatograph mass analysis overestimated the subnormal specimen and diverged + 38 to -33% from the standard assays in the normal range. When similar preparative methods were used, and the results of GLC determined by double isotope analysis, better agreement was obtained (Table 2; B, C). A tendency to higher values was noted in eight out of nine such analyses by GLC, although subnormal levels of steroid were adequately detected. The standard method required four paper chromatographic steps and an oxidation procedure. The GLC methods required three paper, or two paper and one thin-layer step in addition to the GLC procedure, but without oxidation to a secondary derivative. The standard procedure required six days and the GLC method only four days for completion, after a 24 hr pH 1.0 hydrolysis.

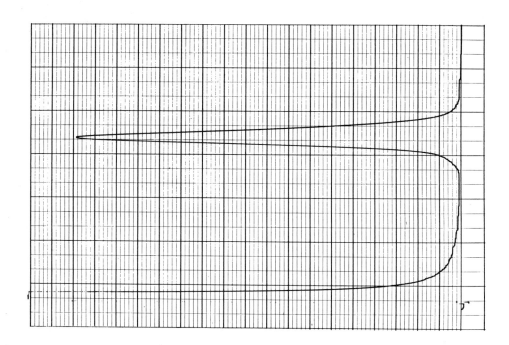

Fig. 1. GLC of aldosterone diacetate on SE-30. Conditions as described in the text.

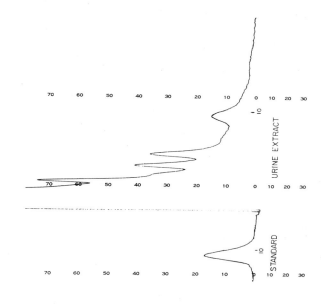

Fig. 2. GLC of aldosterone diacetate from urine after one paper chromatography.

TABLE 1.

| Aldosterone-4-$^{14}$C-diacetate-$^{3}$H | | | Tritium: Carbon-14 |
|---|---|---|---|
| First chromatography, paper | | | 27.0 |
| Second chromatography, paper | | | 25.1 |
| Third chromatography, thin-layer | | | 25.1 |
| Gas chromatography fractions | | | |
| | $^{14}$C CPM-BKG | $^{3}$H CPM-BKG | |
| 1 | 175 | 2262 | 12.95 |
| 2 | 305 | 3825 | 12.55 |
| 3 | 226 | 2823 | 12.50 |
| 4 | 329 | 3975 | 12.10 |
| 5 | 341 | 4273 | 12.55 |
| 6 | 309 | 3946 | 12.75 |
| 7 | 328 | 4325 | 13.20 |
| Total | 2013 | 25429 | 12.63 |
| Theoretical ratio for one acetate: | | | 12.55 |

TABLE 2. Aldosterone secretion rate by gas chromatography. 250 ml urine.

A. Single isotope dilution (2 paper, TLC chromatography).

| No. | Standard assay µg per 24 hr | GLC µg per 24 hr | Difference percent |
|---|---|---|---|
| 1 | 21 | 64 | + 205 |
| 2 | 50 | 63 | + 26 |
| 3 | 118 | 79 | - 33 |
| 4 | 133 | 133 | 0 |
| 5 | 209 | 289 | + 38 |

B. Double isotope dilution (3 paper chromatographies).

| 1 | 5 | 6 | + 20 |
|---|---|---|---|
| 2 | 316 | 388 | + 23 |
| 3 | 1410 | 1980 | + 40 |
| 4 | 49 | 78 | + 59 |

C. Double isotope dilution (1 paper, 1 TLC chromatography).

| 1 | 0 | 9 | -- |
|---|---|---|---|
| 2 | 21 | 35 | + 67 |
| 3 | 51 | 37 | - 28 |
| 4 | 1395 | 2020 | + 15 |
| 5 | 2480 | 3610 | + 46 |

4. <u>Double Isotope Dilution, Using 15 ml of Urine</u>. A more convenient assay for urinary aldosterone levels was studied using 15 ml of urine. The amount of residue is such that chromatography prior to acetylation is not required, and multiple samples can be processed. Since the initial aldosterone content is expected to be 0.1 µg or less in normal subjects, and an average of 25% is lost in each chromatographic step, the detection of millimicrogram amounts is essential for the diagnosis of subnormal values, and quantitation of low normal results. Double isotope dilution was chosen to provide the necessary sensitivity. A tracer of 4-$^{14}$C-aldosterone was added, and derivative formation was done with 100 mc per mM $^{3}$H-acetic anhydride. One mµg of mono-acetate in the isolated fraction was equal to the background count and was considered to be at the limit of detection. Results of analysis using an abbreviated purification, with a single two dimensional thin-layer plate, appears in Table 3, Section A. The normal range is 5-20 µg per 24 hr. Gross overestimation of normal and subnormal values was found in the GLC procedure.

TABLE 3. Double isotope derivative assays: 15 ml urine.

A. One TLC 2 dimensional chromatography prior to GLC.

| No. | Standard method µg per 24 hr | GLC µg per 24 hr | Difference percent |
|---|---|---|---|
| 1 | 0.0 | 7.7 | - |
| 2 | 6.5 | 18 | +192 |
| 3 | 17 | 51 | +200 |
| 4 | 20 | 53 | +165 |
| 5 | 22 | 65 | +195 |
| 6 | 45 | 41 | - 9 |
| 7 | 52 | 78 | + 50 |
| 8 | 60 | 85 | + 42 |

B. One paper and one TLC two dimensional chromatographies.

| | | | |
|---|---|---|---|
| 1 | 1.2 | 1.3 | + 9 |
| 2 | 1.7 | 2.7 | + 59 |
| 3 | 1.8 | 1.9 | + 6 |
| 4 | 4.7 | 5.7 | + 21 |
| 5 | 7.5 | 7.4 | - 1 |
| 6 | 7.7 | 8.3 | + 8 |
| 7 | 7.8 | 8.3 | + 6 |
| 8 | 9.8 | 7.2 | - 27 |
| 9 | 12.3 | 16.1 | - 31 |

The inclusion of one paper and a two dimensional TLC prior to GLC separation achieved acceptable results. The standard and new methods detected low values and variation in the normal range was not unusual. The chromatographic procedures could be completed within 24 hr following acetylation of the extract, instead of 72 hr by the standard method. Recovery of the tracer was 10-25% by either method. Assay specificity was evidenced by the ability to obtain values close to zero in Addisonian urine. The recovery and precision of the method was evaluated by assay of aldosterone standard solution alone, and recovered from urine (Table 4). Linear assay results were obtained for the range 0.04-0.16 µg in the initial sample. Recovery from urine was acceptable for both methods.

TABLE 4. Recovery of aldosterone from urine.

A. Double isotope method using paper chromatography.

| Urine aldosterone mµg | Standard aldosterone mµg | Combined assay mµg | Standard recovery mµg | percent |
|---|---|---|---|---|
| - | 40 | - | 40 | 100 |
| - | 80 | - | 78 | 98 |
| - | 160 | - | 165 | 103 |
| 50 | 40 | 87 | 37 | 93 |
| 50 | 80 | 131 | 81 | 101 |
| 50 | 160 | 206 | 157 | 98 |

B. Double isotope method using thin-layer and GLC.

| Urine aldosterone mµg | Standard aldosterone mµg | Combined assay mµg | Standard recovery mµg | percent |
|---|---|---|---|---|
| - | 40 | - | 38 | 95 |
| - | 80 | - | 82 | 103 |
| - | 160 | - | 150 | 94 |
| 35 | 40 | 79 | 44 | 110 |
| 35 | 80 | 108 | 73 | 94 |
| 35 | 160 | 177 | 142 | 89 |

Reproducibility and variation were tested by replicate analyses of a high level urine, in groups of six aliquots of 15 ml each, by the standard and new techniques. The group mean and range of values were similar by both methods (Table 5). The coefficient of variation, SD percent, was 6.3% for the standard method and 7.0% for the new method.

## DISCUSSION

The analysis of urinary acid-released aldosterone by GLC is subject to inherent problems. The concentration of aldosterone in normal human urine is in the range of 10 µg per l and a reliable method should assay levels as low as 1 µg per l. Milligram quantities of oily residue are obtained from extraction of as little as 15 ml of urine. The presence of multiple impurities is sufficient to obscure the mass records. Removal of these substances requires extensive purification and attendant variable losses of steroid. With the addition of a high specific activity tracer, it has been possible to obtain a mass analysis assay for aldosterone content. This type of analysis is used for measurement of secretion rate of aldosterone when the tracer is injected into the subject under study. Comparison of this method to available double isotope techniques indicates a tendency to overestimation of aldosterone content and lack of sensitivity in the subnormal range, using GLC. These problems could be resolved by analysis of a larger volume of urine and more extensive purification. Since these changes would result in a more cumbersome and time consuming procedure than existing assay methods, further studies were not indicated. Modification of the method to allow processing of double isotope samples did result in a more rapid but less precise assay method. Since some impurities are not readily resolved by the GLC column, the development of increased detector sensitivity alone does not appear to offer a solution. The assay of urine aliquots of 250 ml is currently limited to specimens with normal or elevated aldosterone content when single isotope dilution is used. An approximate secretion rate value may be obtained by double isotope labeling and GLC several days in advance of the standard double isotope procedure by paper chromatography.

TABLE 5. Replicate analyses of urinary aldosterone.

A. Double isotope method using paper chromatography.

B. Double isotope method using thin-layer and GLC.

|   | No. of assays | Range µg per 24 hr | Mean ± SD µg per 24 hr | SD percent |
|---|---|---|---|---|
| A. | 6 | 47-57 | 52.0 ± 3.3 | ± 6.3 |
| B. | 6 | 49-60 | 54.5 ± 3.8 | ± 7.0 |

These studies indicated that GLC analysis required an increased sensitivity of detection, increased specificity in the identification of the steroid in the presence of non-specific contaminants, and a method for protecting the steroid from variable loss during the entire analysis. The measurement of radioactivity of double isotope derivatives of aldosterone in the column effluent provided such results. This procedure represents the use of a radioactivity detection system as the equivalent of the detection cell. The fractions require separate isotope counting because the speed of elution from the column does not allow sufficient counting time for flow counting. The conventional mass detector continues to function as a detector for the steroid fraction. The use of carrier steroid prevents significant adsorption on the column of either the steroid or radioactivity. It was demonstrated that the combined use of paper, thin-layer, and GLC achieved rapid purification. The sensitivity of analysis with tritium labeled acetic anhydride allowed analysis of 15 ml samples of urine which can be readily processed in groups of 16 or more by one technician. The procedure is fully corrected for any loss including the collection of fractions since the ratio of the two isotopes is the basis for calculation. The development of a computer program for data processing has reduced total analysis time and increased the accuracy and efficiency of analysis. The experimental studies of radioactive GLC gave quantitative tests of column performance, identification of structural changes in aldosterone diacetate during GLC, and demonstrated adsorption of small amounts of steroid onto SE-30 columns. The presence of isotope fractionation revealed the potential resolving capacity of the columns and did not limit the assay procedure since total peak collection was used. The isotope measurement system is adaptable to a variety of experimental purposes. Labeled precursor steroids can be isolated and measured after incorporation into aldosterone. The same equipment and methods are available for analysis of other steroids.

The conclusion of this investigation is that an accurate and efficient analysis of aldosterone in human urine has been done by single and double isotope GLC. The single isotope dilution method does not involve collection of column effluent, but requires processing of at least 250 ml or urine. The double isotope dilution method is based on collection of the aldosterone derivative from the column effluent, and is applicable to the analysis of 15 ml of urine. The accuracy and precision of these methods are comparable to the conventional double isotope analysis. The advantages of rapid processing have been confirmed by application to clinical research with decreased expenditure of funds, time, and effort when early results are obtained. As more sensitive and specific mass detection methods become available, it may be possible to improve the use of single isotope dilution methods or to obtain a simultaneous analysis of mass, tritium, and carbon-14 for diverse research purposes.

ACKNOWLEDGEMENT

The technical services of Miss Barbara A. Apple are gratefully acknowledged.

REFERENCES

1. VandenHeuvel, W. J. A. and E. C. Horning, Biophys Res Comm 3: 356, 1960.

2. Meritz, I., J Lipid Res 3: 126, 1962.

3. Kliman, B. and D. W. Foster, Anal Biochem 3: 403, 1962.

4. Wotiz, H. H., I. Naukkarinen, and H. E. Carr, Jr., Biochim et Biophys Acta 53: 449, 1961.
5. Kittinger, G. W., Steroids 3: 21, 1964.
6. Kliman, B. and R. E. Peterson, J Biol Chem 235: 1639, 1960.
7. Karmen, A., L. Guiffrida, and R. L. Bowman, J Lipid Res 3: 44, 1962.
8. Kirschner, M. A. and M. B. Lipsett, J Lipid Res 6: 7, 1965.
9. Peterson, R. E., Recent Progr Hormone Res 15: 231, 1959.
10. Kliman, B., Advances in Tracer Methodology, Vol. II, Plenum Press, New York, 1965 (in press).

# CORTICOSTEROID DISCUSSION

DR. DODSON: I would like to ask Dr. Bailey how extensively he has used his method and whether he has compared its accuracy with other methods and how long a determination takes using his method.

DR. BAILEY: Prior to using GLC we used the Bush method for cortisol and cortisone, namely paper chromatography and the soda fluorescence reaction. Our present results for normal individuals agree with those we used to get, but in our hands the method using GLC is more accurate and reproducible. Using the GLC procedure described previously (Bailey, E., J Endocr 28: 131, 1964) we have studied the excretion of cortisol and cortisone and some of their polar metabolites in 100 or more urine specimens. When the present procedure is in continuous use, the seven corticosteroids can be measured in two urines per day.

DR. VAN DER MOLEN: I would like to ask Dr. Crane about his difficulties in obtaining a linear relationship between the injected amount of steroid and his peak areas. Can you tell me if this might result either from a variation in the artifact formation that occurs during the chromatography as you indicated, or from the injection technique, that you are using?

DR. CRANE: I am sorry I am not able to give you a complete answer as to the cause for this. The defect cannot be in the injection technique since the volume of injection was kept constant as the concentration was varied. I think that since we obtain the "S"-shaped curve in 17-KS and cholestanol it is unlikely that the change in peak height per $\mu$g of the C-21 steroids is due to breakdown of the steroids.

As far as the adsorption to the column is concerned, we have obtained the same non-linear response in our system whether the support was silanized or not. This is the only evidence I can give you to answer your question.

DR. VAN DER MOLEN: May I show two slides? We have done some small things, that are by no means original, but it might be related to this problem. They concern the technique of injecting samples of solutions onto a column using a Hamilton syringe.

We have used two techniques. In the first technique we prefill and pump our needle full with the solution that we want to dispense (as recommended in the manuals for the Hamilton syringes) and subsequently fill the syringe with the volume of solution that we want to apply. In the second technique, we prefill our needle with pure solvent instead of solution. At room temperature (Fig. 1) there was no difference in results between these techniques as far as concerns the linearity of the relationship between the volume that was taken and the amount that was dispensed; before and after the injection the needle will still contain either the solution or the solvent depending on the technique that has been used. However, at high temperatures (Fig. 2), either the solution or the solvent in the needle is flushed out in the chromatograph. Hence, if you pump your needle full with solution before injection, you will always inject the needle volume in addition to the volume that you really want to inject and consequently you will not obtain a straight line relationship going through zero.

Though we would expect a straight line, intersecting the axis at a point representing the needle volume, we often have found a slightly curved line. We presume this to be the result of the inaccuracy in dispensing the smaller volumes. The conclusion drawn from the pictorial representation is also reflected in the standard deviations of the means after converting the obtained results to a standard injection volume.

A. Dispensing of radioactive solution at 22°C

| | cpm/µl after prefilling of needle with | |
|---|---|---|
| | solvent | solution |
| Operator 1: | 4512 s.d. 8.7% n=14 | 6553 s.d. 3% n=10 |
| Operator 2: | 4431 s.d. 7.2% n=14 | 6006 s.d. 3.1% n=10 |

B. Dispensing of a Estrone-TMSi solution at 280°C

| | $cm^2/\mu l$ after prefilling of needle with | |
|---|---|---|
| | solvent | solution |
| Operator 1: | 1.27 s.d. 8.7% n=8 | 1.91 s.d 20% n=8 |
| Operator 2: | 1.46 s.d. 7.5% n=8 | 1.94 s.d. 21% n=8 |

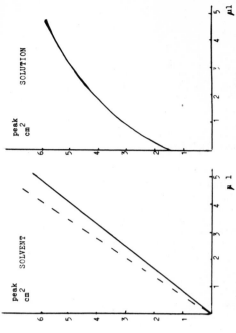

Fig. 2. Reproducibility of injection (5 µl Hamilton syringe). Dispensing of a estrone-TMS solution at 280 C.

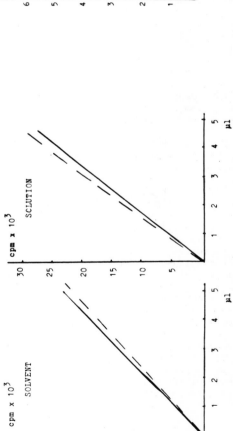

Fig. 1. Reproducibility of injection (5 µl Hamilton syringe). Dispensing of radioactive solution at 22 C.

DR. WOTIZ: I would like to make two comments to Dr. Crane regarding the injection. First, we are somewhat at the mercy of the instrument manufacturer which is rather a pity. There are sample introduction systems available which, in our hands at least, have been quite good. I refer specifically to a system described originally by Tenney and Harris. We have found this method to be highly reproducible and accurate. Unfortunately, the only two manufacturers who made such injectors have discontinued them. We are therefore confined to the Hamilton syringe.

Second, there exists the possibility of washing residual material off the septum. In the particular case you mentioned I wonder if this could be related to the fact that the injector temperature was 350 C, which is really quite high.

DR. CRANE: I am familiar with Dr. Wotiz's mention of this problem at the 1962 Laurentian Hormone Conference, the problem of spurious peaks obtained from an overheated septum. I presented these data because, as far as I can tell, this condition is not very well known.

The septum temperature is important. If the temperature of the septum is in the neighborhood of 300 C, then the silicone rubber septum may damage your column.

We have observed instances in which the column was completely ruined by changing the septum while the column was connected to the flash heater. In the future there may be other types of material used in the manufacture of septa which may be able to introduce substances on the column and cause the same defect or worse. This is the reason why I think this information should be well known.

The topic of injection technique needs further consideration. In regards to flushing of the solution from the microliter syringe, I would like to inquire of Dr. van der Molen, how do you flush the solution in with the solvent? How do you keep the solvent from mixing with the substances that you have in your solution?

DR. VAN DER MOLEN: The technique used is that we refill the syringe two or three times with pure solvent, then adjust it to zero and suck up the solution you want to introduce. We assume, of course, that the needle is just filled with solvent and it may be mixing with the solution.

DR. CRANE: Do you withdraw the solvent and solution out of the needle into the barrel of the syringe before you put the needle into the septum or not?

DR. VAN DER MOLEN: As you say, we just pump solvent into the syringe, adjust the plunger to the zero mark so that the needle is filled with solvent and then draw the solution into the syringe.

DR. CRANE: How long a period of time do you have the needle in the septum before you press the syringe plunger. If you wait too long you may evaporate a portion of your sample before you were ready.

DR. VAN DER MOLEN: Oh, yes, we just penetrate the septum and immediately flush it out.

DR. CRANE: I think that this would introduce the problem of a double peak for your specimen, and that would be the main difficulty in having the needle full of solution. We cannot inject immediately on penetration of the septum because of design of the injection port. The carrier gas flows into the injection port 1 1/4 in behind the front of the septum.

DR. VAN DER MOLEN: I don't understand the question about the double peak.

DR. CRANE: If you delay momentarily after positioning the syringe needle in the injection port before you pressed the plunger you would have introduced by evaporation what was in the needle before you introduced the rest of your sample.

DR. VAN DER MOLEN: I'm sorry I caused this confusion. We first take in the solvent, then pull the solution into the syringe, penetrate the septum and press the punger down <u>immediately</u>.

DR. BROWNIE: I had better clear up the point of contention between Dr. Crane and Dr. van der Molen. Having seen him take up samples into a syringe many times I believe the following procedure was used. First of all, solvent is taken into the needle, the needle is then immersed in the solution to be chromatographed, the required volume of solution is drawn into the syringe, the syringe is removed from the solution and the solution in the syringe is withdrawn further up the syringe barrel thus leaving an air cushion between the solution and the end of the needle.

DR. KLIMAN: In the brochure accompanying the Hamilton syringes, the volume of the needle on a 10 µl syringe is described as containing 0.5 µl. This should be a warning to those who use small injection volumes and emphasizes again the problem of needle volume.

DR. EIK-NES: I have a question for Dr. Bailey. In your article in Journal of Endocrinology, you stated that the sensitivity of assay was better when you converted the adrenocorticoids to their corresponding 17-KS prior to application onto the GLC column. I wonder if you have carried out further experiments along these lines since it currently appears that quantitation of adrenocorticoids by the technique of GLC can only be done by derivative formation of the adrenocorticoids before GLC or by studying the cracking pattern of the adrenocorticoids on the gas chromatograph. Unless the cracking pattern for each steroid is reproducible, the latter method of assay may be quite hazardous.

DR. BAILEY: The conversion of the corticosteroid to its 17-KS derivative by bismuthate oxidation results in an increase in sensitivity which is greater than can be accounted for by differences in molecular weight. This finding was also reported by Kirschner, M. A. and Fales, H. M., Anal Chem 34: 1548, 1962. A more important advantage of the oxidation procedure is that it removes most of the extraneous material from the extract.

DR. WILSON: I am going to take the chairman's privilege of commenting on the elegant work presented by Dr. Bailey. This study was particularly interesting because I once used a similar approach in determining the pattern of urinary 17,20,21-triols and 17,20-diols as the corresponding 17-KS, after oxidation with periodic acid.

The great pitfall of this approach is that any one 17-KS oxidation product may have several precursors. So the prior separation of all such "multiple precursors" is essential for their successful analysis.

Dr. Bailey has presented a list of corticoids, metabolites of cortisol, known to occur in urine, and shown that his TLC system separates all precursors of the same 17-KS. We all know that numerous components in addition to the ones we have thought of, are likely to be present in any fraction of urine extract, particularly from a subject with an endocrine abnormality.

On your slides we see only a few unlabeled peaks, and you say that few unknowns have turned up in your work so far. Nevertheless I would like to list some possible precursors which could interfere in this type of analysis.

1) There are several 21-deoxy $C_{21}$ steroids which have been found in urines from abnormal subjects, whose chromatographic mobility is similar to the less polar cortisol metabolites. These are 21-deoxycortisone, 21-deoxycortisol, pregnane $3\alpha,11\beta,17\alpha,20\alpha$-tetrol. These would give the same ketosteroids on oxidation as the corresponding 21-hydroxycortisol metabolite, so I wonder where they would run on your thin-layer plate.

2) Similarly polar metabolites of 11-deoxycortisol, such as the 3,17,20,21-tetrol also move in the area of some cortisol metabolites. These would give $C_{19}O_2$-17-KS an oxidation. Perhaps some of your unlabeled early GLC peaks represent such metabolites.

3) Metabolites of corticosterone would not give 17-KS, but their oxidation products would show up as GLC peaks.

4) Preformed polar $C_{19}$ steroids, which might chromatograph in the area of cortisone or cortisol and would not be changed by oxidation.

DR. BAILEY: As I have stated in my paper regarding the specificity of the assay it cannot be assumed that in some endocrine abnormalities there will not be steroids present which after the TLC and oxidation, produce peaks under or in the vicinity of those being measured. So far in our studies of urines from normal individuals rheumatoid arthritic patients and women in the third trimester of pregnancy we have no evidence of other steroids being present which might interfere with the assay.

From the list of steroids mentioned by Dr. Wilson, I would expect some to be less polar than cortisone on the TLC system I have used so that if they did occur would not be eluted. The 21-deoxycorticosteroids and 11-deoxycorticosteroids found in adrenogenital syndromes might be present in significant quantities in the unconjugated form in such conditions. I do not know where these would all run in the TLC system but would expect 21-deoxycortisone to run beyond cortisone and so be excluded. 21-deoxycortisol might run in fraction F1 and so give a peak for the 17-KS of cortisol in this fraction where no such peak would normally be present. The 11-deoxycorticosteroids on oxidation would give 17-KS derivatives which would have retention times on the XE-60 phase less than that of 11-ketoetiocholanolone. Corticosterone and derivatives with the same side-chain will give etianic acids on oxidation with sodium bismuthate and these would be removed in the alkali wash.

Do you know Dr. Wilson, if any of these steroids you have mentioned are present in the unconjugated form in urine.

DR. WILSON: It seems to me we must assume that any steroid present in urine could be excreted to a very small extent in the unconjugated form. A sensitive method like yours is extremely revealing and might be very valuable in detecting such metabolites in abnormal urines.

Judging by your charts on normal individuals, there were few, if any, unknown peaks, and the only possible error would be from unknown substances with the same $R_T$, or the same 17-KS from another source.

DR. FALES: We have found that in general the peaks on a gas chromatogram really contained the substances that we had thought they had. It was not too often that there were large concentrations of other components. However, there were frequently large numbers of very small quantities of other components.

DR. KLIMAN: In respect to the tetrahydro-aldosterone tri-acetate I have on occasion run chromatograms on what was represented to be pure material and found there were several peaks eluted and did not follow this further.

The problem may be obtaining pure standard material to work with and in the case of pure labeled tetrahydro-aldosterone, it is available in only a few laboratories.

DR. MENINI: In order to eliminate injection and solvent peak problems, Dr. Norymberski and I developed a technique which allows the introduction of solid analytical samples into GLC.

Figure 3 shows a section view of the top of the modified analytical column surrounded by a preheater.

The top part of standard glass columns (4-5 ft long, 3.5 mm i d ) has been replaced by an evaporation chamber ending with a B 14 ground glass socket. The evaporation chamber fits into the central hole of an aluminum block, secured to the top of the chromatograph housing and heated by a band heater controlled by a variable transformer. A perforated glass tubing is inserted into the evaporation chamber by means of a B 14 cone. The upper part of the cone is connected to two side arms, one for the argon supply and the other for the storage of analytical samples deposited as a solid on small stainless steel gauze rings; the upper side arm is closed with a silicone rubber stopper or better with an expanding rubber seal. When operating conditions are reached, one sample at a time is moved with a magnet towards the vertical tubing and allowed to drop into the perforated tube. The evaporation chamber is kept at 250-260 C to ensure instant vaporization of the analytical sample. Possibly the most valuable part of the technique is the quantitative transfer of the analytical sample onto the gauze ring and this is accomplished as follows: a stainless steel gauze ring is placed in the indentation of a spotting plate made of teflon and a solution of the analytical sample in chloroform is placed in the same indentation. The volume of the solution is not critical but it has been found that volumes ranging from 0.05 to 0.25 ml are convenient and easy to handle quantitatively.

On evaporation of the solvent, a more and more concentrated solution is obtained which finally by capillary action distributes itself on the gauze. Complete evaporation of the solvent leaves the analytical sample on the gauze ready to be introduced in the column. Efficiency of this technique, from the quantitative point of view, has been tested by transferring to gauze rings, by the procedure described, several $\Delta^4$-3-oxosteroids, which were subsequently eluted from the gauze with ethanol. The measured extinction at 240 m$\mu$ accounted for 95-99% of the material used.

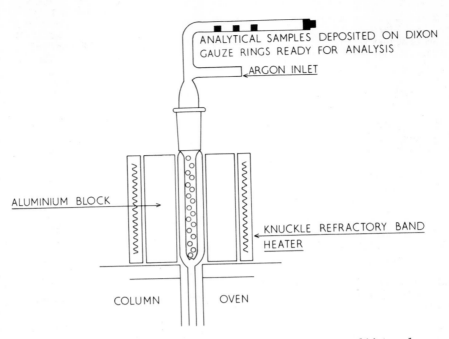

Fig. 3. Section view of top of modified column for solid sample introduction.

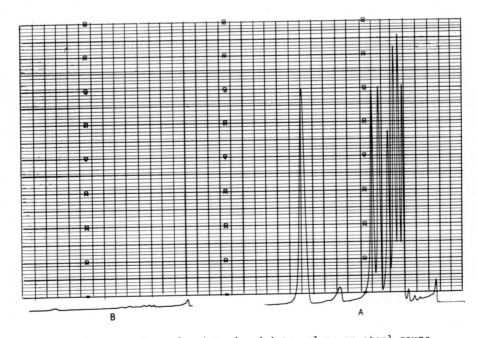

Fig. 4. GLC of samples introduced into column on steel gauze.

In another series of experiments several 17-oxosteroids were first transferred on gauzes, then eluted with ethanol and finally analyzed by Zimmermann reaction. Again the steroids were recovered quantitatively. Further evidence of the efficiency of the transfer is shown in the next slide.

Sample A, consisting of a mixture of polyoxosteroids in 0.2 ml of chloroform was transferred on a stainless steel gauze by the technique described, then the gauze was removed from the indentation in which evaporation took place and the walls of the dimple were washed with chloroform. A second gauze was dipped into this solution which on evaporation gave sample B. Figure 4 shows the gas chromatograms of samples A and B.

Figure 5 shows the actual appearance of the loading port adapted to our instrument. The side arm for the storage of analytical samples is 8 cm long and can accomodate up to 12 samples. The perforated tubing which fits into the evaporation chamber (Fig. 3) is 7 cm long and 10 mm i.d. and can hold 50-60 rings.

Figure 6 shows a variant of the described technique. In this case, the upper part of the cone ends in a stopcock fitted with a teflon key which has a blind hole. When the operating conditions of the instrument are reached the gauze carrying the analytical sample is deposited in the hole, in its vertical position with the open end at the top. On turning the key 180 C the gauze drops into the perforated tube. My experience of nearly three years with this technique indicates that the following points are critical:

1. Solutions in chlorinated solvents give best results for the transfer of the samples on the gauzes. Chloroform, methylene dichloride and ethylene dichloride have been used with equal success; ethanol, methanol and benzene may be used but do not effect a quantitative transfer; this can be obviated by the use of an internal standard.

2. The surface of the dimples in the teflon plates must be perfectly smooth.

3. Very crude extracts leaving on evaporation of the chloroform an oily residue on the dimple may be transferred only partially on the gauze.

4. Towards the end of the evaporation it is best to leave the teflon plates undisturbed. At the final stages of the evaporation the solution is very concentrated and a mechanical separation of a fraction of it from the main portion of the solution may result in substantial loss of material.

The principal merits of the technique may be summarized as follows:

1. Transfer of the analytical sample into the column is quantitative.

2. The solute can be transferred to the gauze and then introduced in the column independently of the volume of the solution with minimal manipulation.

3. Conditions of equilibrium of the system are not disturbed by the introduction of samples.

4. A fraction or the total amount of the material available can be introduced in the column with the same accuracy. This feature is extremely important when the quantity of material available is small.

5. The technique is essentially simple.

MR. THOMAS: Dr. Menini's solid injection system has been automated by two of my colleagues. There is an electrical device which turns a rotor which is encased inside a glass case. The base of the rotor is a teflon ring, perforated with 12 holes. There is an electronic device that turns the teflon rotor around at a given time interval dropping the gauze into the top of the column.

There is one disadvantage with this apparatus in that it needs to be flushed through with the carrier gas for at least an hour before you start GLC. Otherwise you seem to get air pockets that leak through the system.

Fig. 5. Loading port for introduction of solid samples.

Fig. 6. Loading port for introduction of solid samples.

DR. SOMMERVILLE: Can I show two slides very quickly?

We have never used any other technique but the gauze type of solid injection and Fig. 7 shows a simple modification which is based upon Dr. Menini's work and similar to that previously used by Mr. Thomas. It consists of a sleeve of glass which goes into the top of the column which has been widened for this purpose and magnetic gauzes are loaded in the side arm. Apart from the side arm, this is within the column oven and, with our equipment (Pye Panchromatograph), a flash heater is not required.

Figure 8 is taken from our paper in Nature last year (Collins and Sommerville, 1964) and shows the differential trace for progesterone derived from peripheral venous blood. There is, of course, no "Atlantic Ocean" preceding the peak and the trace is indistinguishable from that of authentic progesterone. Finally with regard to reproducibility, I should like to add support to what Dr. Menini has said. When I mentioned that our standards are reproducible perhaps I should write down the figure. If we run a series of standards of 0.1 μg tritiated progesterone, we obtain an integrated radioactivity of 1775 units ± SD 6.3 and the coefficient of variation is 0.04%. These points, I think, add something to the question of a choice between solid and liquid injection.

DR. GOLDFIEN: I would like to ask a technical question and that is, using QF-1 and column temperature of 260 C, how long does the column last? I ask this question because we originally used QF-1 and we had to change our column every three to four weeks and this was using it some place between 190 and 220 C.

DR. KLIMAN: The QF-1 columns are not stable at high temperatures. For testosterone analysis they are used at 205 C. Such columns have to be baked out at a high temperature for considerable time before they can be used with some degree of stability.

DR. WOTIZ: Dr. Kliman, what percentage was the SE-30 phase and by what manner was it prepared?

DR. KLIMAN: This brings up a controversial point. It was 2% SE-30. Dr. Wotiz has reported that aldosterone diacetate is recovered intact with 3% SE-30 but at concentrations lower than this there is degradation of the aldosterone. We haven't experimented with 3% columns because if the diacetate were recovered at higher substrate concentrations the retention time would be increased because of the extra acetate.

It does appear that the transformation of the diacetate on the 2% is essentially quantitative and we found less than 1% of a late eluting material that retained both acetates.

DR. WOTIZ: I do not know how important it would be to debate this point. We do obtain a single major peak for aldosterone diacetate. I am reasonably certain, in view of the infrared data, that the compound we reported is in fact aldosterone diacetate. Because of the apparent critical phase concentration the question of the accuracy of the deposition of stationary phase on the solid support should be considered.

I think one might discuss this whole question of column coatings and their effect on chromatography. There are two different coating procedures which are presently applied in different laboratories. One is the old-fashioned technique of simply dissolving the phase in hot solvent, adding the solid support and stirring until the packing is completely dry. Because of the stirring, fragmentation occurs as an undesirable side reaction.

Nevertheless, one can be certain that if a 3% deposit was prepared for, there will be a 3% coating. Naturally we were interested in better ways of preparing column materials. Dr. Horning sometime ago reported a very elegant filtration technique which prevents fragmentation of the solid support.

However, it has been our experience that the results obtained are highly variable. In six different attempts using exactly the same solution of SE-30 in dichloromethane and the same solid support the filtration technique yielded stationary phases varying in concentrations from 1.8 to 5.6%. Thus, if you used the filtration technique a calculated 2% phase may, in fact, contain much less substrate.

I believe you mentioned that with a 3% phase the retention time would be too long. I think this is overstating the case.

Fig. 7. Modification of Menini's solid sample loading device.

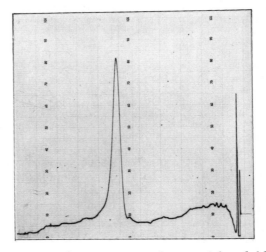

Fig. 8. GLC of progesterone from peripheral blood.

DR. KLIMAN: In answer to the question, I prefer to prepare coatings by the total evaporation technique as you described, applying a weighed amount of substrate to a weighed amount of column packing. When the filtration method was used, the percent coating on the substrate did not necessarily correspond to the percent content of the solution.

DR. VANDEN HEUVEL: We prefer the filtration method to the evaporation deposition approach. In our experience, when you use this filtration coating technique you do observe rather reproducible results.

If you take 20 g of support of the Gas-Chrom or Anakrom type and slurry it with 100 µl of an X% solution of the polymer (1% SE-30 in toluene, for example) you will get about an X% (w/w) coating. If you then take 1 g of the SE-30-coated material and strip off the polymer, you will remove very nearly .01 g. That is our experience. It is reproducible.

DR. KLIMAN: In our earlier experience trying to prepare 1% columns by the filtration technique we found that we had 2% columns because the substrate picked up a considerable volume of toluene solution of SE-30, and if you then evaporated it, at that point you would get a 2%. It may be that with continuous suction and removing as much of this solvent as possible, you might obtain a different percentage coating.

DR. VANDEN HEUVEL: It is important to remove most of the solution from the filter cake by several minutes of suction. The packing at this point should appear to be damp but not wet. In the case of SE-30, one should continue to pull vacuum until the frothy toluene solution ceases to come through the funnel.

DR. VAN DER MOLEN: I would like to make one comment as far as these coating techniques are concerned and I tend to agree with Dr. Vanden Heuvel about the filtration technique. When we started using this technique, we had the same troubles as I believe several other people had. We found rather large variations among our different batches. At this moment I am convinced that these variations resulted mainly from the fact that we did not standardize the times of impregnation and the technique of filtration too well. The day that we started to follow exactly the prescription given by Dr. Horning et al. (Methods of Biochemical Analysis 11: 69, 1963) we obtained constant results.

I would like to ask Dr. Gottfried a question. May I start by saying that I don't have any personal experience with GLC of corticosteroids, so I will not be able to discuss the validity of the presented work on such grounds. But I just want to ask Dr. Gottfried if you could agree purely from a chemical point of view that it might not be ideal, even if you find a constant breakdown of say, 5% or 10%, to inject such compounds on a column. Do you think that it might be possible to do a preliminary oxidation of your compounds as has been done by Drs. Bailey and Kliman in their investigations? I would appreciate if you could elaborate on this point.

DR. GOTTFRIED: I don't know exactly how you would like me to elaborate on this. What I am trying to say is that this can be a useful method for cortisone. A constant 5% breakdown gives you one great advantage. You have two peaks whose retention times are characteristic of the compound in question. This is quite all right, chemically speaking, provided the breakdown is constant. I do accept the fact that it would be preferable to have a compound which does not pyrolyze.

Regarding oxidation, I don't know whether Dr. Bailey has done any work on his oxidation, to find out whether, in fact, it is quantitative at submicrogram levels.

DR. VAN DER MOLEN: You quoted a constant 5% breakdown, but would you agree that this might be constant only because you are using very carefully defined conditions for GLC. An unexpected, or unnoticed change in conditions might result in a different percentage breakdown. We should not advocate this technique as a general technique. I would rather prefer, though this may be a personal preference, to put a stable product onto the column. Would it not be preferable to oxidize your compounds before GLC? I don't like to put something onto a column that is not in its final form. We may have a difference of opinion on this and it may largely be a problem of "feeling" the way we do.

DR. EIK-NES: I don't know if it is advisable to get into the hornet's nest at this point, but since I have done it before I may as well give it a try again. When we are exposing steroids like the adrenocorticoids to GLC and find that the tracings from these experiments contain one or several peaks, shouldn't we exercise some care in the interpretation of data by stating the theoretical plate number for the steroid on the column used? If you have a low theoretical plate number for your column you would not see whether you had one or multiple peaks, and what appeared to be a condition of stability may indeed be a masked picture of thermolability for the chromatographed adrenocorticoid.

I was much excited about Dr. Rosenfeld's publication in *Steroids*. Since the cortols and the cortolones are stable compounds on the GLC column, have you, Dr. Rosenfeld, considered isolating the known adrenocorticoids by conventional techniques, converting these steroids chemically or enzymatically to their corresponding cortols or cortolones and isolate and quantitate these transformation products as their TMS derivatives by GLC? This may lead to a way of doing adequate assay of adrenocorticoids by GLC.

DR. ROSENFELD: Yes, we considered this when we found that the TMS ethers of the cortols and cortolones were quite stable. I agree with Dr. Eik-Nes that enzymatic reduction of the 20 ketone group of cortisol would result in a compound whose TMS ether would be stable in GLC. This would be a worthwhile method to pursue.

DR. VANDEN HEUVEL: I think this work with the TMS ethers is wonderful but Dr. Gottfried, could you explain this 5% breakdown? Some of us are a bit confused.

DR. GOTTFRIED: To start with, the cortisone is converted to a fully silylated derivative. After GLC we always get a peak due to an 11β-silyl ether. When 3 to 6 µg of cortisone-TMS ether is injected, this peak always corresponds to 5% to the amount injected.

DR. GARST: Dr. Kliman, what kind of recovery do you get? Is the percent recovery more or less constant?

DR. KLIMAN: You mean injected into the column? It averages 62% $\pm$ 10%.

DR. GARST: You get that without priming the column?

DR. KLIMAN: The column must be primed frequently.

DR. GARST: We have been using diatoport S as a solid support. We found the aldosterone diacetate which we used as a standard was retained on the column to such an extent that by adding successive equal aliquots we kept getting increasing recoveries which never seemed to reach a plateau and consequently we were unable to use this as a quantitative assay. We are now attempting to use glass beads for the solid support.

I wonder what success you have had with solid supports. According to Dr. Neher, a chromosorb-P catalyzes the decomposition of aldosterol diacetate so that you not only get a free compound along with the diacetate but you also get some other degradation products and these all eluate together and he suggests that gas chrome-P be used instead.

DR. KLIMAN: I would only repeat that frequent priming of the column is necessary. This is also especially true of the other corticosteroids and their derivatives. Occasionally it is a problem even with more stable compounds.

DR. GARST: We never came to the end of the priming.

DR. CRANE: We tried the glass beads enough to know that we have a much larger solvent peak than we would like to have and there was poorer separation of steroids such as cortisol and cortisone.

I would like to ask one question. Why have we not heard much about chromosorb-G?

MISS PATTI: We have tried chromosorb-G with SE-52 and find that it is fairly good for the compounds of interest to us. We have also used SE-52 coated on other supports, such as gaschrom-P, Anachrom ABS and Gaschrom-Z. Of all the supports, chromosorb-G gave us better resolution of the variety of steroids studied. Chromosorb-G is manufactured by Johns-Manville.

To pack a 6 ft column of 5 mm i d, 25 g of chromosorb-G are needed whereas with gaschrom-P only 11.5 g are used. This may be the only disadvantage.

Our interest in this support arose from the report by the company that it could be useful for the separation of polar compounds. It has made possible the detection and separation of several $\Delta^5$-pregnenetriols with SE-52. We are particularly interested in these steroids possessing vicinal hydroxyl groups, namely the 17,20- and 20,21-diols. We have had no problems with the 17,20-diols, however, we have not as yet studied the 20,21-diols.

I do not know the stability of chromosorb-G. We have had a column of SE-52 coated on this support in use for only two months. QF-1 was also investigated with this support. Our data indicated that Anachrom ABS is a much better support for this liquid stationary phase.

DR. KEUTMANN: We have tried both Dr. Horning's and Dr. Wotiz's way of coating the columns and we found one method as good as the other. However, if you stir too fast there is quite a fragmentation in Dr. Wotiz's method.

If you put it into a flask and rotate it then you don't get as much fragmentation as when you stir it.

DR. WOTIZ: Let me just say with respect to the problem of coating that as long as phase concentration and column performance is approximately what is wanted and provided the compound to be analyzed remains stable, the method of column preparation is inconsequential.

As Dr. Keutmann pointed out, if you rotate and evaporate there is a decrease in fragmentation. In our laboratories the fragments or "fines" are removed by re-suspending the material in ethanol and floating them off. Then the alcohol is decanted and the solid dried in an oven.

Lastly, there is a new phase, commercially available, called gas chrom Q, which is an especially hardened support. We find that even after vigorous stirring during evaporation almost no fragmentation occurs.

DR. CRANE: The information from Johns-Manville states that gas chrom-Q is more inert than either the chromosorb-P or the other supports that they have.

DR. VANDEN HEUVEL: The firm that produces this material claims less decomposition of cholesterol using the Q support with the QF-1 liquid phase.

DR. FALES: I am concerned that everybody is getting quite different results with similar phases and nearly similar packing materials.

I think some of this is probably due to differences in preparation of the packing. Some of it is due to the subtle differences in techniques such as injection. Some of it is undoubtedly due to variation in the equipment that is used. If a piece of equipment has a hot spot or a cold spot of course you will get different results. It is most important to try for better and better columns even if it breaks your heart to throw away a column that is moderately good.

DR. TAIT: I would like to ask Drs. Kliman and Lipsett a question.

Using double isotope methods and paper and thin-layer separation techniques it is rather easy to obtain blank values of one nanogram. It is, however, rather difficult using these methods to reduce this blank further. I would like to ask if it is recommended that a gas rather than an old-fashioned step at the end of the procedure would be more efficient to do this.

DR. KLIMAN: For aldosterone the problems produced by gas chromatography probably equal the problems that are solved by GLC.

In general one can say that the methodology is not much improved. Specificity, accuracy, and precision are not improved, but operating time is reduced for a limited number of samples. The advantage seems to be in making some type of preliminary patient diagnosis and determining whether to proceed with very extensive research studies or whether to drop the study at an early stage.

It has been my impression from studies with testosterone acetate that radioactive reagent blank material, whatever it is, does not follow testosterone acetate successfully through GLC. The blanks with the method using double isotope gas chromatography generally are anywhere from zero to one-third of a m$\mu$g in the product at the final stage of analysis.

I recommend GLC for testosterone acetate derivatives and I believe Dr. Lipsett agrees.

DR. LIPSETT: I wish we could do as well as Dr. Kliman in getting rid of the non-specific contaminants. Our blanks still average 0.02 µg, occasionally 0.03 µg just running water through the procedure. I would guess that as one uses more chromatographic systems following acetylation the blank will be reduced even further.

If Dr. Kliman is getting down to zero I think he has proved this.

DR. KLIMAN: We use one additional paper chromatography step before GLC and this probably accounts for the blank difference.

DR. TAIT: After using this additional paper chromatographic step how much excess tritium do you have at that point?

DR. KLIMAN: We have gone up to six chromatographic steps on paper and TLC without obtaining a similar blank. Prior to GLC the reagent blank is 3 to 10 mµg.

# MEASUREMENT OF URINARY STEROIDS BY GAS CHROMATOGRAPHY. $3\alpha$, 17-DIHYDROXYPREGNANE-20-ONE AND $C_{21}$ TRIOLS

R. S. Rosenfeld, Institute for Steroid Research, Montefiore Hospital and Medical Center, New York

It is the purpose of this report to present GLC methods for the estimation of $3\alpha,17$-dihydroxypregnane-20-one (17-OHP), pregnane-$3\alpha,17,20\alpha$-triol (PT), 5-pregnene-$3\beta,17,20\alpha$-triol ($\Delta^5$PT) and several isomeric $\Delta^5$-pregnenetriols in partially purified urinary extracts. Elevated levels of these compounds are frequently associated with hyperactivity of the adrenal cortex. They are difficult to estimate by the alternative techniques such as measurement of acetyldehyde after oxidation of the side chain or by the formation of sulfuric acid chromogens. In our laboratories PT and $\Delta^5$PT are routinely measured by GLC and satisfactory agreement has been obtained with the more laborious procedure of measurement as acetaldehydrogenic steroids.

## EXPERIMENTAL

*Isolation and measurement of compounds.* Fractions containing 17-OHP, PT, $\Delta^5$PT or isomeric $\Delta^5$-pregnenetriols were obtained from urine by methods described by Fukushima and his associates (1,2). Chromatography of the extract containing 17-OHP was carried out on Whatman No. 1 paper (18 x 118 cm) at 24 C for 48 hr. The stationary phase was methanol:water (3:1) and the mobile phase was 2, 2,4-trimethylpentane:toluene (3:1). The area containing 17-OHP was located by a pilot strip (46-50 cm from origin). For the five $C_{21}$-triols, the non-ketonic fractions of urine were chromatographed on paper in the system, 2,2,4-trimethylpentane:toluene:methanol:water (3:5:4:1) and the appropriate areas were located with reference compounds. In this system, PT (93-98 cm) moved well ahead of 5PT (65-73 cm) while 5-pregnene-$3\alpha,16\alpha,20\alpha$-triol was located between them. Both $20\beta$-hydroxy compounds, 5-pregnene-$3\beta,16\alpha,20\beta$-triol and its $3\alpha$ epimer were much more polar (20-30 cm). The areas containing the compounds were cut from the paper and extracted with warm methanol. Alternatively, for PT and $\Delta^5$PT, a portion of the non-ketonic extract was applied in a narrow band to a 20 x 20 cm glass plate coated with silica gel G and the chromatogram was developed for 20 min in the system ethyl acetate:cyclohexane (7:3). Location of the appropriate areas and extraction was carried out by usual procedures.

GLC analyses of 17-OHP were made only on the underivatized compound. PT and $\Delta^5$PT were measured as the free compounds or after conversion to the TMS ether derivatives; the three pregnanetriols containing a $16\alpha$-hydroxy group were measured only as TMS derivatives. For the free compounds, a portion of concentrated extract from the chromatograms was dissolved in 1:1 chloroform-methanol and an aliquot injected into the column with a 10 μl syringe. TMS derivatives were prepared in the following way: paper extracts were filtered through a celite pad to remove debris and, after concentrating the filtrate, about 0.1 mg of cholestanol was introduced as an internal standard. The mixture was dissolved in approximately 0.4 cc of pyridine to which was added 40 μl of hexamethyldisilazane and 8 μl of trimethylchlorosilane. The solution was sealed with a teflon cap and remained overnight. After removal of the reactants at about 60 C with a stream of nitrogen, the residue was mixed with 0.2 cc of chloroform and a portion was injected in the column.

*Gas chromatography.* For the underivatized 17-OHP, PT or $\Delta^5$PT, analyses were carried out with a 1.8 m x 5 mm glass column packed with SE-30 (3% by wt) supported on 100-140 mesh Gas Chrom P. The column was maintained at 235 C and during the determinations, the argon pressure was 30 psi; an ionization detection system with a radium source was used. Analytical samples of the compounds to be measured were employed as reference standards and the standard curves shown in Fig. 1 were constructed in the usual way (3, 4). Relative retention times as compared with cholestane (7.8 min) were 0.93 for 17-OHP, 1.08 for PT and 1.10 for $\Delta^5$PT. In addition, approximately 1 mg of each standard was chromatographed, collected, and examined by infrared spectrometry. After acetylation, the PT and $\Delta^5$PT eluates showed spectra identical with pregnane-$3\alpha,17,20\alpha$-triol-3,20 diacetate (chloroform, 1150-800 cm$^{-1}$) and 5-pregnene 3β,17,20α-triol-3,20 diacetate (carbon disulfide, 1200-900 cm$^{-1}$) respectively. The 17-OHP eluate afforded a spectrum identical with 3α,17-dihydroxy-17β-methyl-D-homoetiocholane-17α-one (chloroform, 1150-800 cm$^{-1}$).

TMS ethers of the $C_{21}$-triols were chromatographed under identical conditions except that the column temperature was 226 C. Authentic samples of PT, 5PT, 5-pregnene-3α,16α,20α-triol, 5-pregnene-3α,16α,20β-triol, and 5-pregnene-3β,16α,20β-triol were derivatized and the standard curves are shown in Fig. 2. Relative retention times of TMS ether derivatives as compared with cholestane (8.4 min) were 1.30 (PT), 1.55 (5PT), 1.13 (5-pregnene-3α,16α,20α-triol), 1.22 (5-pregnene-3α,16α,20β-triol) and 1.63 (5-pregnene 3β,16α,20β-triol).

## RESULTS AND DISCUSSION

Fractions containing 17-OHP were sufficiently purified by paper chromatography so that GLC of the material extracted from the appropriate area afforded tracings as in Fig. 3. Values for 24 hr urinary excretion of 17-OHP by this method ranged from 0-0.3 mg in six subjects with apparently normal adrenals and were 2.1, 7.1, and 8.9 mg in three patients with adrenocortical hyperactivity. The formation of 3α,17-dihydroxy-17β-methyl-D-homoetiocholan-17α-one during GLC is quantitative and does not affect the accuracy of the determination. Fukushima and co-workers obtained this D-homo compound by heating 17-OHP above its melting point (5).

Estimation of urinary PT and 5PT usually involves quantitation of appropriate fractions eluted from columns of paper chromatograms (6-12) and the methods have been reviewed (9). The most specific appears to be the acetaldehydogenic method of Cox (6). In 26 determinations where the GLC method was compared with the Cox procedure, PT values ranged from 0.75 to 25.2 mg per day in subjects with a variety of clinical conditions and averaged about 10% lower than corresponding values obtained by the acetaldehydrogenic method. In 21 analyses for 5PT by both methods, values ranged from 0.27 to 33.5 mg per day by GLC averaging 5% lower than by the Cox procedure. The daily excretion of PT in subjects with normal adrenocortical function ranged from 0.13 to 1.0 mg which is in reasonable agreement with GLC data reported by Kirschner and Lipsett (13). The daily excretion of 5PT in similar subjects varied from 0.10 to 0.91 mg. These data were obtained by analyses of fractions examined as the underivatized material or as the TMS derivatives containing an internal standard; there were no significant differences in the values. In our hands, however, measurement of the TMS derivatives of PT and 5PT has proven superior. Whereas about 5-8 μg of underivatized steroid gave an optimum peak in the gas chromatogram (requiring about 100-200 μg of PT or 5PT in the paper or TLC extract for ease in dilution and removal of aliquots for injection), from 1-2 μg of the corresponding TMS ether afforded an equivalent response. In addition, the derivatives appeared to be less susceptible to vagaries of column performance. Therefore derivatization of the extracts prior to injection is the method of choice. Figure 4 shows typical chromatograms of fractions containing PT; 4a was run without derivatization while 4b is a chromatogram of the TMS ethers containing cholestanol TMS as the internal standard. Figure 5 illustrates corresponding chromatograms of 5PT extracts. Since quantitation of PT and 5PT by GLC is rapid, convenient and in acceptable agreement with alternate techniques, it is used in these laboratories for the routine measurement of these triols.

With regard of the three $\Delta^5$-pregnenetriols containing a 16α-hydroxy group, GLC of the appropriate extracts from the paper chromatograms permits ready identification and quantitation of the individual compounds. The two 20β-hydroxy compounds epimeric at C-3 frequently migrate together but are easily separated from the less polar 5-pregnene 3α,16α,20α-triol. GLC afforded clear resolution of the compounds, since the relative retention times of 5-pregnene-3α,16α,20β-triol-TMS ether and its 3β-hydroxy epimer were 1.22 and 1.63 respectively. 5-pregnene-3α,16α,20α-triol runs between 5PT and PT from which it may not always be cleanly separated; however, the shorter relative retention times of its TMS ether (1.13) compared with 1.30 and 1.55 for PT-TMS ether and 5PT-TMS respectively allowed facile analysis by GLC. Figure 6 shows a chromatogram of a paper eluate containing 5-pregnene-3α,16α,20α-triol after TMS ether formation.

## STEROID STANDARDS FOR GAS CHROMATOGRAPHY

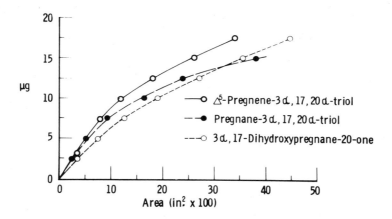

Fig. 1. Quantity vs peak area for underivatized compounds 17-OHP (0- - -0), 5PT (0———0), PT (0 - - 0-). Reproduced by permission *J Chromatog*.

## TRIMETHYLSILYL ETHER DERIVATIVES

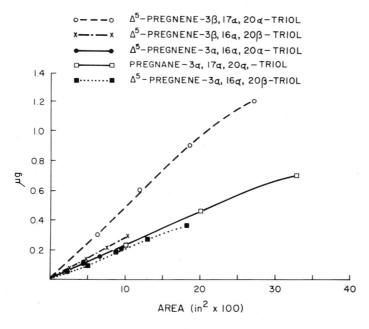

Fig. 2. Quantity vs peak area for TMS ethers. 5PT (-0--0-), 5-pregnene-3β,16α,20β-triol ( x---x ), 5-pregnene-3α,16α,20-triol (●——●), PT (□——□), 5-pregnene-3α,16α,20β-triol (■·······■).

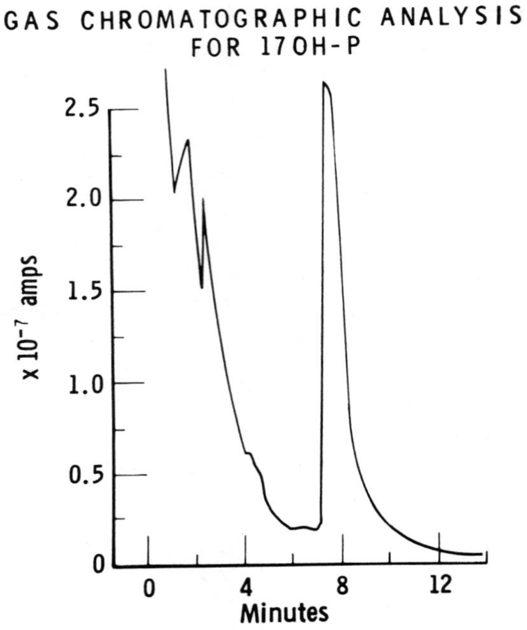

GAS CHROMATOGRAPHIC ANALYSIS
FOR 17OH-P

Fig. 3. Gas chromatographic analysis for 17-OHP.  Reprinted by permission
J Chromatog.

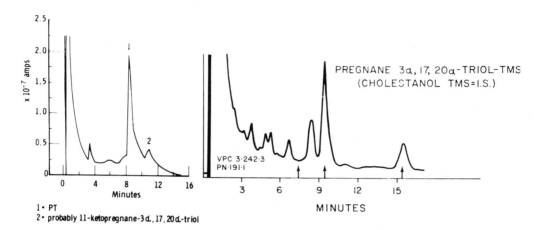

Fig. 4. GLC analysis for PT. 4a: underivatized. 4b: TMS ethers (PT-TMS = 9.3 min, cholestanol-TMS = 15.2 min).

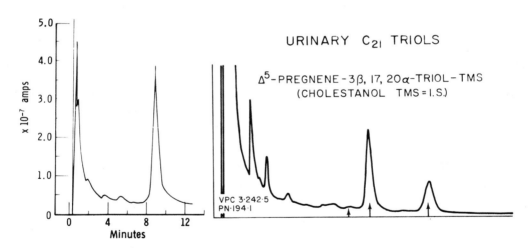

Fig. 5. GLC analysis for 5PT. 5a: underivatized. 5b: TMS ethers (5PT-TMS = 10.3 min, cholestanol-TMS = 15.2 min).

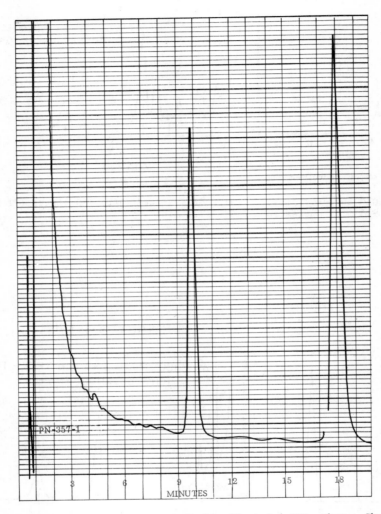

Fig. 6. GLC analysis for 5-pregnene-3α,16α,20α-triol-TMS ether. Cholestanol-TMS ether = 18 min.

The convenience and rapidity of GLC analysis for 16-hydroxy $C_{21}$ steroids are advantages over colorimetric determinations where $3\alpha$ and $3\beta$ epimers may not be readily differentiated and over enzymatic analyses where separate determinations are required for each epimer. These 3,16,20-triols may have significance in studies concerning progesterone and adrenocortical hormone metabolism as well as in estrogen biosynthesis.

## SUMMARY

GLC on SE-30 has been used to quantify $3\alpha$,17-dihydroxypregnane-20-one, pregnane-$3\alpha$,17,$20\alpha$-triol and 5-pregnene-$3\beta$,17,$20\alpha$-triol in eluates of paper chromatograms of urinary extracts. While the 3,17,20-triols may be measured either as free compounds or as the trimethylsilyl ethers, derivatization is preferable. In addition, 5-pregnene-$3\alpha$,$16\alpha$,$20\alpha$-triol, $5\Delta$-pregnene-$3\beta$,$16\alpha$,$20\beta$-triol, and 5-pregnene-$3\alpha$,$16\alpha$,$20\beta$-triol may be measured by this technique.

## REFERENCES

1. Fukushima, D. K., T. F. Gallagher, and O. H. Pearson, J Clin Endocr 20: 1234, 1960.
2. Fukushima, D. K., H. L. Bradlow, L. Hellman, B. Zumoff, and T. F. Gallagher, J Clin Endocr 21: 765, 1961.
3. Rosenfeld, R. S., M. C. Lebeau, S. Shulman, and J. Seltzer, J Chromatog 7: 293, 1962.
4. Rosenfeld, R. S., M. C. Lebeau, R. D. Jandorek, and T. Salumaa, J Chromatog 8: 355, 1962.
5. Fukushima, D. K., S. Dobriner, M. S. Heffler, T. H. Kritchevsky, F. Herling, and G. Roberts, J Am Chem Soc 77: 6585, 1955.
6. Cox, R. I., J Biol Chem 234, 1693, 1959.
7. Wilson, H., M. B. Lipsett, and D. W. Ryan, J Clin Endocr 21: 1304, 1961.
8. Stern, M. I., J Endocr 16: 180, 1957.
9. Bongiovanni, A. M., and W. R. Eberlein, Anal Chem 30: 388, 1958.
10. Nowaczynski, W., E. Koiw, and J. Genest, J Clin Endocr 20: 1503, 1960.
11. DeCourcy, C., J Endocr 14: 164, 1956.
12. Cox, R. I., Biochem J 52: 339, 1952.
13. Kirschner, M. A., and M. B. Lipsett, Steroids 3: 277, 1964.

# ANALYSIS OF URINARY PREGNANEDIOL AND PREGNANETRIOL BY GAS-LIQUID CHROMATOGRAPHY

Marvin A. Kirschner and Mortimer B. Lipsett, Endocrinology Branch, National Cancer Institute, Bethesda, Md.

The measurement of urinary pregnanediol and pregnanetriol are of clinical importance. Methods used previously have generally depended on the formation of sulfuric acid chromogens after purification of the steroid (1-3). The color reaction with sulfuric acid is non-specific and for pregnanediol, sensitivity is poor. We therefore have used GLC for the measurement of pregnanediol and pregnanetriol in urine extracts.

## GAS-LIQUID CHROMATOGRAPHY

A dual-column gas chromatographic system designed by the Glowall Corp., Glenside, Pa., was used in these studies. The columns consisted of a matching pair of glass spirals 6 ft x 3.4 mm i
One column was packed with Gaschrome P, 60-80 mesh coated with 1% SE-30, and the other was similarly packed with either 2% XE-60 or 1% NGS, resulting in a "polar" and "non-polar" set of columns either of which could be used by altering the effluent lead to the detector. All glassware and supports were presiliconized with dimethyldichlorosilane, after the method of Haahti (4). The detector consisted of a Lovelock Argon Ionization Detector, with radium foil operating at 850 C. Peak areas were estimated as the product of peak height and retention volume. Standard curves were linear from 0.5 to 2.0 µg. From 2.0 to 4.0 µg the curve was linear, but with a slightly greater slope. It was therefore considered essential to select proper dilutions of the urine fractions so that peak areas would fall between the upper and lower limits of the standard curve.

## METHODS

One-tenth of a 24 hr urine collection is incubated with 500 units per ml of Ketodase® overnight and extracted twice with an equal volume of anhydrous ether. The ether extract is washed with 1N sodium hydroxide, and water until neutral, and evaporated under reduced pressure.

The dried extract is dissolved in acetone, and applied across 3-5 cm on a 0.375 mm silica gel plate (Silica Gel G, Brinkman Co., Great Neck, N. Y.). Generally at least two samples could be applied to a standard 20 x 20 cm plate. Twenty µg of pregnanetriol standard are applied at each end of the plate, and the chromatogram is developed for 45 min in the system benzene:ethyl acetate (40:60). After drying in air, the plate is sprayed with 0.1% solution of Rhodamine 6G in ethanol and examined under UV light, where the standards and urinary metabolites appear as orange lines on the yellow background. The zone opposite the pregnanetriol standard is eluted with 5 ml acetone, dried under an air stream, taken up in a proper dilution of acetone, along with an appropriate marker, for introduction into the column. If pregnanetriol is to be converted to the TMS ether before GLC, the dried thin-layer eluate is dissolved in 0.2 ml chloroform and 0.2 ml hexamethyldisilazane and 5 drops of trimethylchlorosilane are added. The mixture is incubated at 56 C for 1/2 hr, dried under a stream of dry air and the TMS ethers taken up in hexane for introduction into the SE-30 column. On occasion, some SE-30 columns have been found to adsorb highly polar steroids such as pregnanetriol. With these columns, the TMS ethers are useful since they chromatograph well on XE-60 at 215 C with an excellent area response.

For determination of pregnanediol, the dried ether extract is acetylated with 10 drops acetic anhydride and 5 drops pyridine, and incubated in a capped vessel overnight at 25 C or for 3 hr at 56 C. Excess reagents are removed under a stream of dry air, and the acetylated extract is applied to a thin-layer plate with a pregnanediol diacetate reference standard. The plate is developed in the thin-layer system, benzene:ethyl acetate (80:20) for 45 min. The zone opposite the standard is eluted with acetone, and aliquots taken for analysis on SE-30.

## RESULTS

The effect of acetylation of the pregnanediol before TLC and GLC is demonstrated in Fig. 1. GLC of pregnanediol after TLC resulted in a chromatogram with several interfering peaks. If, however, the crude urine extract was acetylated, prior to TLC and GLC, the pregnanediol diacetate is shifted away from background impurities, making accurate quantitation possible as well as enabling use of a larger aliquot. The majority of the $C_{21}O_2$ and $C_{21}O_3$ steroids would be separated from pregnanediol diacetate after TLC (Table 1), with the exception of allopregnanediol-diacetate, pregnenediol diacetate and pregnanolone acetate. Subsequent GLC on SE-30 (Table 1) permits measurement of pregnanediol diacetate without significant interference from these other $C_{21}O_2$ acetates. If appreciable amounts of allopregnanediol were present, it would appear as a shoulder on the descending limb of the pregnanediol peak.

### Validity of the Methods.

Specificity. The urinary steroids were identified by their $R_f$'s in the thin-layer system and relative retention times in GLC. Retention times of these steroids differed by less than 2% from those of authentic standards. When authentic steroids were added to urine, the appropriate peaks were augmented and no new peaks were seen.

Sensitivity. Specific studies were not conducted to outline the sensitivity of the method. However, from experience with many analyses, the excretion of 50 μg per 24 hr of these metabolites can be accurately measured using aliquots of 10-20% of a 24 hr sample.

Precision. To establish precision for pregnanetriol measurements, a single 24 hr urine was incubated with Ketodase®, extracted with ether, then divided into 8 equal aliquots, each of which was processed through TLC and GLC. The coefficient of variation was 3% at a 24 hr excretion level of 0.55 mg. To establish a precision estimate for the pregnanediol determination, 8 equal aliquots of a 24 hr urine were separately carried through the entire procedure, starting with enzyme incubation. At a level of excretion of 1.9 mg per 24 hr, the coefficient of variation was 5%.

Accuracy and Recovery. In Table 2 are listed pregnanediol values found in each of five separate urines. To duplicate aliquots were added the equivalent of 0.5 mg per 24 hr of pregnanediol, and recoveries are indicated. Eight urines were similarly studied for pregnanetriol. Four samples were analyzed by direct injection of the thin-layer eluate on 1% SE-30, while the other four were converted to TMS ethers and analyzed on 2% XE-60. As noted, either GLC method results in excellent recoveries.

We reported previously (5) that measurement of urinary pregnanediol by GLC gave results considerably lower than those obtained using the sulfuric acid chromogen following a single column chromatography. We have subsequently analyzed the pregnanediol diacetate fractions obtained using the Klopper procedure (1). Table 3 presents these results and it is apparent that in almost every case the values obtained from GLC are considerably less than those obtained with the Klopper procedure. When the pregnanediol diacetates taken from the columns were chromatographed on SE-30 the following tracings were obtained. The pregnant subjects had a single peak, and one of these is shown in Fig. 2. However, in the normal subject and the patients with amenorrhea and ovarian cysts, multiple peaks were obtained (Fig. 2 and 3).

Urinary Pregnanediol and Pregnanetriol in the Normal Subject. The excretion of pregnanediol in the normal male averaged 0.41 mg per 24 hr. In the normal menstruating woman, pregnanediol excretion during the proliferative phase averaged 0.55 mg per 24 hr, and during the luteal phase increased to an average of 4.0 mg per 24 hr. The excretion of pregnanetriol averaged 0.89 mg per 24 hr in men and 0.77 mg per 24 hr in women. In two normal women, followed through four menstrual cycles, pregnanetriol excretion doubled the luteal phase concomitant with the rise in pregnanediol excretion.

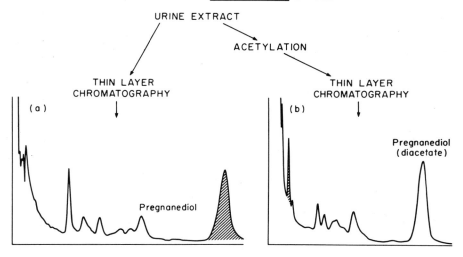

Fig. 1. Comparison of gas chromatograms of pregnanediol and pregnanediol diacetate.

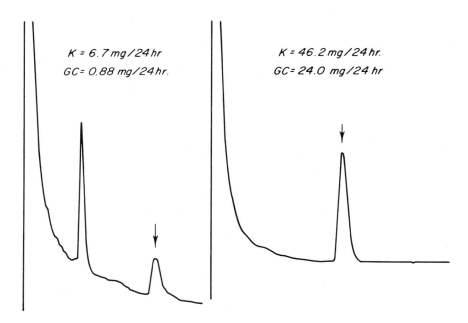

Fig. 2. Gas chromatograms of pregnanediol diacetate fractions after the second column chromatography of the Klopper method. Arrow points to pregnanediol diacetate peak. Values for pregnanediol excretion calculated by Klopper method and GLC are shown. Left panel - patient with ovarian cyst; right panel - pregnancy.

TABLE 1. Chromatographic characteristics of acetates.

| Trivial name | $R_f$ Thin-layer* | Relative Retention time GLC** |
|---|---|---|
| Androsterone acetate | 0.39 | 0.53 |
| Etiocholanolone acetate | 0.39 | 0.53 |
| Dehydroepiandrosterone acetate | 0.39 | 0.55 |
| Pregnanediol diacetate | 0.71 | 1.32 |
| Allopregnanediol diacetate | 0.71 | 1.49 |
| Pregnanolone acetate | 0.62 | 0.83 |
| Pregnenediol diacetate | 0.64 | 1.44 |
| Pregnenolone acetate | 0.57 | 1.00 |
| 17-Hydroxypregnanolone acetate | 0.39 | 1.17 |
| 17-Hydroxypregnenolone acetate | 0.39 | 1.29 |
| Pregnanetriol diacetate | 0.43 | 1.92 |
| 20β-hydroxyprogesterone acetate | 0.44 | 1.23 |

*Thin-layer system, benzene:ethyl acetate (80:20).

**Retention times relative to cholestane on 1% SE-30.

TABLE 2. Recovery studies.

| Pregnanediol mg per 24 hr | | | | Pregnanetriol mg per 24 hr | | | | Pregnanetriol mg per 24 hr (as TMS on XE-60) | | | |
|---|---|---|---|---|---|---|---|---|---|---|---|
| det'd. | added | expected | found | det'd. | added | expected | found | det'd. | added | expected | found |
| 0.32 | 0.50 | 0.82 | 0.86 | 1.90 | 0.50 | 2.40 | 2.50 | 0.21 | 0.50 | 0.71 | 0.78 |
| 0.57 | 0.50 | 1.07 | 1.09 | 1.33 | 0.50 | 1.83 | 1.83 | 0.80 | 0.50 | 1.30 | 1.45 |
| 0.23 | 0.50 | 0.73 | 0.72 | 1.50 | 0.50 | 2.00 | 2.16 | 0.35 | 0.50 | 0.85 | 0.85 |
| 0.53 | 0.50 | 1.03 | 1.02 | 1.15 | 0.50 | 1.65 | 1.63 | 0.50 | 0.50 | 1.00 | 1.00 |
| 2.30 | 0.50 | 2.80 | 2.70 | | | | | | | | |

TABLE 3. Comparison of GLC of pregnanediol diacetate fractions obtained by Klopper method with sulfuric acid chromogen.

| Patient | Diagnosis | Urinary pregnanediol mg per 24 hr | |
| --- | --- | --- | --- |
| | | Klopper | GLC |
| At | Pregnancy | 22.0 | 11.7 |
| Sl | " | 30.0 | 28.8 |
| Go | " | 32.8 | 19.6 |
| Di | " | 46.2 | 24.0 |
| Vi | Ovarian cyst | 6.7 | 0.88 |
| Sp | Normal | 3.2 | 0.2 |
| Fr | Amenorrhea | 0.34 | 0.14 |

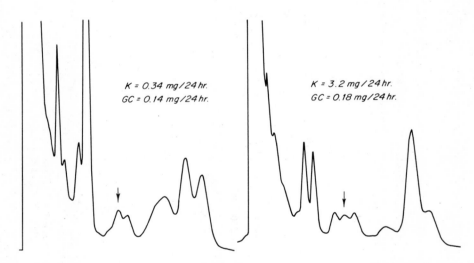

Fig. 3. Gas chromatograms of pregnanediol diacetate fractions after the second column chromatography of the Klopper method. Arrow points to pregnanediol diacetate peak. Values for pregnanediol excretion calculated by Klopper method and GLC are shown. Left panel - proliferative phase urine; right panel - luteal phase urine.

## DISCUSSION

The analysis of pregnanetriol by this scheme presented no undue problems, since preliminary TLC adequately reduced background contamination and served to isolate this metabolite from most other steroids found in urine. However, when the urine was treated with β-glucuronidase and followed by solvolysis, the pregnanetriol fractions were considerably more contaminated with background impurities. Thus the accuracy of pregnanetriol measurement can be increased by using the glucuronide fraction only.

The scheme presented here for analysis of urinary pregnanediol is more complex when compared to others reported. VandenHeuvel et al. (6) reported good separation of pregnanediol-TMS ether from the $C_{19}O_2$-17-ketosteroids on the phase, XE-60, and suggested that androsterone, etiocholanolone, dehydroepiandrosterone and pregnanediol could all be measured in the same sample. Although we used the same phase and conditions, we could not obtain comparable results with urine extracts and decided that pregnanediol should be isolated from the $C_{19}O_2$-17-ketosteroids before the TMS ethers were formed.

There are many reports of the measurement by GLC of urinary pregnanediol in pregnancy and simple solvent extractions of the hydrolyzed urines are sufficient. This is due, of course, to the large excretion of pregnanediol so that other steroids are diluted in the aliquots used. When these relatively simple procedures were extended to the measurement of pregnanediol in the urine of normal subjects (7, 8) the chromatograms demonstrated many poorly separated peaks of similar height and area among which the pregnanediol peak could not always be clearly differentiated from the background, and the precision of such methods was inadequate.

Acetylation of the crude extract and TLC prior to GLC resulted in elution of pregnanediol free of contaminants, and free of metabolites of progesterone, or closely related compounds. In several studies of pregnanediol excretion (7-10) these separations were not obtained and the pregnanediol peaks may have included such metabolites as allopregnanediol, pregnenediol, pregnenolone or 11-hydroxyandrosterone.

There has not been a systematic comparison of the method for urinary pregnanediol using column chromatography and sulfuric acid chromogen formation with the GLC method for pregnanediol. Jansen (11) and Cox (12) have observed several lower values by analysis with GLC. We have noted consistently lower values for pregnanediol excretion when measured by GLC as compared to those measured by the Klopper procedure. This discrepancy requires further study but might be explained by the relatively less specific nature of chromatogenic reactions with sulfuric acid, such that isomers and impurities would lead to higher values. This seems to be borne out by the greater discrepancy found at lower levels of pregnanediol excretion. The tracings of Fig. 3 are a graphic demonstration of the many impurities that are present in the pregnanediol-diacetate eluate in the Klopper method.

The excretion of pregnanetriol is comparable to that obtained by several different methods (2, 3, 13) as well as that reported by Rosenfeld et al. (14) using GLC of paper chromatogram eluates.

Measurement of the excretion of several other metabolites of progesterone can be made by the combined use of TLC and GLC. It appears that pregnenetriol can be analyzed as its TMS ether from the appropriate thin-layer eluate on SE-30, and pregnenediol can be measured as its diacetate along with pregnanediol. Since many of these minor metabolites are often present in quantities of 0.01 mg per 24 hr or less, further preliminary purification may be required.

## SUMMARY

Methods are described for the measurement of urinary pregnanediol and pregnanetriol in the normal subject. The use of GLC increases the specificity and accuracy of the analyses of these steroids.

## ACKNOWLEDGEMENT

The samples of pregnanediol diacetate obtained after the second column chromatography of the Klopper procedure were kindly provided by Dr. Claude Moyer, Department of Obstetrics and Gynecology, Albert Einstein College of Medicine, New York.

## REFERENCES

1. Klopper, A., E. A. Michie, and J. B. Brown, <u>J Endocr</u> 12: 209, 1955.

2. Bongiovanni, A. M. and W. R. Eberlein, <u>Anal Chem</u> 30: 388, 1958.

3. Fotherby, K. and D. N. Love, <u>J Endocr</u> 20: 157, 1960.

4. Haahti, E. O. A., <u>Mercatorin Kirjapaino</u>, Turku, Finland, 1961.

5. Kirschner, M. A. and M. B. Lipsett, <u>Steroids</u> 3: 277, 1964.

6. VandenHeuvel, W. J. A., B. G. Creech, and E. C. Horning, <u>Anal Biochem</u> 4: 191, 1962.

7. Turner, D. A., G. E. Seegar-Jones, I. J. Sarlos, A. C. Barnes, and R. Cohen, <u>Anal Biochem</u> 5: 99, 1963.

8. Patti, A. A., P. Bonanno, T. F. Frawley, and A. A. Stein, <u>Obst & Gynec</u> 21: 302, 1963.

9. Seegar-Jones, G. E., D. Turner, I. J. Sarlos, A. G. Barnes, and R. Cohen, <u>Fertility & Sterility</u> 13: 544, 1962.

10. Patti, A. A., P. Bonanno, T. F. Frawley, and A. A. Stein, <u>Acta Endocr</u> 42: Suppl 77, 1963.

11. Jansen, A. P., <u>Clin Chim Acta</u> 8: 785, 1963.

12. Cox, R. I. J., <u>J Chromatog</u> 12: 242, 1963.

13. Cox, R. I. J., <u>J Biol Chem</u> 234: 1693, 1959.

14. Rosenfeld, R. S., M. C. Lebeau, R. D. Jandorek, and T. Salumaa, <u>J Chromatog</u> 8: 355, 1962.

# THE APPLICATION OF GAS-LIQUID CHROMATOGRAPHY TO THE MEASUREMENT OF PLASMA PROGESTERONE

Alan Goldfien, M. Edward Yannone*, Darrell B. McComas and Carolina Braga, Departments of Obstetrics and Gynecology, Medicine and the Cardiovascular Research Institute, University of California School of Medicine, San Francisco.

*Part of this work was carried out while a U.S.P.H.S. trainee in the Department of Obstetrics and Gynecology.

This work was supported in part by NIH Research Grants HD 00640, H2685 and Training Grant 5TI RD-6.

Difficulties encountered in the assay of progesterone in biological fluids have hampered advances in our understanding of its secretion and metabolism. Investigators have, therefore, sought more sensitive, specific and accurate methods. A double derivative isotope technique capable of estimating the small amounts of progesterone present in normal human plasma has been described (1) and used to measure plasma progesterone levels throughout the normal menstrual cycle (2).

Methods utilizing GLC for the measurement of steroids have been described (3, 4) but in these methods quantification was limited to amounts which can be measured by traditional colorimetric and spectrophotometric technics. Although detectors capable of responding to minute amounts of steroid hormones have been developed, the presence of a variety of non-specific contaminants may interfere with the measurement of submicrogram quantities of steroids in biological fluids.

It has been possible by the use of preliminary purification by solvent partition and paper chromatography and by the use of a dry injector which allows quantitative transfer of a sample to the gas chromatograph (5) to develop a method capable of measuring the amounts of progesterone found in the plasma of normal and pregnant women (6). The following report describes this method with modifications.

## METHOD

*Materials and Apparatus.* Reagent grade solvents are employed throughout the procedure and redistilled before use. Chromatographic quality methylene chloride is used instead of diethyl ether as described in the original procedure. Standard solutions of progesterone are prepared by dissolving progesterone (Elite Chemical Company), purified by repeated recrystallization from ethanol and water, in 95% ethanol. Progesterone-7-$^3$H with a specific activity of 25 mg per mg (New England Nuclear Corp.) is purified by chromatography and an ethanolic solution containing 15,000 cpm (about 1 nanogram) per 10 µl is prepared. The phosphor solution is prepared by dissolving Liquifluor® (Pilot Chemicals) in freshly distilled toluene. GLC is done on a Research Specialties Company Model 600 chromatograph or Model 60-10 compact gas-liquid chromatograph equipped with a hydrogen flame ionization detector. Except where noted the data were obtained using a 28 in U-shaped stainless steel column, i d 2 mm packed with Gas Chrom CLH, 60/70 mesh, coated with 1% XE-60 (Applied Science Lab., Inc.) and conditioned overnight at 230 C in nitrogen. Analyses are performed at a column temperature of 195 C using nitrogen carrier at 24 psi. The vaporization chamber is kept at 280 C and samples are introduced through a rubber diaphragm with a solids injector held in the flash heater for one min. The solids injector described by McComas and Goldfien (5) has been modified so that in the extended position the spiral tip protrudes 3/4 instead of 3/8 in beyond the bevel of the needle (Fig. 1).

A Model 880 Autoscanner (Vanguard Instrument Co.) is used to locate progesterone-$^3$H on paper chromatograms and liquid scintillation counting is done in a liquid scintillation spectrometer (Packard Instrument Co.) at 1100 v.

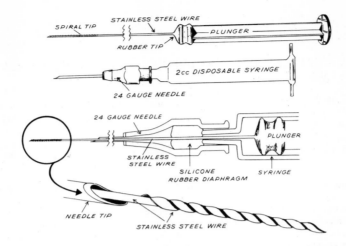

Fig. 1. The solids injector used for introducing the plasma extract into the gas chromatograph. The spiral tip protrudes 3/4 in beyond the base of the needle when extended.

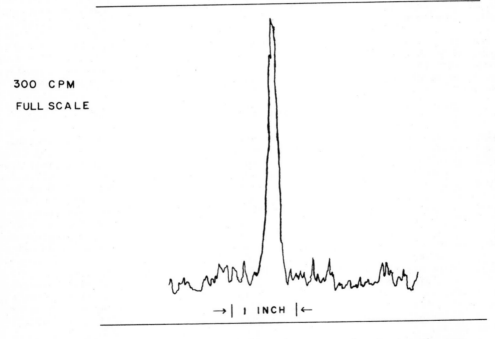

Fig. 2. Scan of paper strip before chromatography showing the concentration of the sample at the origin achieved by using ascending chromatography for application of samples to the paper.

Procedure. Plasma, obtained by the centrifugation of heparinized blood is stored at -15 C until used. Fasting samples are used when possible in order to facilitate purification by reducing the quantity of lipid present.

Ten µl (15,000 cpm) of the standard solution of progesterone-$^3$H are mixed with each plasma sample at the time of analysis and to each of two counting vials which are held for subsequent counting. Plasma is then extracted by gently shaking for ten min with one volume of methylene chloride. After centrifugation and transfer of the solvent to a clean test tube, the extraction is repeated. The combined extracts are washed successively with 1/10 volume each of NaOH, water, 0.3N acetic acid and water. The solvent is then evaporated to dryness by filtered air and the residue dissolved in 5 ml of warm 70% methanol.

The samples are then spun in a refrigerated centrifuge at -15 C for 60 min at 6,500 rpm. The liquid phase is carefully removed from the precipitated solids to a clean test tube and shaken with an equal volume of heptane. The heptane phase is re-extracted twice with 0.5 volume of 70% methanol, and the combined aqueous-methanol extracts are reduced to approximately 30% of the original volume by evaporation. The remaining liquid, which is predominantly water, is then extracted with 5 ml of methylene chloride.

The methylene chloride extract is taken to dryness and concentrated in the tip of a 10 x 75 mm test tube. The residue is then applied to 5 mm strips of paper (Whatman No. 1) for chromatography. Ascending chromatography reduces time required for spotting and is performed as follows. The residue is dissolved in four drops of methanol:benzene (1:1). The paper is dipped into the tube until the solution is absorbed. It is then removed and allowed to dry and the entire procedure is repeated two more times. Application of the extract in this manner results in its concentration on the paper at the level of the opening of the test tube. It is possible to spot a large number of chromatograms simultaneously with a minimum of effort using this method. Figure 2 shows a scan of a sample spotted in this manner. The spotting tubes reveal no appreciable residual radioactivity following this procedure. After equilibration for two hr in an atmosphere of 70% methanol-heptane the chromatograms are developed with heptane for two to four hr. In this system, the $R_f$ value for progesterone is 0.68. The paper is dried and the labeled progesterone located by means of a radioactivity chromatogram scanner. The radioactive portion of the strip is cut and eluted into special conical vials by adding one drop of water followed by 1 ml of 100% methanol. The eluate is then evaporated to dryness.

The residue is dissolved in 99 µl of 95% ethanol by thoroughly stirring with the needle of a 10 µl Hamilton syringe. Nine µl of the solution are removed and placed in a counting vial. The remaining solvent is evaporated and the residue transferred by four successive 10 µl washes to the insert of the solids injector. Each series of 1 µl droplets is allowed to evaporate on the insert for several min before addition of the next. Following completion of the transfer of the conical vial and the µl syringe are washed with methanol and the washings placed in another counting vial.

The injection of plasma extracts into the gas-liquid chromatograph is preceded and followed by the injection of progesterone standards. The progesterone peak areas (peak height x width at half height) are plotted against the corresponding mass and used as a standard calibration curve. The amounts of progesterone in the plasma extracts are then determined from this curve. After the addition of phosphor solution (15 ml) each vial is counted in the liquid scintillation counter.

The following formula is used for determining the concentration of progesterone in the initial sample.

$$\text{Progesterone (µg per 100 ml plasma)} = \frac{100 \; SM}{P \; (10 \; C - V)}$$

C = cpm of progesterone-$^3$H in the 1/11 aliquot of the final extract.

V = cpm of progesterone-$^3$H remaining in the 10 µl syringe and vial after transfer of the plasma extract to the solids injector.

S = cpm of progesterone-$^3$H added to each plasma sample as an internal indicator.

M = Micrograms of progesterone corresponding to peak on the gas chromatogram.

P = Volume of the plasma sample in ml.

## RESULTS AND DISCUSSION

Although it is possible under the conditions described to regularly detect two millimicrograms of a progesterone standard, it is necessary to have approximately seven mµg to insure reproducible results in samples extracted from biological materials. The original sample, therefore, should contain at least 15 to 20 mµg of progesterone to allow for recoveries of 30 to 40%.

In order to study the relationship between sample size, retention time and peak area, 34 samples containing 3 to 130 mµg of progesterone were injected. The relationship between mass and retention time is plotted in Fig. 3. The relationship between sample size and peak area is illustrated in Fig. 4 in which the data from the same experiment and four additional samples are plotted. The linear relationship observed in this range is also observed in the range of 0.1 to 1 µg. Representative tracings obtained from the injection of extracts of three plasma samples are shown in Fig. 5.

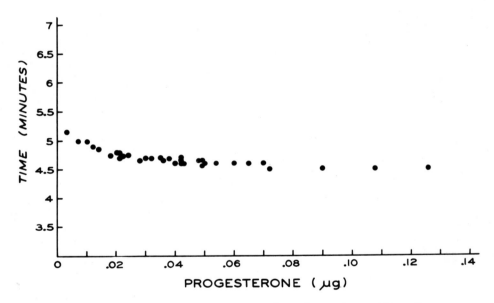

Fig. 3. The relationship between retention time and quantity of progesterone injected is illustrated. This study was done on a 28 in column packed with an alkyl cyanosilicone coated support which was obtained as a sample from Panta Industries, Berkeley, California. Reprinted with permission (6).

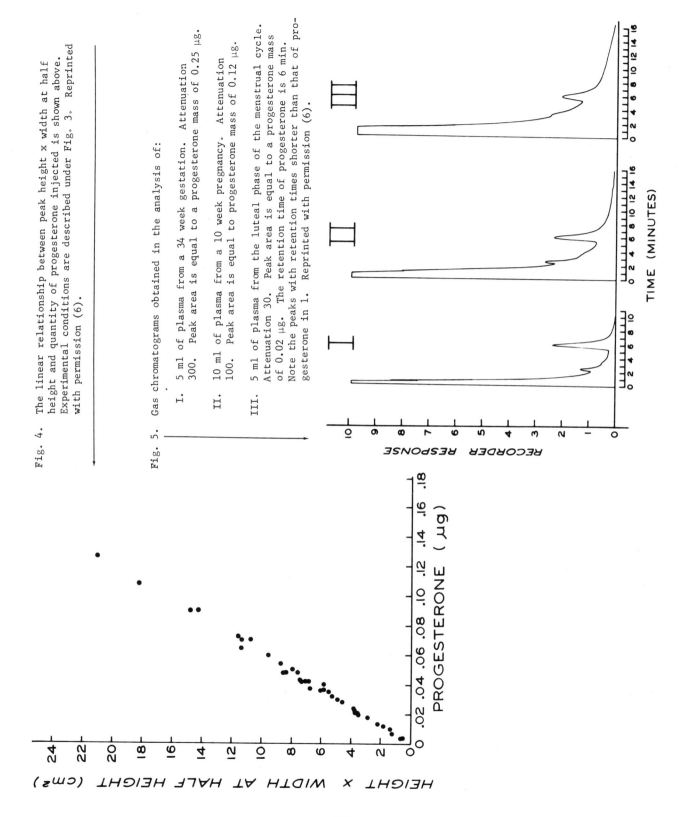

Fig. 4. The linear relationship between peak height x width at half height and quantity of progesterone injected is shown above. Experimental conditions are described under Fig. 3. Reprinted with permission (6).

Fig. 5. Gas chromatograms obtained in the analysis of:

I. 5 ml of plasma from a 34 week gestation. Attenuation 300. Peak area is equal to a progesterone mass of 0.25 μg.

II. 10 ml of plasma from a 10 week pregnancy. Attenuation 100. Peak area is equal to progesterone mass of 0.12 μg.

III. 5 ml of plasma from the luteal phase of the menstrual cycle. Attenuation 30. Peak area is equal to a progesterone mass of 0.02 μg. The retention time of progesterone is 6 min. Note the peaks with retention times shorter than that of progesterone in 1. Reprinted with permission (6).

Since the injected sample may contain sufficient steroid to produce a greater than full scale deflection, we have found a foldover zero suppression switch to be of value (7). This device effectively extends the range without altering the attenuation of the amplifier.

The specificity of this method depends on the ability of the paper and GLC systems utilized to isolate progesterone. The extensive studies of Zander (8) demonstrate the usefulness of the heptane-70% methanol system in isolating progesterone from other known steroids. The only additional peaks which we have found in the eluates of the progesterone area of these chromatograms are those noted in Fig. 5. No further peaks were found when extracts of larger samples of plasma were chromatographed on longer columns packed with 1% SE-30 and 1% QF-1.

The precision and accuracy of the method were determined by analyzing specimens of pooled male plasma with added progesterone. The results of these analyses appear in Tables 1 and 2. In the range of 0.024 to 0.115 µg the average deviation is approximately 10% of the mean. In order to determine the reproducibility of the assay, duplicate plasma specimens from 15 patients were analyzed. The results of these determinations are recorded in Table 3. The average deviation from the mean of duplicate samples in this experiment was 6%, SD $\pm$ 4.7%.

TABLE 1. Analysis of pooled male plasma with varying amounts of added progesterone showing precision of the method.

| Sample volume in ml | Number of samples | Calculated content of progesterone in µg | Measured content of progesterone in µg Mean $\pm$ SD |
|---|---|---|---|
| 5 | 5 | 0.115 | 0.113 $\pm$ 0.011 |
| 1 | 7 | 0.108 | 0.096 $\pm$ 0.012 |
| 1 | 5 | 0.024 | 0.021 $\pm$ 0.0018 |

TABLE 2. Analysis of duplicate aliquots of pooled male plasma with varying amounts of added progesterone showing accuracy of the method.

| Calculated content of progesterone in µg | Mean measured content of progesterone in µg |
|---|---|
| 0.017 | 0.012 |
| 0.024 | 0.021 |
| 0.051 | 0.045 |
| 0.45 | 0.40 |
| 0.65 | 0.64 |
| 0.108 | 0.096 |
| 0.115 | 0.113 |
| 1.15 | 1.05 |
| 2.15 | 2.04 |

TABLE 3. Comparison of duplicate analyses in samples containing a wide range of progesterone.

| Sample size | Source of sample | Measured concentration in μg per 100 ml | | Percent deviation from the mean |
|---|---|---|---|---|
| | | Sample I | Sample II | |
| 1 ml | 39 weeks gestation | 13.8 | 13.4 | 1.47 |
| 3 ml | 38 weeks gestation | 12.9 | 14.8 | 6.86 |
| 3 ml | 37 weeks gestation | 16.0 | 17.0 | 3.03 |
| 3 ml | 36 weeks gestation | 20.5 | 20.1 | 0.99 |
| 5 ml | 33 weeks gestation | 13.0 | 15.0 | 7.14 |
| 4 ml | 29 weeks gestation | 10.0 | 9.3 | 3.63 |
| 6 ml | 29 weeks gestation | 10.0 | 12.0 | 9.09 |
| 4 ml | 27 weeks gestation | 7.2 | 5.4 | 14.29 |
| 4 ml | 24 weeks gestation | 7.8 | 7.8 | 0.00 |
| 4 ml | 20 weeks gestation | 4.7 | 5.4 | 6.93 |
| 4 ml | 20 weeks gestation | 6.8 | 8.3 | 9.93 |
| 5 ml | 17 weeks gestation | 4.6 | 4.5 | 1.10 |
| 15 ml | 8 weeks gestation | 3.4 | 3.0 | 6.25 |
| 15 ml | Luteal phase Menstrual cycle | 2.3 | 2.7 | 8.00 |
| 20 ml | Anovulatory hirsute Female | 0.26 | 0.19 | 15.56 |

Plasma samples (7-10 ml) were obtained from normal ovulating women and analyzed for progesterone. The results of these analyses are shown in Fig. 6. Although we have few levels available for comparison, those illustrated appear to be at the lower end of the range of those reported by Woolever (2). Levels measured in nine anovulatory women lie between 0.15 and 0.3 µg per 100 ml. These levels appear to be higher than those found in the proliferative phase in the normal subjects, an observation consistent with the values previously reported in anovulatory women (2).

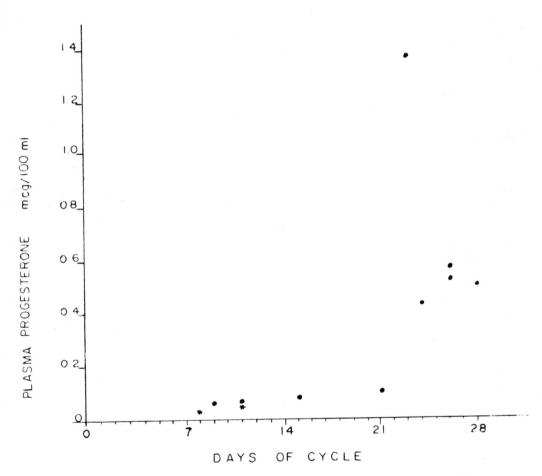

Fig. 6. Progesterone levels in plasma obtained from normal, ovulating women. Levels are plotted on the proper day of the cycle. The asterisks indicate that the values are less than those plotted.

This technique has been applied to several studies and the details of these will be reported separately. Measurements of plasma progesterone during pregnancy in 76 patients indicate a progressive rise from levels of approximately 2 µg per 100 ml at eight weeks to between 10 and 20 µg per 100 ml at term (9). In preliminary measurements the production rate of progesterone by the rat ovary has been found to vary from 0.2 mµg per min in proestrus to as high as 20 mµg per min in late diestrus (10). The production rate of progesterone by the dog adrenal has also been found to vary from 0.01 to 0.09 µg per 100 ml per min in the resting animal.

Recent experiments suggest that the cyclohexane:methanol:water (100:100:10) system may be substituted for heptane:methanol:water system for the initial separation. This is of special value when simultaneous estimates of the androgens are desired.

## SUMMARY

A method is described for the measurement of progesterone in human plasma. After preliminary purification, the extracts are subjected to paper chromatography. The portion of the chromatogram containing the progesterone is then injected into the GLC column using a solids injector. Samples containing as little as 10-20 millimicrograms can be analyzed with good accuracy and precision. The applications of the method to measurement of plasma progesterone levels in samples obtained from pregnant women and from normal ovulating and anovulatory women are illustrated.

The method achieves a degree of accuracy and reproducibility comparable to more complicated methods.

## ACKNOWLEDGEMENT

We wish to acknowledge the valuable assistance of Shirley Gullixson, Barbara Borgers and Marie McCabe.

## REFERENCES

1. Woolever, C. A. and A. Goldfien, Inter J Appl Rad Isotopes 14: 163, 1963.

2. Woolever, C. A., Am J Obst Gynec 85: 981, 1963.

3. Wotiz, H. H. and H. F. Martin, Anal Biochem 3: 97, 1962.

4. Kumar, D., J. A. Goodno, and A. C. Barnes, Nature (London) 195: 1204, 1962.

5. McComas, D. B., and A. Goldfien, Anal Chem 35: 263, 1963.

6. Yannone, M. E., D. B. McComas, and A. Goldfien, J Gas Chromatog 2: 30, 1964.

7. Hamilton, L. H., Anal Chem 34: 445, 1962.

8. Zander, J. and H. Simmer, Klin Wochschr 32: 529, 1954.

9. Yannone, M. E. and A. Goldfien, in preparation.

10. McDonald, D. M. and A. Goldfien (abstract, in press).

# DETERMINATION OF PROGESTERONE IN HUMAN PERIPHERAL BLOOD USING GAS-LIQUID CHROMATOGRAPHY WITH ELECTRON CAPTURE DETECTION

H. J. van der Molen and D. Groen, Department of Obstetrics and Gynecology, State University, Utrecht, The Netherlands

Isolation and identification of progesterone in human peripheral blood by chemical criteria has been achieved for plasma of pregnant women (1) and plasma of normal women during the menstrual cycle (2). Several methods have been described in the past for the estimation of progesterone in plasma of females during pregnancy (3, 4). The lack of sensitivity of these methods has, however, until recently prevented the adequate estimation of plasma progesterone in early pregnancy and during the menstrual cycle.

Short and Levett (5) using a sulfuric acid fluorescence technique could measure as little as 0.05 μg of progesterone. The lack of specificity of the fluorescence reaction at the lower limits of sensitivity makes it very likely that plasma progesterone values recorded by this method (1.1-2.3 μg per 100 ml plasma during the luteal phase of the cycle) are too high. Recently, Heap (6) described another sulfuric acid fluorescence assay for progesterone; after enzymatic conversion of progesterone to 20β-hydroxypregn-4-ene-3-one, this compound could be estimated fluorometrically with a sensitivity of 3-5 mμg. Results for human plasma obtained with this method have not been published. Sommerville et al. (7) using a modification of an earlier developed method, based on the UV absorption of progesterone-bis-thiosemicarbazone, could measure amounts as low as 0.1 μg. They reported values of 1-4 μg per 100 ml plasma during the luteal phase of the cycle. Woolever and Goldfien (8) have described a double isotope derivative method. They were able to measure 0.1 μg of progesterone and estimated (9) that plasma levels for normal women ranged between 0-0.53 μg per 100 ml plasma during the preovulatory and 0.6-2.1 μg per 100 ml plasma during the postovulatory phase of the menstrual cycle. In a pool of male plasma, the level was 0.66 μg progesterone per 100 ml (8). Riondel et al. (10) used a different isotope derivative method and were able to measure as little as 0.002 μg progesterone. In normal women the mean progesterone level was 0.12 μg per 100 ml plasma during the follicular phase and approximately 1.0 μg per 100 ml plasma during the luteal phase. In male plasma they found values of 0.20-0.30 μg per 100 ml plasma.

GLC for the estimation of progesterone was first applied to the determination of progesterone in human myometrium (11) and the identification of progesterone in human pregnancy plasma (12). GLC methods for the assay of plasma progesterone have been developed by Yannone et al. (13), who used the flame ionization detector and Collins and Sommerville (14) who used the argon ionization detector. The sensitivity of these methods is in the order of 0.01-0.02 μg progesterone.

It has been shown previously (15, 16) that electron absorption capacity of steroids may be increased following chloroacetate formation. We (17) have synthesized and studied the characteristics of a series of steroid-chloroacetates. All these chloroacetates could be quantitated using the electron capture detector after GLC with a sensitivity in the order of nanograms or less. The application of this principle to a method for the estimation of testosterone in human peripheral blood using GLC has been described (16). The 20-keto group of progesterone may easily be reduced by a specific 20β-hydroxysteroid dehydrogenase prepared from Streptomyces hydrogenans (18).

This report will deal with the application of the electron capture detection, of the chloroacetate derivative, of enzymatically reduced progesterone (20β-hydroxy-4-pregnene-3-one) as a means of estimating the small amounts of progesterone present in peripheral blood of normal men and women.

## MATERIALS

*Solvents and Reagents.* Solvents were obtained from BDH (Analar reagent) or Merck and purified as described previously (16, 17).

Special care had to be taken for the purification of <u>benzene</u>. After evaporation of the amount of normal distilled p.a. benzene, that is used for the extraction of silica gel after the last TLC, it appears that this benzene contains impurities that give a strong background signal with the electron capture detector. This could be prevented by repeatedly washing the benzene with concentrated sulfuric acid and water, followed by distillation twice through a 100 cm Vigreux column.

<u>Silica gel</u> GF254 (Merck, according to Stahl, containing UV fluorescent indicator) was used without any additional purification.

The solutions that were used for the enzymatic conversion of progesterone to 20β-hydroxy-4-pregnene-3-one (20-OHP), were all prepared with water twice distilled from an all-glass apparatus:

1. <u>Phosphate buffer pH 5.2 containing EDTA</u>

   0.15 M phosphate buffer pH 5.2 is made up of 97.5 ml of 0.15 M $KH_2PO_4$-solution and 2.5 ml of 0.15 M $Na_2HPO_4$-solution. To 100 ml of the buffer solution is added 100 mg ($2.7 \times 10^{-3}$M) EDTA (disodium salt of ethylene-diamino-tetraacetic acid).

2. <u>Tris-(hydroxymethyl-aminomethane) buffers</u>

   a) 0.1 M pH 8.1 without EDTA
   50 ml of a 0.1 M solution of tris-hydroxymethyl-aminomethane (Boehringer, Mannheim) and 26 ml 0.1 N HCl are mixed and diluted to 100 ml.

   b) 0.005 M pH 8.2 with EDTA
   A 0.005 M buffer solution is prepared by dissolving 300 mg Tris in 500 ml distilled water; the pH is adjusted (using a pH meter) by addition of 0.01 HCl. To 100 ml of the buffer solution is added 100 mg EDTA to make a final concentration of $2.7 \times 10^{-3}$M.

3. <u>Coferment solution</u>

   5 mg DPNH (reduced diphosphopyridine nucleotide = NADH - reduced Nicotinamide-Adenin-Dinucleotide) obtained from Boehringer (Mannheim) is dissolved in 3 ml 0.1 M Tris-buffer pH 8.1 without EDTA. This solution is stored at 4 C and prepared freshly at least once a week.

4. <u>Enzyme dilution</u>

   A concentrated crystal suspension of 20β-hydroxysteroid dehydrogenase prepared from <u>Streptomyces hydrogenans</u> is obtained from Boehringer (Mannheim). This suspension is stable for many months when stored at 4 C. A small volume of the concentrated suspension (5 mg protein per ml) is diluted before use with 4 vol 0.005 M Tris-buffer pH 8.2 with EDTA. Stored at 4 C over a period of 2-3 months, we have never observed a loss of activity of the diluted enzyme solution.

<u>Scintillation fluid</u> containing 4 g PPO and 40 mg POPOP per 1 toluene. PPO (2,5-diphenyloxazole) and POPOP (1,4-bis-2-(5-phenyloxazoyl)-benzene) were scintillation grade obtained from Nuclear Enterprises (Edinburgh).

<u>Progesterone</u> and <u>20β-hydroxy-4-pregnene-3-one</u> were obtained from Steraloids Inc. (Pawling, N. Y.) and repeatedly crystallized from organic solvents. Progesterone had a mp of 128 C and 20-OHP had a mp of 174 C. UV absorptions ($E_{241}$ mu) and infrared spectra were identical with those reported in the literature.

<u>Testosterone-chloroacetate</u> and <u>20β-hydroxy-4-pregnene-3-one-chloroacetate</u> were prepared and purified as reported elsewhere (16, 17).

$^3$<u>H-progesterone</u> was obtained from New England Nuclear Corp. with a quoted specific activity of 10 mC per mmole (30,000 dpm per 0.0005 µg). It was purified by chromatography in the TLC system benzene:ethylacetate (1:1), cyclohexane:ethylacetate (1:1) and benzene:ethylacetate (4:1). After mixing of the purified $^3$H-progesterone with unlabeled progesterone, the mixture was taken through the method for the estimation of plasma progesterone. Specific activity as estimated at different stages in the method remained constant, indicating the purity of the purified $^3$H-progesterone. A solution of 10,000 cpm per ml (0.0005 µg per ml) in methanol was stored at 4 C.

$^3$H-20β-hydroxy-4-pregnene-3-one was prepared from the purified $^3$H-progesterone using the enzymatic reduction as described hereafter. After purification through several TLC's, an aliquot of the $^3$H-20-OHP was mixed with the pure unlabeled steroid and taken through several thin-layer and paper chromatographies. The specific activity remained constant through these procedures.

$^3$H-20β-chloroacetate-4-pregnene-3-one was prepared from purified $^3$H-20-OHP using the chloroacetylation method as described hereafter. After purification through several TLC's an aliquot was mixed with authentic unlabeled steroid and taken through several chromatographic procedures. The specific activity remained constant through all these steps.

<u>Glassware</u> used for the determination was soaked overnight in chromic acid, rinsed with tap water, soaked overnight in a detergent (RBS 25), rinsed with tap water, soaked overnight in diluted hydrochloric acid, rinsed 10 times with tap water and 10 times with deionized water, and finally allowed to dry at room temperature. The 2 ml tubes used to collect samples for GLC were siliconized using a 5% solution of dimethyldichlorosilane in benzene. This had to be done in order to prevent losses of the small (0.01-0.001 µg) samples through absorption onto the glass.

## METHODS

TLC was carried out on 20 x 20 cm glass plates with silica gel $GF_{254}$ as described previously (16). Since we were often able to detect traces (0.002-0.01 µg per 2 ml solvent) a free steroid or steroid-chloroacetates in the solvent after development of the plates, the tanks used for the TLC were cleaned and freshly prepared for each chromatography.

GLC was carried out with an F&M model 400 Biomedical Analyzer. For the detection of the steroid-chloroacetates we used an electron capture detector, that was operated with a pulsating potential gradient (54 v per cm) of 0.75 usec duration and a pulse interval of 5, 15, 50 or 150 usec. The pulse interval used throughout this investigation was 150 usec. Columns for GLC were prepared as described by Horning et al. (19). Even with the very small amounts (0.01-0.001 µg) of steroid used, this technique gave good results without any apparent loss of samples or peak tailing. U-shaped glass columns (90 cm long, 0.4 cm i d ) filled with 1% XE-60 coated on Gas-Chrom P (80-100 mesh) were employed.

The column temperature was kept at 215-220 C, with the detector at 200 C and the flash heater at 250 C. Pure Nitrogen (velocity: 75 ml per min) was used as carrier gas; Argon, containing 10% methane (velocity: 225 ml per min) was added as a pure gas to the carrier gas stream just prior to entering the electron capture detector.

Standards and unknown samples were introduced onto the column dissolved in benzene using a graduated 10 µl Hamilton syringe (all injection volumes were kept close to 5 µl).

The recorded peaks were evaluated using triangulation, calculating peak height x width of the peak at half peak height.

Assay of radioactivity was done using a Nuclear Chicago model 725 liquid scintillation spectrometer. Samples were evaporated in glass vials; after dissolving the residues in 0.1 ml ethanol, 10 ml scintillation fluid was added. Efficiency for $^3$H-counting was approximately 35%. Enough counts were accumulated to give a counting accuracy of 1-2%. The background varied from 35-40 cpm; the amount of radioactivity in the aliquots of the plasma samples after substraction of the background gave approximately 600-800 cpm. Quenching corrections were not necessary.

## ESTIMATION OF PROGESTERONE IN HUMAN BLOOD

The main stages in the method are:

1. Addition of $^3$H-progesterone to plasma.

2. Ether extraction of alkaline plasma.

3. TLC isolation of free progesterone.

4. Reduction of progesterone with 20β-hydroxysteroid dehydrogenase.

5. Chloroacetylation of 20-OHP.

6. TLC of 20-OHP-chloroacetate.

7. Addition of testosterone-chloroacetate as internal standard.

8. Sampling for counting of radioactivity.

9. GLC with electron capture detection.

The method in detail:

A solution of approximately 10,000 cpm $^3$H-progesterone is taken to dryness in 50 ml glass tubes; 10 ml plasma is added to each tube. Appropriate standards for counting of the added radioactivity and water blanks, containing $^3$H-progesterone only, are included in each series.

After addition of 0.25 ml 20% sodium hydroxide solution the samples are extracted 6 times with 15 ml diethyl ether; the pooled ether extracts are washed twice with 5 ml distilled water. After evaporation and concentration of the ether extracts under nitrogen, the residues are chromatographed on silica gel plates in the TLC system benzene:ethyl acetate (2:1). The silica gel areas in the lanes of the samples corresponding to those of authentic progesterone are scraped off and extracted with 95% methanol (3 times with 1 ml). The combined extracts are taken to dryness and concentrated in the tip of a conical 15 ml centrifuge tube.

Reduction. The residues are dissolved in one drop of ethanol. After addition of 0.5 ml of 0.1 M phosphate buffer pH 5.2, 0.03 ml of the coferment solution and 0.03 ml of the diluted enzyme solution, the contents of the tube are thoroughly mixed and incubated for 2 hr at 37 C. Following the incubation, 1 ml distilled water is added and the solution is extracted with ethyl acetate (4 times with 1 ml). The combined ethyl acetate extracts are evaporated and concentrated under nitrogen.

Chloroacetylation. The residues after reduction are dried for 2-3 hr in a vacuum desiccator. To the dried residues is added 0.5 ml of a solution of monochloroacetic anhydride in tetrahydrofuran (10 mg per ml) and 0.1 ml pyridine. Reaction is carried out overnight in a desiccator. After addition of 1 ml of distilled water, the chloroacetate is extracted with 3 times 1 ml ethyl acetate. The combined ethyl acetate extracts are washed once with 1 ml 6 N HCL, twice with 1 ml distilled water and subsequently evaporated to dryness. The residues after chloroacetylation are chromatographed on silica gel plates in the system benzene:ethyl acetate (6:1). Areas in the sample lanes corresponding with those of authentic 20-OHP-chloroacetate are scraped off. Extraction is done using a water:benzene partition. One ml of benzene is added to the silica gel, after mixing, 0.2 ml distilled water is added, mixing is repeated and centrifugation gives a separation of the benzene and the water layers. This extraction is repeated 3 times. The benzene layers are transferred to 2 ml siliconized conical centrifuge tubes and dried down after each extraction.

The residues in the micro centrifuge tubes are dissolved in 1 ml of methanol containing testosterone chloroacetate (0.01 μg per ml for blanks and plasma samples obtained from males and females during the first half of the cycle or from amenorrheic non-pregnant females; 0.04 μg per ml for plasma samples obtained from females during the second half of the cycle). The contents of the tubes are thoroughly mixed; one tenth is taken for counting of radioactivity, the remainder is taken to dryness and concentrated in the tips of the tubes. The concentrated extracts are dissolved in benzene (10 μl when 0.01 μg internal standard is added; 20 μl when 0.04 internal standard is added).

Selecting an adequate sensitivity setting of the detector and recorder, 5 μl of each tube is injected into the gas chromatograph. Areas of the testosterone chloroacetate and 20-OHP-chloroacetate peaks are measured by triangulation. Quantitation is performed by comparison with peak areas of standard solutions containing 0.010 or 0.005 μg per 5 μl of both testosterone- and 20-OHP-chloroacetate, that are injected under the same conditions.

The micrograms of progesterone (P) in the samples can be calculated as:

P = R x C x U x A x 0.80 µg

$R = \dfrac{\text{cpm }^3\text{H-progesterone initially added to plasma sample}}{10 \times \text{cpm }^3\text{H in aliquot prior to GLC}}$

$C = \dfrac{\text{peak area (cm}^2\text{) of 0.01 µg testosterone chloroacetate}}{\text{peak area (cm}^2\text{) of 0.01 µg 20-OHP-chloroacetate}}$

$U = \dfrac{\text{peak area (cm}^2\text{) of 20-OHP-chloroacetate in sample}}{\text{peak area (cm}^2\text{) of testosterone chloroacetate in sample}}$

A = µg testosterone chloroacetate added as internal standard

$0.80 = \dfrac{\text{molecular weight of progesterone}}{\text{molecular weight of 20-OHP-chloroacetate}} = \dfrac{314}{393}$

To obtain the amount of progesterone in the plasma sample, the mass (µg) of the $^3$H-progesterone, that was initially added to the plasma and is measured in the water blank assay, has to be subtracted from the values found for the samples.

## RESULTS AND DISCUSSION

<u>The quantitative estimation using electron capture detection</u>. The applicability of the described method for the estimation of plasma progesterone depends for a large part on the reliability of the quantitative estimation using the electron capture detector following GLC.

The calculation of the amount of progesterone in a sample, hinges on the presumptions that:

1) The sensitivity of the electron capture detector is high enough to detect the small amounts of progesterone isolated from peripheral blood;

2) The response of the electron capture detector as a function of the amount of injected steroids is linear for: a) different amounts of 20-OHP-chloroacetate, b) different amounts of testosterone chloroacetate.

3) The behavior of the added internal standards ($^3$H-progesterone and testosterone chloroacetate) is the same as that of the compounds under investigation (progesterone and 20-OHP-chloroacetate).

<u>Detector sensitivity</u>. Figure 1 illustrates the high sensitivity of electron capture detection of the chloroacetate derivative as compared to the electron capture or flame ionization detection of free steroids and flame ionization detection of the chloroacetate derivative under identical condition.

0.001 µg of the pure chloroacetates may easily be measured with good precision (± 10%). Several factors, however, may influence the absolute sensitivity of the detector. If, for example, the silicone rubber septum through which the samples are injected onto the column is replaced, the detector sensitivity decreases severely and is gradually restored over a period of two or three days if the carrier gas flame is maintained. To assure a continuous sensitive detector, we always condition the septum before use in a stream of nitrogen in another gas chromatograph at 240 C.

Whereas we routinely use pure nitrogen as a carrier gas and Argon containing 10% methane as purge gas, a 2-3 fold increase in sensitivity is obtained (Fig. 2) if Argon containing 10% methane is used as both carrier and purge gas.

If complete residues of plasma samples containing very small (0.001 µg) amounts of progesterone are injected, impurities in the final residue, that broaden the solvent front, are the limiting factor for an accurate quantitation of the small amounts.

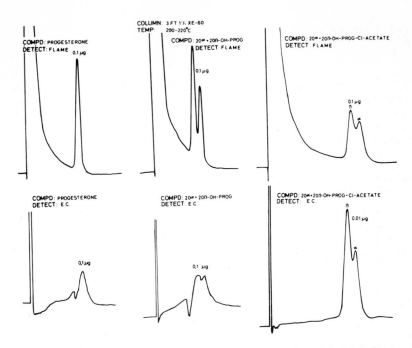

Fig. 1. Electron capture and flame ionization detection of free steroids and the chloroacetate derivatives of 20α- and 20β-hydroxy-4-pregnene-3-one.

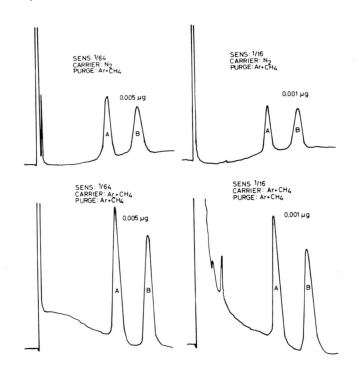

Fig. 2. GLC tracings of mixtures of pure testosterone chloroacetate (A) and 20β-hydroxy-4-pregnene-3-one-chloroacetate (B) at different sensitivity settings using different carrier gasses.

For practical purposes a sensitive detector allows multiple injections of small aliquots of the final residues, thus increasing the precision of gas-chromatographic quantitation.

<u>Linearity of detector response</u>. Figure 3 shows typical calibration curves ($cm^2$ peak area vs µg) obtained after exposing different amounts of testosterone chloroacetate and 20-OHP-chloroacetate to the electron capture detector following GLC on the XE-60 column. We always observe a straight line relationship with amounts in the range of 0.001-0.200 µg.

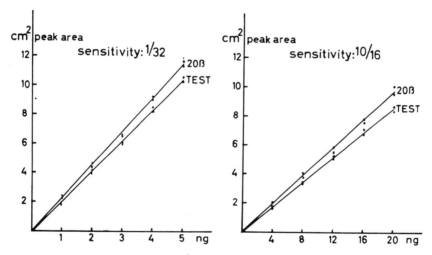

Fig. 3. Calibration curves ($cm^2$ peak area vs nanograms) of pure 20β-hydroxy-4-pregnene-3-one-chloroacetate and testosterone-chloroacetate.

Behavior of internal standards. The final check on the purity of $^3$H-progesterone is the constancy of the specific activity after mixing with authentic unlabeled progesterone and exposing this mixture to several chromatographic procedures. We do not have any indications that the $^3$H-progesterone added to plasma does not behave similar to plasma progesterone.

The addition of testosterone chloroacetate as an internal standard before GLC is needed to correct for: 1) losses of 20-OHP-chloroacetate before it reaches the detector, and 2) variation in sensitivity of the electron capture detector.

Through whatever steps (dilution, transfer, adsorption onto glass) standard solutions containing testosterone-chloroacetate and 20-OHP-chloroacetate were taken before GLC, their peak ratio after GLC remained constant; within the limits of the pure standards that were immediately exposed to GLC.

The results in Table 1, taken from calibration curves (all reported with the same attenuator settings) over a period of several months, clearly illustrate that variation of detector sensitivity strongly influences absolute peak areas whereas the ratio of testosterone-chloroacetate to 20-OHP-chloroacetate peak areas remains constant.

TABLE 1. Spontaneous variation in detector sensitivity (attenuation 1/32) over a period of several months for a mixture (0.005 μg per 5 μl) of 20β-hydroxy-4-pregnene-3-one-chloroacetate and testosterone-chloroacetate.

| Peak area (cm$^2$) Test-Cl-Acet. | Peak area (cm$^2$) 20β-OH-Cl-Acet. | Ratio[x] |
|---|---|---|
| 6.72 | 7.20 | 1.07 |
| 8.26 | 10.12 | 1.22 |
| 6.56 | 7.70 | 1.17 |
| 11.76 | 13.30 | 1.13 |
| 9.64 | 11.20 | 1.16 |
| N = 29 | Mean | 1.15 |
| | SD | 0.044 (4%) |

[x] Ratio = $\dfrac{\text{peak area (cm}^2\text{) 20β-hydroxy-4-pregnene-3-one-chloroacetate}}{\text{peak area (cm}^2\text{) testosterone-chloroacetate}}$

For daily routine we inject different amounts (0.001-0.010 μg) of standard solutions containing equal amounts (0.005 and 0.010 μg per 5 μl) of testosterone- and 20-OHP-chloroacetate. The results in Table 2 demonstrate the constancy of the ratio of the peak areas.

Specificity of the method. Table 3 shows the behavior in the chromatographic systems of several steroids, that might be present in peripheral plasma. After TLC of the free steroids in the system benzene:ethylacetate (2:1), a "progesterone" area between $R_f$ 0.50 and $R_f$ 0.70 is eluted. Of the steroids that might be in this area none is expected in the "20β-hydroxy-4-pregnene-3-one-chloroacetate" area ($R_f$ 0.45-0.65) after TLC in benzene:ethylacetate (6:1) and following the specific 20β- reduction and chloroacetylation. The specific retention behavior of steroid-chloroacetates on the XE-60 column as shown in the last column of Table 3 adds further to the specificity of the method for the estimation of plasma progesterone.

TABLE 2. Reproducibility of quantitation of pure testosterone-chloroacetate and pure 20β-hydroxy-4-pregnene-3-one-chloroacetate.

| N | Ratio | SD | SE |
|---|---|---|---|
| 10 | 1.12 | 0.066<br>4.9% | 0.018<br>1.6% |
| 9 | 1.08 | 0.054<br>5.0% | 0.018<br>1.7% |
| 7 | 1.12 | 0.055<br>4.9% | 0.021<br>1.9% |
| 9 | 1.12 | 0.065<br>6.0% | 0.022<br>2.0% |
| 9 | 1.09 | 0.066<br>6.0% | 0.022<br>2.0% |

N = Number of estimations

$$\text{Ratio} = \frac{\text{peak area 20β-hydroxy-4-pregnene-3-one-chloroacetate}}{\text{peak area testosterone-chloroacetate}}$$

TABLE 3. Thin-layer chromatography and gas chromatography of steroids.

| Steroid | Thin-layer chromatography | | Gas chromatography |
|---|---|---|---|
| | $R_f$ in benzene: ethylacetate (2:1) | $R_f$ after chloroacetylation in benzene:ethylacetate (6:1) | On 1% XE-60 Relative[x] retention of chloroacetates |
| Pregnanediol | 0.18 | 0.91 | 39.0 |
| Testosterone | 0.26 | 0.42 | 21.4 |
| 20α-hydroxy-4-pregnene-3-one | 0.32 | 0.47 | 33.2 |
| 20β-hydroxy-4-pregnene-3-one | 0.32 | 0.52 | 29.7 |
| 17α-hydroxyprogesterone | 0.33 | 0.12 | -- |
| Etiocholanolone | 0.35 | -- | -- |
| Androsterone | 0.40 | 0.55 | 9.3 |
| 3β-hydroxy-5-pregnene-20-one | 0.43 | 0.70 | 15.2 |
| Dehydroepiandrosterone | 0.44 | 0.59 | 10.9 |
| 4-androstene-3,17-dione | 0.45 | 0.18 | -- |
| 3α-hydroxy-5β-pregnane-20-one | 0.46 | 0.71 | 13.0 |
| Estradiol | 0.57 | 0.82 | 45.6 |
| Progesterone | 0.59 | 0.25 | -- |
| Estrone | 0.88 | 0.56 | 17.5 |

[x] Retention time relative to that of 5α-cholestane = 1.

When water blanks or plasma samples from ovariectomized adrenalectomized women (with or without addition of 10,000 cpm $^3$H-progesterone) were taken through the method, omitting the addition of testosterone-chloroacetate internal standard, the gas chromatographic tracings showed no peaks in either the 20-OHP-chloroacetate area or the testosterone-chloroacetate area.

In the final gas chromatograms of more than a hundred different unknown plasma samples we have not seen any peak that might interfere with the quantitative evaluation of either the testosterone-chloroacetate or the 20-OHP-chloroacetate peaks.

The absence of any measurable 20-OHP-chloroacetate peak in the residue of water blanks containing only $^3$H-progesterone proves the specific activity of the $^3$H-progesterone to be as high as expected (10,000 cpm per 0.0005 μg).

Recoveries. The recovery of the method would theoretically be of no importance as long as enough radioactivity and chloroacetate is left to permit a reliable quantitation. A high recovery, however, will allow more than one measurement by GLC, thus improving the precision of estimation.

If the time of incubation with 20β-hydrogenase is increased, recoveries tend to decrease. Eighty percent $^3$H-progesterone may be recovered as free $^3$H-20β-hydroxy-4-pregnene-3-one following TLC after reduction under the conditions (2 hr at 37 C) described in the method. Table 4 illustrates that the recovery of radioactivity through the complete method is strongly decreased after incubating during 16 hr instead of 2 hr.

TABLE 4. Recovery of $^3$H from $^3$H-progesterone up to gas chromatography using different times of incubation with 20β-dehydrogenase.

| Incubation | | 2 hr 37 C | 16 hr 37 C |
|---|---|---|---|
| Recovery | range | 44.6 - 57.8% | 5.9 - 23.8% |
| | mean | 48.9% | 16.7% |
| | n | 22 | 24 |
| | SD | 5.5% | 5.6% |

Another important factor that may influence the recovery is the method of extraction after the last TLC. To obtain an extract for GLC that is completely free of silica gel, a water-benzene partition has to be used. If the water is first added to the silica gel followed by extraction with benzene, recoveries are low (as demonstrated by the recoveries of pure $^3$H-20-OHP-chloroacetate in Table 5. If, however, the benzene is first added and mixed with the silica gel, followed by addition of water, an equally clean extract is obtained but recoveries are much higher.

Routinely the recovery of radioactivity up to the GLC step is in the order of 50% (mean for 96 estimations: 51.4%, range: 38.6 - 72.3%).

The use of siliconized tubes to collect samples before the final GLC prevents the absorption of the small amounts of chloroacetates onto the glass. Using this procedure, 80% of the residues may be easily transferred onto the column. Though this may vary according to quality of the glass, we have observed with unsiliconized tubes losses up to 70% that could be explained only by adsorption of the substances onto the glass.

TABLE 5. Efficiency of elution of $^3$H-20β-hydroxy-4-pregnene-3-one-chloroacetate from silica gel after thin-layer chromatography.

BENZENE-WATER PARTITION

| Sequence of addition: | | 1) 0.2 ml water<br>2) 1 ml benzene | 1) 1 ml benzene<br>2) 0.2 ml water |
|---|---|---|---|
| Recovery | range | 18.5 - 37.9% | 46.4 - 85.0% |
|  | mean | 26.9% | 65.0% |
|  | n | 9 | 9 |
|  | SD | 10% | 10% |

<u>Sensitivity of the method</u>. 0.010 μg of pure testosterone added to water is recovered in good yield as can be detected easily (Table 6). It is possible to make a reliable (10% precision) quantitation of 0.001-0.002 μg pure 20-OHP-chloroacetate. Considering the recoveries of radioactivity up to GLC sampling for counting of radioactivity (10% loss) and completeness of transfer of a sample onto the column for GLC (80%), the total recovery of substances reaching the electron capture detector amounts to from 30-35%. This would theoretically limit the sensitivity (accepting that 0.001-0.002 μg may be quantitated with good precision) to approximately 0.002-0.005 μg progesterone per sample (0.02 μg per 100 ml plasma). These figures are in the order of the smallest amounts that we have actually been able to measure in plasma samples.

TABLE 6. Determination of progesterone added to water.

| | Progesterone in 10 ml water | | |
|---|---|---|---|
| Added (μg) | 0.020 | 0.050 | 0.100 |
| Found (μg) | 0.018 | 0.053 | 0.111 |
| | 0.024 | 0.051 | 0.095 |
| | 0.016 | 0.055 | 0.107 |
| | 0.020 | 0.051 | 0.105 |
| | 0.028 | 0.061 | 0.102 |
| | 0.018 | 0.063 | 0.113 |
| | 0.017 | 0.050 | 0.110 |
| | 0.019 | 0.043 | 0.104 |
| | | 0.043 | |
| | | 0.045 | |
| Mean | 0.020 | 0.051 | 0.106 |
| SD | 0.003 | 0.007 | 0.006 |
| | 15.0% | 13.7% | 5.8% |

<u>Accuracy and precision</u>. From the results in Table 6 and 7 may be concluded that pure progesterone added to water or plasma in amounts ranging from 0.01-0.1 μg is recovered in good yield.

The precision of the complete method depends mainly on the precision of the quantitation using GLC. The results in Tables 1 and 2 have shown that the precision of the determination of peak ratios for the pure chloroacetates is in the order of 5-6%. Whereas generally an aliquot containing between 0.003 and 0.015 μg of the chloroacetates is exposed to GLC, this precision would be expected to be independent of the absolute amounts of progesterone present in the original sample. The standard deviation for GLC quantitation of samples containing amounts large enough to permit multiple injections is in the same order as those of the pure standards (Table 8).

TABLE 7. Determination of progesterone in plasma.

|  | μg progesterone in | | | |
|---|---|---|---|---|
|  | 10 ml plasma A | 10 ml plasma A + 0.010 μg added | 10 ml plasma B | 10 ml plasma B + 0.050 μg added |
| Found (μg) | 0.019 | 0.025 | 0.011 | 0.057 |
|  | 0.022 | 0.027 | 0.008 | 0.057 |
|  | 0.015 | 0.020 | 0.006 | 0.055 |
|  | 0.016 | 0.028 | 0.005 | 0.063 |
|  | 0.021 | 0.023 |  |  |
| Mean | 0.019 | 0.027 | 0.007 | 0.059 |
| SD | 0.0026 | 0.0034 | 0.0025 | 0.0047 |
|  | 13.5% | 12.5% | 35.0% | 8.0% |
|  | Recovery (μg) | 0.008 | Recovery (μg) | 0.052 |

TABLE 8. Multiple gas-chromatographic quantitations of residues of the same (10 ml) plasma samples.

| Sample | μg progesterone per 10 ml plasma in | | | | | | | |
|---|---|---|---|---|---|---|---|---|
|  | A | B | C | D | E | F | G | H |
| Found (μg) | 0.0034 | 0.0052 | 0.042 | 0.045 | 0.072 | 0.106 | 0.155 | 0.199 |
|  | 0.0038 | 0.0068 | 0.053 | 0.048 | 0.076 | 0.107 | 0.168 | 0.168 |
|  |  |  | 0.048 | 0.050 | 0.078 | 0.092 | 0.159 | 0.192 |
|  |  |  |  |  |  | 0.111 | 0.146 | 0.188 |
|  |  |  |  |  |  | 0.107 | 0.148 | 0.205 |
|  |  |  |  |  |  | 0.100 |  |  |
| Mean | 0.0036 | 0.0060 | 0.048 | 0.048 | 0.075 | 0.104 | 0.155 | 0.186 |
| SD | 0.0004 | 0.0014 | 0.003 | 0.003 | 0.004 | 0.008 | 0.009 | 0.014 |
|  | 11% | 23% | 6% | 6% | 5% | 7.7% | 5.8% | 3.5% |

With unknown plasma samples, however, the residue that is subjected to GLC may contain non-specific impurities that may result in a solvent front broader than that observed with standards. This tends to result in peaks on a sometimes sloping background, which makes the evaluation using triangulation of peaks for samples containing small amounts less precise.

From the results in Tables 6, 7 and 8 and other serial estimations (also including plasma testosterone estimations) it appears that the precision of a single estimation of small amounts (0.010 μg or less per sample; 0.1 μg per 100 ml plasma) may be as low as 10-40%. The precision of amounts higher than 0.010 μg per sample is in the order of from 10-15%. The precision at high values (0.1 μg per sample) may be as high as 5-6%. In the case that the final samples contain amounts high enough to perform multiple GLC quantitations, the precision may still be improved. Considering the recoveries obtained under the conditions used in our laboratory multiple injections can be successfully applied to samples containing 0.020 μg (0.2 μg per 100 ml plasma) or more, progesterone per sample.

*Plasma values.* With the described method we found in plasma of nine normal males progesterone concentrations ranging from 0.020-0.049 μg per 100 ml. In plasma of normal women the concentration of progesterone varies greatly. Preovulation values ranged from 0.01-0.079 μg per 100 ml plasma; post-ovulation values were found between 0.60 and 1.99 μg per 100 ml plasma. Figure 4 demonstrates the gas chromatograms of the 20-OHP standard and the 20-OHP obtained from plasma during the follicular and luteal phases. Figure 5 shows a typical example of the different stages of the menstrual cycle.

<u>Conclusions</u>. Provided that the utmost care is taken in analyzing these small amounts of progesterone, the relative simplicity of the method allows the handling of a rather large number of samples within a reasonable time period. Excluding the purification of solvents and cleaning of glassware, a skilled technician can analyze 20 plasma samples for both testosterone and progesterone within a 5-day working week. This may well place this and other methods using the same principles among the techniques that can be used for routine endocrine investigations.

The reliability of the method compares well with those of the most sensitive double isotope derivative techniques (10). Though the sensitivity is as high as those of the isotope techniques, the precision at low concentrations (10-40% for concentrations of 0.10 µg or less per 100 ml) may be lower than those (23% at the 0.05 µg per 100 ml level) reported by Riondel <u>et al</u>. (10) for a double isotope technique.

Though it is difficult to compare the specificity of methods at the lower levels of sensitivity, it is striking that the plasma progesterone values that we observe in plasma of males and females before ovulation ( 0.02-0.08 µg per 100 ml) tend to be lower than values (0.10-0.30 µg per 100 ml) reported for comparable subjects.

The principle of electron capture detection of chloroacetates for the determination of plasma progesterone is the same as that used previously (16) in a method for the estimation of plasma testosterone. This principle is of course not limited to these steroids and may be applied equally well to the determination of other steroids. We are currently estimating in our laboratory progesterone, $20\alpha$-hydroxy-4-pregnene-3-one, $20\beta$-hydroxy-4-pregnene-3-one, androstenedione and testosterone in a single plasma sample. Following addition of appropriate labeled steroids to the plasma sample and a preliminary paper chromatographic separation of the free steroids, progesterone is converted to $20\beta$-hydroxy-4-pregnene-3-one and androstenedione is converted to testosterone; following the formation of chloroacetates and addition of appropriate internal standards for GLC, quantitation is performed using electron capture detection.

## SUMMARY

A method for the estimation of progesterone in human peripheral blood is described. After converting progesterone to $20\beta$-hydroxy-4-pregnene-3-one the monochloroacetate derivative is prepared. Nanogram amounts of this derivative can be detected using electron capture detection following GLC. The precision of the method ranges from 10-40% for samples containing less than 0.01 µg, from 10-15% for samples containing more than 0.01 µg and may be as high as 5-6% for samples containing 0.1 µg or more. Plasma progesterone values in male plasma samples ranged from 0.02-0.05 µg per 100 ml. Values found in plasma of women ranged from 0.01-0.08 µg per 100 ml before ovulation and from 0.60-1.99 µg per 100 ml after ovulation.

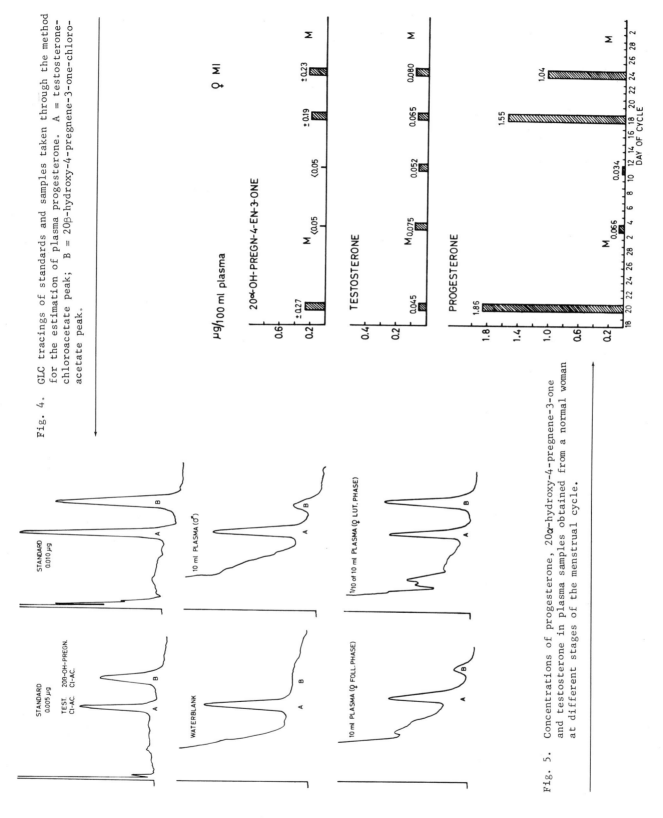

Fig. 4. GLC tracings of standards and samples taken through the method for the estimation of plasma progesterone. A = testosterone-chloroacetate peak; B = 20β-hydroxy-4-pregnene-3-one-chloroacetate peak.

Fig. 5. Concentrations of progesterone, 20α-hydroxy-4-pregnene-3-one and testosterone in plasma samples obtained from a normal woman at different stages of the menstrual cycle.

## REFERENCES

1. Zander, J., *Nature* 174: 406, 1954.

2. van der Molen, H. J. B., Runnenbaum, E. E. Nishizawa, E. Krinstensen, T. Kirschbaum, W. G. Wiest, and K. B. Eik-Nes, *J Clin Endocr* 25: 170, 1965.

3. Short, R. V., in Gray, C. H. and A. L. Bachrach, *Hormones in Blood*. Chapter 13, Progesterone. Academic Press, New York, 1961.

4. Zander, J., in Dorfman, R. I., *Methods in Hormone Research*, Vol. 1, Chapter 3, Progesterone, Academic Press, New York, 1962.

5. Short, R. V. and I. Levett, *J Endocr* 25: 239, 1962.

6. Heap, R. B., *J Endocr* 30: 293, 1964.

7. Sommerville, I. F., M. T. Picket, W. P. Collins, and D. C. Denyer, *Acta Endocr (Kbh)* 43: 101, 1963.

8. Woolever, C. A. and A. Goldfien, *Intern J Appl Rad* 14: 163, 1963.

9. Woolever, C. A., *Am J Obst & Gynec* 85: 981, 1963.

10. Riondel, A., J. F. Tait, S. A. S. Tait, M. Gut, and B. Little, *J Clin Endocr* 25: 229, 1965.

11. Barnes, A. C., D. Kumar, and J. A. Goodno, *Am J Obst & Gynec* 84: 1207, 1962.

12. Futterweit, W., N. L. McNiven and R. I. Dorfman, *Biochim Biophys Acta* 71: 474, 1963.

13. Yannone, M. E., D. B. McComas, and A. Goldfien, *J Gas Chromatog* 2: 30, 1964.

14. Collins, W. P. and I. F. Sommerville, *Nature* 203: 836, 1964.

15. Landowne, R. A. and S. R. Lipsky, *Anal Chem* 35: 532, 1963.

16. Brownie, A. C., H. J. van der Molen, E. E. Nishizawa, and K. B. Eik-Nes, *J Clin Endocr* 24: 1091, 1964.

17. van der Molen, H. J., D. Groen, and J. H. van der Maas, in preparation.

18. Henning, H. D. and J. Zander, *H. S. Zeitschr physiol Chem* 330: 31, 1962.

19. Horning, E. C., W. J. A. VandenHeuvel, and B. G. Creech, in Glick, D., *Methods of Biochemical Analysis*, Vol. XI, Chapter 2, Separation and Determination of Steroids by Gas Chromatography, Interscience Publishers, New York-London, 1963.

# THE ANALYSIS OF PLASMA PROGESTINS BY GAS CHROMATOGRAPHY

I. H. Carlson, A. J. Blair and R. K. Meyer, Department of Zoology, University of Wisconsin, Madison, Wisconsin.

This study was supported by Grants A-5068 and 5-1AM-5240 of the National Institutes of Health, and by the Ford Foundation.

## INTRODUCTION

The use of GLC as an analytical technique for the identification and quantitation of progestational steroidal hormones has liminations, but nonetheless, is a most useful analytical procedure. Some progestins, such as 17α-hydroxy-4-pregnene-3,20-dione become rather intimately associated with the column constituents and consequently are poor candidates for GLC analysis. However, even though the progestins in question can be adequately characterized and quantified by GLC, other considerations, such as sample preparation and the variability of measurement, as a result of sample preparation, must be made. Considerable evidence exists to indicate the necessity for preliminary chromatographic separation of progestins from contaminating substances found in whole blood or serum samples, prior to GLC analysis. Realizing the importance of these considerations, we have developed a method that has allowed us to analyze the progestin content of ovarian venous blood and systemic blood of the laboratory rat.

## MATERIALS AND METHODS

(a) <u>Organic extraction</u>.

Whole heparinized blood, 6 ml, was immediately diluted with several volumes of 0.9 N saline. The blood cells were removed by centrifugation, washed with saline, recentrifuged, and supernatants combined. To the combined supernatant, 70 ml of cold acetone was added. This mixture was allowed to stand for 12 hr at 40 C. The protein precipitate was removed by filtration through glass wool, and washed several times with cold acetone. The filtrate was evaporated under reduced pressure to constant volume. The remaining aqueous phase was extracted three times with five vol of methylene chloride. The organic phase was dissolved in chloroform and reduced to a volume of 1.0 ml.

(b) <u>Preliminary chromatography</u>.

The solution of the dry extract in chloroform was diluted with Skellysolve-C 1:1, v/v. This solution was then subjected to a gradient elution system depicted in Fig. 1. The 1.0 cm diameter column was prepared with silica gel, 100 mesh, and the column packed to a height of 5 cm followed by wetting the column with a solution of Skellysolve-C and chloroform 1:1 v/v. Care was taken to exclude all of the air spaces in the column. The column effluent was collected in 4 ml aliquots, each of which was evaporated to dryness under air. The fractions were dissolved in 80 μl of chloroform and applied to TLC plates.

Generally all undesirable lipids were removed in the first four fractions. However, if plasma samples in excess of 30 ml were extracted, some lipid did appear in other fractions. Lipid contamination was adequately removed by the TLC separation. If might appear as though gradient elution is a tedious task, but we believe that the 20 min spent per sample is worthwhile. One caution should be stated. The method of packing, rate of elution, and length of column, should be kept as uniform as possible. We have found that different technicians seem to have different individual techniques that result in slightly different progestin elution patterns.

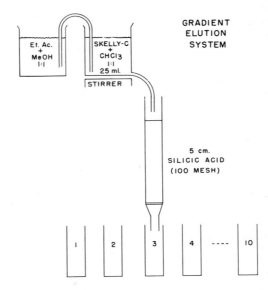

Fig. 1. Gradient elution system for removal of lipids from plasma extract.

Chromatographic separation of progestins was accomplished by TLC using the system chloroform: ethyl acetate, 13:1, v/v. The thin-layer plate was dried with a hand dryer, and scanned under UV light at 2400 A in a "Chromatovue" model C-3 scanner. UV light absorbing zones corresponding to progesterone and 20α-hydroxy-4-pregnene-3-one were eluted with methanol:ethyl acetate (1:1) with a suction collection device that allowed subsequent filtering of the aspirated silica gel on a standard glass plate. A second TLC step using the same solvent system and "white" plates was used for the purpose of removal of fluoresceine, reduction of contaminants in the eluate of the progesterone zone, or for further chromatography of acetylated components of the eluted zones. This will be referred to as the two-step thin-layer system.

Acetylation of hydroxyl containing compounds was accomplished using overnight incubation at 37 C of 0.25 ml of a 20% solution of acetic anhydride in benzene, diluted with an equal volume of pyridine.

(c) *Gas-liquid chromatography.*

GLC analyses of progestins was accomplished with a Barber-Colman gas chromatograph, model 10, equipped with a $^{90}$Sr argon ionization detector. In all cases, glass U-shaped columns of 3 i d and 6 ft in length, were used. Most routine analyses were accomplished with 90-100 mesh glass bead columns with a liquid phase loading of 0.2% w/w. The liquid phases employed most frequently were QF-1, XE-60, and SE-30. The glass beads were coated with the liquid phase by the method outlined by Nogare and Juvet (1). The glass bead columns were conditioned at a column temperature of 240 C, for a period of 12 hr or more, depending upon column bleed rate. The column temperature was set initially at 230 C for purposes of column evaluation. The flow rate was adjusted so that a pressure drop across the column was 30 psi, with the usual column effluent flow rate of from 50-63 ml per min. The cell bath was set at 235 C and the flash heater was maintained at 230 C.

Column evaluation consisted of determining the retention time and peak shape of a suitable standard steroid. For progestin analyses 1.0 μg of testosterone in 1.0 μl was employed. Further adjustment of the column temperature was made so that the retention time of testosterone was not less than 4 min and not more than 8 min. The injection of 1.0 μl containing 1.0 μg of a progesterone standard was made with a subsequent adjustment of the column temperature if the retention time was greater than 16 min. In the case of a broad peak the column was rejected. Column selectivity was determined by employing a mixture of 20α-hydroxy-4-pregnene-3-one and 20β-hydroxy-4-pregnene-3-one. If the particular column could resolve these isomers with a retention time difference of over one min, the column was deemed useful for qualitative analyses.

Quantitative analyses of progestins by GLC required the inclusion of an internal standard in the final progestin extract for the purposes of evaluating the column losses of the progestin. Testosterone was employed as the internal standard. Prior to any attempt at progestin quantitation, a standard curve of the progestin in question was plotted against peak height. Further reference determinations were made throughout the analyses to ascertain the loss of sensitivity of the detection system by injecting known amounts of the progestin in question and referring to the previously prepared standard curves.

(d) *Recovery studies.*

Recovery studies employing $^{14}$C-progesterone were carried out for the organic extraction, gradient elution, TLC, GLC, column effluent collection, as well as for the overall sample preparation for GLC analysis. Column effluent collection of labeled steroids was accomplished with a Packard Gas Chromatography Fraction Collector. The trapping of the steroidal material was accomplished with small glass vials loosely filled with methylene chloride saturated cotton wool. Fractions were collected before, during, and after the appearance of a peak on the recorder chart. Collection times were three min in each case. Elution of the steroidal material from the cotton wool was accomplished with methanol. Radioactivity was measured in a Packard Tri-Carb Liquid Scintillation spectrometer.

All solvents employed throughout the analysis were distilled prior to use.

## RESULTS AND DISCUSSION

The percent recovery and observed variability of the various samples prepared for GLC are presented in Table 1. The employment of acetone precipitation of the plasma protein prior to organic extraction substantially reduced the incidence of emulsion formation. Not only does this facilitate extraction but it also reduces the variability caused by the difficulty of uniform extraction of steroidal materials from the emulsion interface. Plasma samples in excess of 30 ml have been successfully extracted yielding equally high recoveries with minimum variability.

TABLE 1. Progestin extraction and chromatography recovery studies.

| Treatment | Recovery of $^{14}$C-progesterone (SD) |
|---|---|
| Organic extraction of whole blood | 94.5% (3.7) |
| Gradient elution | 91.0% |
| Thin-layer chromatography | 92.0% (6.3) |
| Overall recovery prior to GLC | 74.5% (7.6) |
| GLC column effluent recovery | 54.6% (3.7) |

TLC has revolutionized the preparation of samples for GLC analysis. However, the problem of co-chromatography of some steroidal compounds as a result of rather limited solvent travel may lead to complications. We have observed that cholesterol seriously interferes with the GLC analysis of progesterone. Since cholesterol is present in blood extracts to a larger extent than progesterone, its removal is imperative. This has been accomplished by our group, with what we refer to as the "two-step TLC" procedure. This procedure entails an initial thin-layer separation of the gradient elution eluate on a "white" thin-layer plate, followed by acetylation of the progesterone zone eluate, and a final thin-layer separation on a second "white" thin-layer plate. This results in the total removal of cholesterol from the extract to be subjected to GLC analysis. The 20-hydroxy-4-pregnene-3-one is freed from contaminating substances that seriously affect the successful analyses of these compounds by GLC, and therefore can be chromatographed.

The choice of the type of column for GLC of progestins by our group has been influenced by many factors. Glass bead columns of the type we employ with low liquid phase loading make it possible to analyze progestins at lower column temperature than similar length columns packed with silanized diatomaceous earth (Table 2). While glass bead columns may be slightly less efficient in terms of number of theoretical plates, the selectivity of liquid phases such as XE-60 and QF-1 more than compensates for this factor. Glass bead columns appear also to require less conditioning than the conventional columns. Furthermore, glass beads require, at most, acid washing to remove impurities that may be present. Our experience has been that glass bead columns can be reproduced more easily than diatomaceous earth columns.

If we consider Fig. 2 it is apparent that a 0.2% XE-60, 6 ft glass bead column can resolve the two isometric progestins, 20α-hydroxy-4-pregnene-3-one and 20β-hydroxy-4-pregnene-3-one, with short retention times and low column temperature of 210 C. We emphasize, however, that progesterone and 20α-hydroxy-4-pregnene-3-one are initially separated in the thin-layer step thereby avoiding the problem of poor resolution of these compounds by an XE-60 column (Table 2). Analysis of progesterone and 20α-hydroxy-4-pregnene-3-one by an 0.2% XE-60 column, in extracts of rat ovarian venous and peripheral blood yields chromatograms as illustrated in Fig. 2, 3, and 4. These tracings indicate that the extracts are relatively free from major contaminants that could otherwise interfere with plasma progestin analysis.

For the purpose of planning experiments, we have found it desirable to assess the types and quantities of plasma progestins in extracts prepared by our method. The employment of testosterone as an internal standard along with quantitation of progestins by referral to calibration curves prepared at the same time of analysis as shown in Fig. 5 and 6 have been found to be adequate. The method of plotting peak height against concentration is not only adequate, but also allows rapid calculation of the concentration in biological extracts.

We feel that it is apparent by this time that the plasma samples, which are to be analyzed for progestin content, must be adequately prepared prior to GLC. While this requirement would appear to mean a considerable expenditure of time on the part of the investigator, one should remember this demand is not unique to GLC. Furthermore, the preparation of the sample prior to GLC analysis, coupled with the inherent high resolution of GLC enables the investigator to be highly certain of his analytical conclusions, a feature not common to other methods of analysis.

TABLE 2. Retention times of progestins, standards, and extracts.

| Compound | TLC $R_f$ | Retention time in min | | |
|---|---|---|---|---|
| | | 0.2% SE-30* | 4.0% XE-60** | 0.2% QF-1† |
| Progesterone std. | 0.49 | 16.1 | 29.2 | 14.5 |
| Progesterone rat | 0.49 | 16.1 | 29.2 | 14.5 |
| 20α-ol std. | 0.23 | 18.1 | 28.2 | 11.0 |
| 20α-ol rat | 0.23 | 18.1 | 28.2 | 11.0 |
| 20β-ol std. | 0.23 | 16.3 | 25.0 | 8.5 |
| 20β-ol rat | 0.23 | 16.3 | 25.0 | 8.5 |
| 20α-ol acetate std. | 0.67 | 25.5 | | |
| 20α-ol acetate rat | 0.67 | 25.5 | | |
| 20β-ol acetate std. | 0.63 | 24.0 | | |
| 20β-ol acetate rat | 0.63 | 24.0 | | |

\* On 90-100 mesh glass beads, at 210 C, flow rate 56 ml per min.

\*\* On 60-80 chromosorb P, at 250 C, flow rate 53 ml per min.

† On 90-100 mesh glass beads, at 200 C, flow rate 56 ml per min.

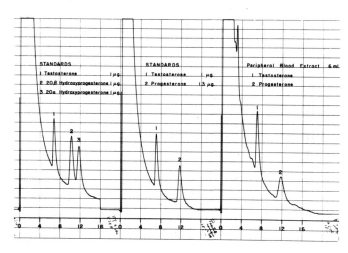

Fig. 2. Chromatogram of standard progestins and progesterone extracted from 6 ml of rat peripheral blood on day 6 of pregnancy.

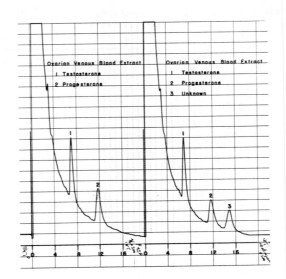

Fig. 3. GLC recording of progesterone extracted from 2.5 ml ovarian venous blood of day 6 pregnant rats.

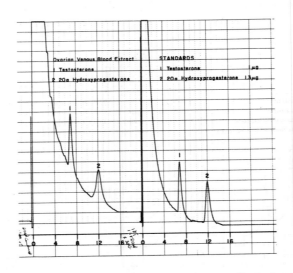

Fig. 4. GLC recording of standard and extracted 20α-hydroxyprogesterone from ovarian venous blood of a day 6 pregnant rat.

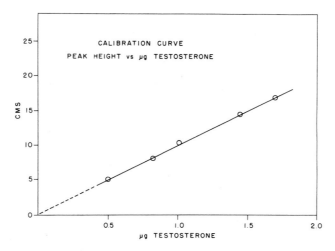

Fig. 5. Calibration curve of peak height vs μg progesterone as reference for correction factor calculated by observed peak height of internal standards of testosterone (0.2 %XE-60).

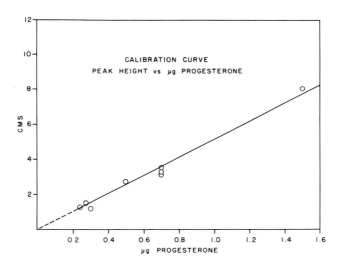

Fig. 6 Calibration curve of peak height vs μg progesterone. Lower points plotted during progestin analysis indicating slight loss of sensitivity (0.2% XE-60).

The major drawbacks of column losses and low detector sensitivity are being overcome by the availability of new detectors and column support materials. Furthermore, it is now possible to purchase at a modest price very adequate GLC analytical systems that are considerably more versatile than the equipment first marketed.

REFERENCES

1. Nogare, S. D. and R. S. Juvet, Gas-liquid Chromatography, Chapter II, p. 45, Interscience Publishers, New York-London, 1962.

# DISCUSSION OF PROGESTERONE, $C_{21}$-DIOLS AND TRIOLS

DR. SOMMERVILLE: I should like to report upon recent developments in our work upon the quantitative determination of progesterone in human plasma by TLC and gas-liquid radiochromatography. The following modifications have been introduced since our report upon the method (Collins and Sommerville, Nature 203: 836, 1964).

1. Modification of the Pye Panchromatograph: in the published method we referred to the fact that, occasionally, major losses occur due to a faulty seal between the base of the column and the molecular entrainer (flow-splitter) but we showed that these could be corrected in terms of the integrated radioactivity of the internal standard. This defect has now been overcome by removing the molecular entrainer and replacing it by a direct connection (swagelok). Difficulties due to fusion of the contents of the combustion tube have been reduced by lowering the temperature to 700 C. The main modification to which I have already referred has been for the determination of tritium and the complete system was illustrated in the slide which I showed in yesterday's session.

a) <u>Retention time.</u> In the original method a column of 1% cychlohexane-dimethanol succinate on Gas Chrome P (60-80 mesh) was run at 210 C and the retention time of progesterone was 72 min. In view of the purity of the final product, the retention time has been reduced to 32 min by using a column temperature of 225 C. Thus the time involved in performing four plasma assays in parallel (ether extraction of alkaline plasma; two-dimensional TLC and gas-liquid radiochromatography) is 3 1/2 hr.

b) <u>Calculation in terms of integrated radioactivity of $^3$H-progesterone.</u> In the published method 0.25 µg progesterone-4-$^{14}$C was added to each sample. Using a tritiated standard it is possible to reduce the amount of internal standard very considerably. As mentioned previously, 0.1 µg tritiated progesterone (425 mc per mM) yielded a high and reproducible number of integrator units (1775 $\pm$ SD 6.3). I should like to emphasize this modification as the results obtained with the tritiated standard indicate that the few values which we published for male plasma were overestimates and the results in plasma from peripheral venous blood collected during the follicular phase of the menstrual cycle by the modified method are now in the range 0.1-0.2 µg per 100 ml plasma. I am glad to find that these results are compatible with those reported by Dr. van der Molen. I should add that we obtain similar results in the luteal phase of the menstrual cycle or in pregnancy - whether the internal standard is labeled with $^{14}$C or tritium.

DR. GOLDFIEN: I was not well acquainted with your technique, Dr. Sommerville, and it may be simpler than ours. I am glad you agree with the results in general. I just wanted to comment about the levels of progesterone obtained by Dr. Carlson. In Fig. 1 are shown progesterone levels in rat ovarian vein obtained by Donald McDonald, a fourth year medical student who is working in my laboratory. He has cannulated the ovarian vein tying off the blood supply so that he might consider what might be the production rate of that ovary. There is some difficulty because you cannot be sure you haven't altered the flow.

In the estrus period, where Dr. Carlson quoted concentrations from a 10-30 µg per 100 ml, our values range from 15-31 and that is quite close. These are only three animals which I am talking about now.

In late diestrus, where his levels were about 50-150 µg per 100 ml, ours ranged from 30-60 µg per 100 ml. In the earlier phase, before cornification, we find very little progesterone, about 1 µg per 100 ml.

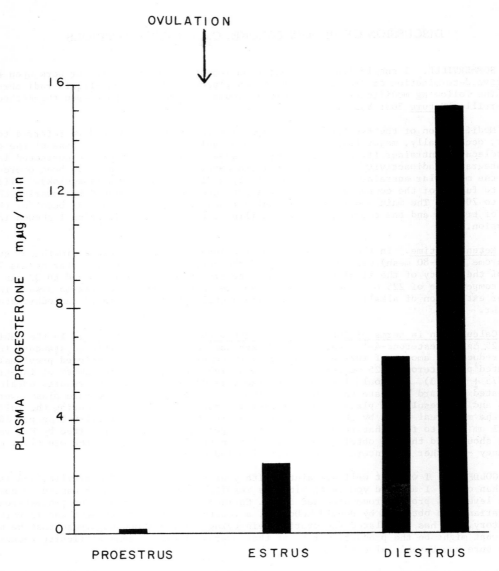

Fig. 1. Mean production rate of progesterone by the ovary (3 animals).

DR. VAN DER MOLEN: I would like to rectify an error and ask Dr. Goldfien a question. I have been told that I was wrong in assuming that nobody had estimated estrogens in normal plasma. Dr. Kroman has actually been doing this (Clin Chim Acta 9: 73, 1964). As you have shown us, Dr. Goldfien, you are able to measure a few thousandths of a microgram using a flame ionization detector. We know this is not a single observation and this sensitivity has also been claimed recently for the argon ionization detector. My own experience with the latter has not been as good.

DR. GOLDFIEN: The sensitivities we reported were obtained with a flame ionization detector. The sensitivity using the strontium ionization detector in our hands was approximately an order of magnitude different.

DR. ROSENFELD: Dr. van der Molen, I have two questions: first, about what length of time is tritiated progesterone stable in methanol solution? Second, do you find any 20-hydroxyprogesterone in the plasma?

Admittedly, you and Dr. Carlson are dealing with different animals, but I thought you might have some information on this.

DR. VAN DER MOLEN: First, answering your question about the radioactive progesterone. We have obtained this compound from New England Nuclear Corp., purified it by several TLC and paper chromatographies and store it as a concentrated stock solution in benzene in the freezer. For practical use in the plasma progesterone method we make up dilutions in methanol, that contain approximately 10,000-15,000 cpm per ml. From these dilutions we check the specific activities after addition of labeled progesterone and taking the mixture through the method. If I calculate correctly each single batch of diluted solution will last five to six weeks and we have never experienced difficulties with this internal standard.

To answer your second question, as I showed in the last flow sheet for the estimation of plasma steroids we are actually using a hexane/65% methanol paper chromatography system for this separation. It may be worthwhile to tell how we arrived at the estimation of 20$\alpha$- and 20$\beta$-hydroxy-4-pregnene-3-one in plasma. Yesterday I explained some of the reasons why we switched from cholesterol-chloroacetate to 20$\beta$-hydroxy-4-pregnene-3-one-chloroacetate as an internal standard for GLC during the estimation of plasma testosterone. The latter compound can be detected with approximately the same sensitivity in the electron capture detector as testosterone-chloroacetate, whereas detection of cholesterol chloroacetate is approximately 20 times less sensitive. In doing so we observed difficulties when we started analyzing plasma samples from women in the post-ovulatory period of their menstrual cycle. We could not obtain a clear 20$\beta$-hydroxy-4-pregnene-3-one-chloroacetate peak in these samples, there was always a shoulder or a clear additional peak with the retention time of 20$\alpha$-hydroxy-4-pregnene-3-one-chloroacetate. We know that 20$\alpha$- and 20$\beta$-hydroxy-4-pregnene-3-one do not separate too well from testosterone during TLC of the steroids. We analyzed this further and switched to paper chromatography for a satisfactory separation of testosterone, 20$\alpha$- and 20$\beta$-hydroxy-4-pregnene-3-one. Figure 2 gives some typical analytic results in one subject throughout the menstrual cycle. All I can say about the specificity of this method is that the presumptive 20$\alpha$- and 20$\beta$-hydroxy-4-pregnene-3-one behave chromatographically (in the free form in TLC and paper chromatography and following chloroacetylation in TLC and GLC) like the authentic steroids.

May I come back to a question I tried to ask before? I believe that Dr. Carlson stated that he separated 20$\alpha$ and 20$\beta$ reduced progesterone with TLC. We have always been forced to use paper chromatography. I wonder if you really obtained separation and whether the small inflection in your 20$\beta$-hydroxy peak might have been due to the 20$\alpha$ compound?

DR. CARLSON: I did not mean to imply that we were separating 20$\alpha$ and 20$\beta$-ol by TLC. However, as one of my slides indicated, one can resolve these isomers with an XE-60 column. I did make reference to a slight shoulder on the leading edge of a tracing of 20$\alpha$-ol from a plasma extract, as being a trace of 20$\beta$-ol. This is a common occurrence when one of the compounds is present in very small amounts. However, in plasma from rats during late pregnancy one can obtain tracings indicating both 20$\alpha$ and 20$\beta$ as distinct fractions. From studies of derivative formation with further GLC of these fractions, we are convinced that we have isolated 20$\alpha$ and 20$\beta$-ol.

DR. ROSENFELD: Two comments to Dr. Carlson. First about the life of thin-film glass bead columns; how long do they last? Secondly, it seems to me that if one would use a defatting procedure in these analyses, like, for example, a three funnel partition between petroleum ether and 90% methanol, you could very nicely separate cholesterol and other lipids from most of the steroids which have been discussed. The steroids, of course, are all soluble in the methanol phase.

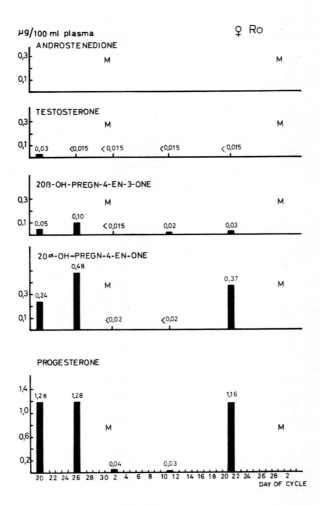

Fig. 2. Plasma levels of androstenedione, testosterone, progesterone, 20α-hydroxyprogesterone and 20β-hydroxyprogesterone throughout the menstrual cycle.

DR. CARLSON: With respect to the column, I grant you that a QF-1 column, if operating continuously at 225 C for any length of time, has a short life of from three to four weeks. On the other hand, an XE-60 column will remain functional with little change in retention times, for several months. Similarly, an SE-30 column will last considerably longer than both of the former. However, one does not have to keep the column in the machine all the time. One can remove the column after an analysis, thereby further increasing the life span of the column. One of the major factors affecting column life is the build-up of pyrolized material which causes broadening of the peaks. We have found this to be a limiting factor rather than the loss of liquid phase. In order to avoid this undesirable problem one must prepare clean samples by adequate preliminary chromatography. The suggestion by Dr. Rosenfeld regarding the petroleum ether-methanol partition is a good one, which we have considered.

I would like to point out that the acetone precipitation of plasma proteins allows one to isolate circulating gonadotrophins, the levels of which can be correlated with the steroid levels found in the same plasma sample.

DR. VAN DER MOLEN: I have some comments on the remarks that Dr. Goldfein made. This has been brought up in the past and will probably be brought up in many future meetings in discussions about extraction of steroids from plasma.

I do believe that the extraction of alkaline plasma is a good extraction and gives very clean extracts. However, you should not add your alkali before lunch and try to extract after lunch. You will find a jelly that is difficult to extract and will even give troubles in removing it from the tube. It has been proven many times in the past and we have also found this extraction as good for testosterone as for progesterone and could find no indications of decomposition or appreciable losses. But, again, you should extract immediately, or at least within the next half hour following additions of the alkali.

DR. EIK-NES: I was wondering about Dr. Carlson's experience with fluoresceine and low recovery of steroids from thin-layer plates. Have you tried the luminescent chemical (index 609) available from DuPont? We have used this material in all our TLC over the past two years and we have not seen that this material interfered with elution of steroids from thin-layer plates, GLC of steroids or liquid scintillation counting of steroids.

Finally, with regard to extraction of plasma, I do not think this is too much of a problem even when extracting plasma at an alkaline pH. I would like to put in a plug for the use of the lowest polar solvent you can get by knowing, admittedly, the partition coefficient for the steroid you are looking for. Thus, I don't feel that ethyl acetate should be used as a standard solvent for all steroid extractions. The best initial extract is the one of lowest weight but still containing better than 95% of the steroid to be estimated. Extraction of a plasma steroid of low polarity with a solvent of high polarity will remove the steroid most efficiently but may also give excess weight of the crude extract. This tends to increase the number of purification steps needed prior to column, paper, TLC or GLC.

DR. CARLSON: The loss of internal standard steroids in fractions that were eluted from fluoresceine treated plates was our first indication of difficulty with this material. In retrospect, the fluoresceine we used was a mixture suggested by the Brinkman people in one of their early brochures. However, I feel this may be the reason for the losses we observed, for, since that time, newer products have been developed, as Dr. Eik-Nes indicates, and may not cause on-column losses of steroids.

DR. FUTTERWEIT: In terms of the defatting procedure I wonder if, in a simple procedure, a small column might not be employed.

DR. TAIT: I would like to ask Dr. Sommerville a question about the errors of his method. The remarks apply to all methods including our own for plasma aldosterone where one has to add large amounts of indicator and errors are consequently escalated.

In his paper in Nature he added 0.2 $\mu g$ of indicator. Taking the combined error of his measurement of radioactivity and mass as 6% as he quoted, this would lead to a minimum standard error of the estimate of 0.012 $\mu g$ per sample or 0.06 $\mu g$ per 100 ml for a 20 ml plasma sample. This would lead to an error of 60% for progesterone estimates in the follicular phase. This arises because one is estimating 0.02 $\mu g$ per 20 ml by subtracting 0.20 from 0.22 $\mu g$.

I realize that the indicator added has now been lowered to 0.1 µg. This would still make for a large error estimated in the follicular phase.

DR. SOMMERVILLE: I think that Dr. Tait is quite right about the $^{14}C$ method as we described it in Nature and, as I said before, the use of a relatively large amount of $^{14}C$ labeled standard introduces an error in the determination in male plasma and to a less extent in the follicular phase of the cycle. In Nature we gave results with 0.1 µg $^{14}C$ internal standard and there was a coefficient of variation of 2.9%, whereas, as I have emphasized in this meeting, with tritium the coefficient of variation is reduced to 0.04%.

Reference was made to the use of silica gel and I should mention that we use a silica gel defatting column in our in vitro work as a preliminary to gas-liquid radiochromatography.

DR. TAIT: The combined error of mass and radioactive estimates would still be 3%. This would lead to a minimum standard error of the estimate of 3% of 0.1 µg. It is still a considerable error due to the addition of a large amount of indicator.

DR. SALHANICK: May I ask Dr. Carlson a question? What was the blood level of progesterone that you found in the rat and how does this compare with the human.

DR. CARLSON: The level of progesterone in rat blood plasma appear to be, perhaps 10 times as great as those found in the human. Of course this comparison depends upon the stage of the estrous cycle and the menstrual cycle.

DR. WILSON: Metabolites such as 5-androstene-3β,17β-diol, 5-pregnene-3β,2α-diol and 5-pregnene-3β,17α,20α-triol have been particularly difficult to determine by conventional means.

I would like to show some approximate values which we obtained by GLC on relatively crude urine extract fraction. The surprising amount of information gained makes a good illustration of the value of GLC as a rapid screening procedure for following fractionations.

The urines were first divided into a glucuronide fraction, obtained by continuous ether extraction at pH 5 after incubation with β-glucuronidase, and a sulfate fraction, similarly extracted after acidifying the residual urine to 1N $H_2SO_4$. Eluates from a partition column, containing groups of steroids, were then divided into ketonic and non-ketonic fractions. Each of these was treated with digitonin to obtain the 3β-hydroxy steroids, and these were analyzed by GLC using a 3% SE-30 column.

Figure 3 shows the results on the non-ketonic 3β-hydroxy fraction of a subject with an interstitial cell carcinoma of the testis. As some peaks unquestionably represent more than one substance, the values are only approximate. Nevertheless, the peaks are very well defined when one considers that only group separations had been performed.

Until recently, we always thought of $\Delta^5$-3β-hydroxy metabolites as being in the sulfate fraction. Now it can be seen that three such steroids were excreted as glucuronides. Although 5-pregnenetriol has previously been found as a glucuronide, we believe this is the first demonstration of 5-androstenediol and 5-pregnenediol glucuronides.

Figure 3 also shows that the same components were found in much larger amounts in the sulfate fraction. The very high excretion levels of the tumor patient are evident.

Figure 4 contrasts these levels with the much lower values found in a pool of urine from two normal men. Again, a little of each diol was excreted as the glururonide, whereas 80% of the 5-pregnenetriol was in this fraction.

Figures 5 and 6 show tracings from the corresponding 3β-hydroxy ketonic fractions. Titers given for dehydroepiandrosterone (DHEA) would include epiandrosterone which has the same $R_T$ on SE-30.

The tumor patient had about 22% of his total DHEA in the glucuronide moiety, compared to 10% for the normal men. This patient also showed a large clean peak in column eluate B2, not seen in the controls. This has been fairly well identified as 16α-hydroxy-DHEA. It too was excreted to the extent of 30% as the glucuronide.

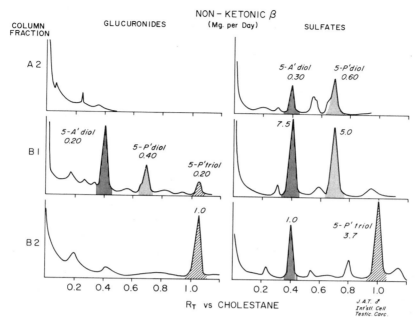

Fig. 3. GLC tracings of 3β-hydroxy non-ketonic fractions of successive column eluates from the urine of a subject with an interstitial cell tumor of the testis. Peaks which are shaded and labeled corresponded exactly with the $R_t$ of pure reference standards. Other peaks represent unknown materials. The figures are mg per day calculated from peak heights.

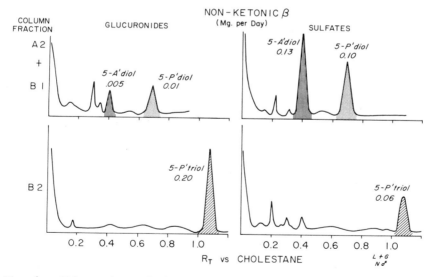

Fig. 4. GLC tracings of the 3β-hydroxy non-ketonic fractions from the pooled urine of two normal men.

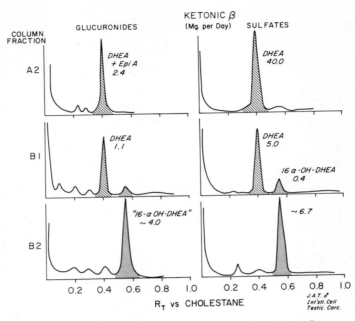

Fig. 5. GLC tracings of the 3β-hydroxy ketonic fractions from a subject with an interstitial cell tumor of the testis.

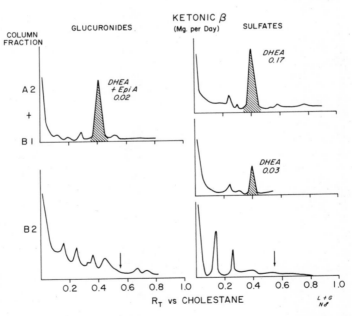

Fig. 6. GLC tracings of the 3β-hydroxy ketonic fractions from the pooled urine of two normal men.

This type of analysis could not be performed on most of the corresponding 3α-hydroxy fractions owing to the far more numerous components. However, we believe that such approximate qualitative and quantitative analyses of suitably clean fractions, is an important use of GLC. Besides giving an idea of what compounds are present, it also yields invaluable information for planning the next purification step, such as how wide a papergram or thin-layer plate is needed, and what reference standards should be run in parallel.

Moreover, as seen here in the discovery of 16α-hydroxy-DHEA, the GLC tracing may provide unexpected leads for further experiments.

DR. VAN DER MOLEN: I appreciate that Dr. Kirschner found considerable differences between the pregnanediol values obtained using Klopper's method and GLC. As we all know, there was for some mysterious reason no Allen correction included in the original Klopper method. It has been demonstrated subsequently that, calculating your values from the peak absorption only, may result in erroneous high values. One of the questions I would like Dr. Kirschner to answer is, whether he applied any such corrections or not. Another point I would seriously question is about the recoveries of the two methods that have been used. I understand that these methods are in use at different places. Whereas no internal standards are included in either method, the recoveries might differ and offer another explanation for the observed differences.

The method for the determination of urinary pregnanediol in urine as developed by Klopper et al. (J Endocr 12: 209, 1955), is among the ones most widely used. Some doubt has been expressed, however, as to the specificity of the sulfuric acid reaction for the quantitative estimation of the prenanedioldiacetate in the final extract of low titer urines, that might contain interfering aspecific chromogens (Lipp, G., Acta Endocr (Kbh) 33: 501, 1960; van der Molen, H. J., "Infrared Spectrometry of Steroids" Thesis, Utrecht, 1961).

One measure to overcome possible interfering absorptions of aspecific chromogens is the application of the Allen correction for the quantitative evaluation of the sulfuric acid coloration. Non-specific chromogens, that do not adhere to the presumptions underlying the Allen correction, might still lead to erroneous results.

We have compared for a large series of urines the results obtained using the sulfuric acid reaction and the results using GLC with flame ionization detection for the quantitation of pregnanedioldiacetate isolated from the urine samples according to Klopper's method.

Residues after the final chromatography of the pregnanedioldiacetate were dissolved in 1 ml chloroform: 1/10th of the sample was taken for GLC and evaporated to dryness. The remainder was evaporated and used for the sulfuric acid reaction (measuring at 380, 420, and 460 mµ using the Allen correction).

To the samples for GLC were added appropriate amounts of 5α-cholestan-3-one as internal standard and aliquots were injected onto a 6 ft 1% SE-30 column using the flame ionization detector for detection (Fig. 7).

The SE-30 column was chosen in order to obtain a maximal separation between the pregnanediol-diacetates and from the pregnanolone-acetates. The precision of quantitation using GLC is in the order of 5%.

The overall accuracy of the method was calculated (standard deviation = $\frac{(diff^2)}{2N}$ from the differences in results between duplicate urine samples. No difference was observed between the accuracy of quantitation using GLC or the sulfuric acid reaction (Table 1).

It appears (Fig. 8) that, at low concentrations, a large number of results obtained after GLC quantitation, are significantly lower than results of the same residue after quantitation with the sulfuric acid reaction. This indicates the presence in the final residue of impurities that may lead to too high results when the sulfuric acid reaction is used.

Besides its specificity, GLC quantitation has the great advantage that the sensitivity of flame ionization detection is so much higher than the sensitivity of the sulfuric acid reaction (0.05-0.10 µg vs 10-20 µg) that a rather small aliquot of the 24 hr urine can be analyzed to make an accurate quantitation of the amount of pregnanediol present.

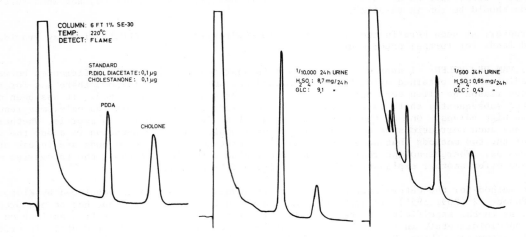

Fig. 7.  Gas chromatographic tracings of pregnanedioldiacetate and cholestanone standards and urine samples.

Column: 6 ft 1% SE-30 on 80/100 mesh Gas Chrom P.
Column temp: 210 C; detector temp: 250 C; injection port temp: 230 C; carrier gas: nitrogen, 75 ml per min.

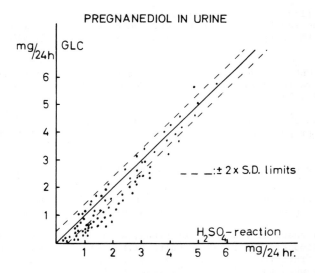

Fig. 8.  Correlation between the amounts of pregnanediol measured in the same urine samples, respectively using the sulfuric acid reaction and GLC.

TABLE 1. Accuracy of the estimation of pregnanediol in urine, analyzing 1/20th of 24 hr urine (range: 0.10-5.53 mg per 24 hr).

| Method of quantitation | Number of duplicate analysis | SD mg per 24 hr |
|---|---|---|
| $H_2SO_4$-reaction | 64 | 0.19 |
| Gas chromatography | 68 | 0.20 |

DR. KIRSCHNER: In the comparative study, the Allen correction was not applied. A single reading was taken at 420 mµ. Regarding an internal standard, we have not found this necessary for routine use. We assume that our hydrolytic yields are the same as others, and have shown previously that recoveries from TLC is virtually quantitative. Since our GLC aliquots are generally 0.5-2.0 µg, I doubt that minimal adsorption by the column is significant with such large quantities. The fact that authentic pregnanediol added to urine is recovered quantitatively, seems to support these assumptions.

DR. WOTIZ: A couple of years ago we published a method for the determination of pregnanediol in crude extracts and although this is by no means as refined as Dr. Kirschner's procedure we find that we can analyze not only pregnancy urines, but, also urines from man and ovariectomized women.

I would like to express agreement with the statement made by Dr. van der Molen. We have compared the Klopper procedure with GLC of crude extracts on a 3300 plate SE-30 column. Comparison of these data shows that at high levels of excretion we obtain slightly greater values with GLC. In the proliferative phase GLC obtains considerably lower figures. By and large the lowest levels we normally find are of the order of 150 to 180 µg.

According to the publication by Kirschner and Lipsett, the average pregnanediol excretion of males was 0.41 mg per 24 hr. Using our method on urine from nine males we found a mean value of 0.46 mg. In the proliferative phase, Dr. Kirschner reports a value of 0.55 mg. In 100 determinations we found 0.52 mg. In the luteal phase Dr. Kirschner reported 4.0 mg, we also get a mean value of exactly 4 mg.

Cox reported that the values he obtained using a preliminary column chromatogram followed by GLC, when compared to those obtained by the Klopper procedure showed little difference. This suggests that the discrepancy reported by Dr. Kirschner is due to a defect in the application of the Klopper procedure.

DR. KIRSCHNER: In our GLC system the relative retention time of allopregnanediol diacetate was 1.49 vs 1.32 for pregnanediol diacetate. The separation factor of 1.13 is not good enough to allow quantitations of the allopregnanediol. It appears in the tail of the pregnanediol diacetate peak. Since there is usually much less of this metabolite in urine, I doubt if it significantly contributes to the pregnanediol peak. I would not want to rely on this separation factor, however, to measure the allopregnanediol in urine.

In the five pregnancy urines reported by Cox, there was a good agreement between GLC and Klopper determinations. In the single follicular phase urine the GLC value was 0.2 mgm per day vs 0.5 by sulfuric acid method. In two of the four comparative studies of Jansen, the GLC values were significantly lower. If the differences between pregnanediol values estimated by GLC and the Klopper method were due to non-specific contaminants in the latter, it could be predicted that these differences would be greater at lower levels of excretion and relatively less important at the higher levels. Dr. van der Molen's comparisons seem to bear out these predictions.

DR. VAN DER MOLEN: I am happy with the views that have been expressed. Especially because they might explain these rather unexpected differences and may stress again, as have been demonstrated in several publications, that you may make tremendous errors by not applying the Allen correction.

It must be worthwhile to show some pictures that we prepared before we started this investigation in order to obtain an impression about the separation of progesterone metabolites during GLC. We did not try to put all eight different pregnanediols in these pictures, because they would have looked even more crowded than they do!

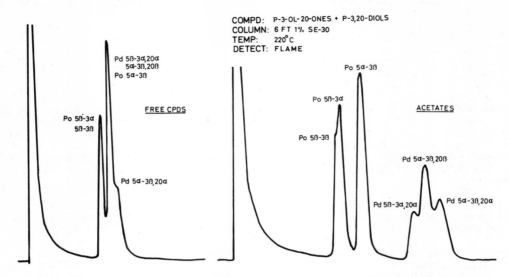

Fig. 9. Separation of pregnane-3-ol-20-ones (Po) and pregnane-3-,20-diols (Pd) and their acetates on a 6 ft 1% SE-30 column.

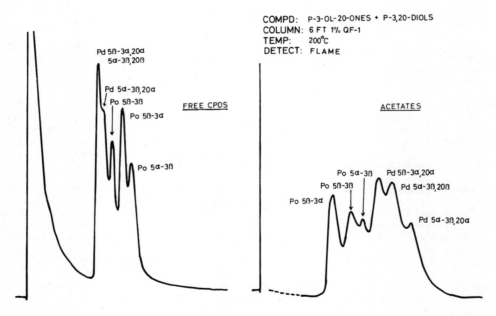

Fig. 10. Separation of pregnane-3-ol-20-ones (Po) and pregnane-3,20-diols (Pd) and their acetates on a 6 ft 1% QF-1 column.

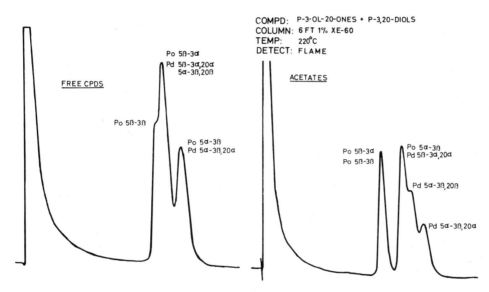

Fig. 11. Separation of pregnane-3-ol-20-ones (Po) and pregnane-3,20-diols (Pd) and their acetates on a 6 ft 1% XE-60 column.

Actually, because we were mainly interested in separating 5β-pregnane-3α, 20α-diol-diacetate from 5β-pregnane-3α-ol-20-one-acetate we preferred the SE-30 column for our studies.

DR. BROWNIE: Dr. Kirschner has presented the estimate of coefficient of variation for pregnanediol determination by GLC.

I wonder if he could give us an estimate of the coefficient variation at the level of 50 μg a day which he suggested was the lower level that could be handled by this method?

DR. KIRSCHNER: We do not have sufficient data to give that estimate.

DR. ROSENFELD: With reference to the separation of 5α and 5β-compounds on SE-30, although these are $C_{19}$'s and not $C_{21}$'s we had a problem of separating 3α-hydroxyetiocnolanolone from 3β-hydroxyandrosterone in a TLC eluate. Figure 12 shows the TMS ether of such an extract. The peak at 11.9 min is the 5β-compound and the 14.8 min peak is the 3β-hydroxyandrosterone derivative. They seem to separate quite well.

DR. SOMMERVILLE: The Allen correction was not included in Dr. Klopper's method as published in 1954 but has been used in many departments, including my own, for the past 10 years. It is especially important to use the correction when determining urinary pregnanediol in male subjects. Could I ask Dr. Wotiz for a mean value by his method in male urine?

DR. WOTIZ: 0.46 mg per 24 hr.

DR. SOMMERVILLE: In a paper which we published on Klinefelter's syndrome (Giorgi and Sommerville, <u>J Clin Endocr</u> 1963) our pregnanediol levels using the Allen correction were $1.26 \pm 0.50$ mg per 24 hr in healthy men.

The other point which I should like to mention is this important matter raised by Dr. Tait about the permissible level of labelled internal standard. I have now discussed this matter with Dr. Tait who agrees that I may put on record that his criticism would be answered by the use of less than 0.05 μg tritiated progesterone. We obtain a significant integration with as little as 0.01 μg tritiated progesterone and so there should be no difficulty in following his valuable suggestion.

Finally, I should like to add a word about the application of GLC to the determination of plasma pregnanediol. In my department a method is being developed for the determination of several progesterone metabolites in human plasma and this involves two-dimensional TLC and the use of tritiated pregnanediol kindly supplied by Dr. Erlich of Chicago. Figure 13 shows the result of an assay in late pregnancy plasma (the aliquot corresponds to 1 ml plasma). The main peak on the left is pregnanediol diacetate. This was obtained with a 1% CDMS column and separation of the $C^5$ stereoisomers can be obtained with other phases (e.g., QF-1).

DR. EIK-NES: Since we are exchanging historical notes, I had a prepublication of Dr. Klopper's method in my laboratory where somebody had written "at low levels use the Allen correction."

I wonder, Dr. Wilson, if you have done more work on the steroid glucuronosides you discussed a while ago?

DR. WILSON: All that we have done I mentioned briefly, that is that we found the compounds in the fractions that were extracted by continuous ether after hydrolysis with β-glucuronidase. This extraction is open to some criticism in that it has been found that some sulfates are split during that procedure. We were aware of this fact and therefore we examined some urines which had been extracted by the Kellie-Wade procedure.

The glucuronides were separated from the sulfates on a TLC system and the glucuronides fraction so obtained was hydrolyzed with β-glucuronidase and the beta fraction obtained. We again found the same peaks.

DR. ENGEL: The comments I have to make may, perhaps, be more appropriate for tomorrow's session. Since they touch upon matters that we will have to discuss tomorrow morning, I should like to get them off my chest now in the expectation that by tomorrow each of the arguments I propose will be effectively refuted by the experts in the field.

One of the problems that we face when we try to make up our minds whether to use GLC or one

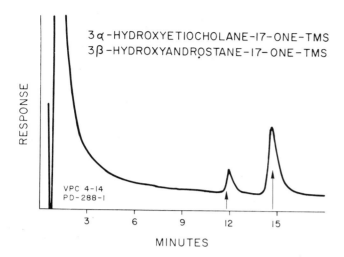

Fig. 12. Separation of TMS ethers of the 3α- and 3β-hydroxy-etiocholanolone-17-one.

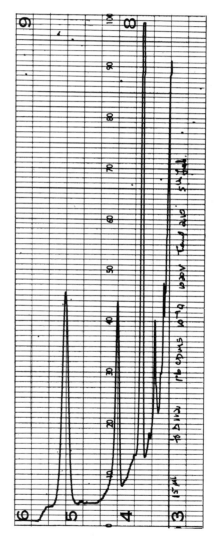

Fig. 13. Gas chromatogram of pregnanediol diacetate from pregnancy plasma. Peak 5 is pregnanediol diacetate.

of the more conventional methods of analysis is whether there is, in fact, a saving of time.

When one considers the fact that if it takes approximately 15 min for a peak to emerge from the gas chromatogram, this establishes a limit to the number of determinations that can be done in a single day. A battery of gas chromatograms is not a solution. Automation of sample introduction and digital readout of the integrated plot might help to alleviate this difficulty but still we are dealing with a situation in which one sample at a time has to go through the instrument.

We can compare this situation with the more conventional methods where one can set up 10, 20, or even 30 tubes and sit in front of the spectrophotometer and read them. To be sure, the preliminary processing must be taken into consideration as a factor in determining which way we are going to jump.

The attributes of GLC are so well known that we need not dwell upon them; among them are sensitivity and resolving power. When we wish to use GLC for quantitative analysis, it seems to me that we deal with three kinds of situations in which we have a greater or lesser need for these two attributes, sensitivity and resolution.

At one extreme we deal with multicomponent mixtures where we are trying to obtain an analysis of a group of compounds isolated from a more complex mixture in such a way that to the best of our ability, we have removed extraneous compounds. Clearly, GLC is a powerful tool since one is utilizing both the resolution and the sensitivity of the method plus specificity or lack of specificity of the detector depending upon what system is employed.

One sees also advantages in the type of situation that Dr. Kirschner described this afternoon. This is the case in which it is not convenient to exclude all but the compounds to be measured by prior treatment. Here one can distinguish between the one compound one desires to measure and the other compounds that are difficult to remove except by the high resolving power of GLC.

The third situation was discussed at some length this morning. This is the one in which the purification is so extensive that a single peak in addition to solvent appears on the gas chromatogram. Here, in fact, the thousand theoretical plates of the gas chromatogram are used solely and exclusively for separating a volatile solvent from a steroid or a steroid derivative.

It is possible that in this situation we should use a different instrument which one might name a "flame ionization balance" or an "electron capture balance"? It would consist of a solid injection system plus a detector coupled to an amplifier and recorder and would dispense with a column. It seems to me that this would be a useful analytical tool and I wonder if the apparatus manufacturers have considered building such an instrument for the specific purpose of measuring single compounds when the important criterion is the high sensitivity that can be achieved with the electron capture or flame ionization detector or some similar instrument.

DR. WILSON: Just one quick answer over the top of my head. You don't know whether there is going to be one peak there until you put it through the column.

I might say that in agreement with what Dr. Engel says, other things being equal if you wish to do colorimetric determinations such as the Zimmermann reaction, you can do 40 or 50 unknowns plus standards in two hours. It would take days to do that by GLC. We certainly are well aware of this.

DR. WOTIZ: I would like to reply to Dr. Engel's comments. First of all, let us consider the problem of the laboratory charged with carrying out a large number of assays of relative simplicity, such as the determination of pregnanediol or pregnancy estriol. While it is certainly restrictive to be able to analyze only one extract at a time, the period of approximately 15 min per assay would permit about 25 determinations a day. While this is far less than the number of final colorimetric quantifications which can be made per day, it does include a major portion of the purification prior to measurement. In fact, either of these two determinations can be carried out in about three hr, allowing as many as 60 analyses per week or slightly better. While I don't advocate the use of a battery of gas chromatographs, two reliable instruments could in fact increase this productivity significantly without increasing the labor force.

Another situation to be considered is the analysis of steroids present in fairly large concentrations (mg per 24 hr) but requiring the separation of several individual components from each other as well as from unknown contaminants. For example, the analysis of 17-ketosteroids or 17-ketogenic steroids or the fractionation of several pregnancy estrogens. This generally requires a relatively simple preliminary clean-up such as described by Drs. Bailey, Kirschner, Rosenberg and others. Generally speaking, I think it would be fair to say that such procedures are often more rapid than equivalent methods despite the fact that only one injection at a time can be made. The significant time saving occurs at the earlier phases of the assays. Moreover, one must point out, that the often decried non-specificity of the GLC detectors can be an advantage here. On one hand, one may detect components of interest not previously known and on the other, any drugs, dietary constituents or other substances, which may interfere with an established assay procedure are readily detected in the recorder tracing. This further offers the opportunity to inspect final results even long after readings are taken - something which generally cannot be done when colorimetric readings are recorded by hand.

One need also point out that the non-specific detectors offer the advantage of seeing substances not otherwise readily observed, such as cortol and cortolone. To make up, at least in part, for this often deplored lack of detector specificity there are other contributing factors which must be considered, such as the specific separation of individual components by paper or thin-layer chromatography, often involving more than one such step, followed by GLC involving perhaps 2500-4500 theoretical plates. Since this high efficiency separation can be accomplished in but a few min, for critical assays GLC on two or three columns may be carried out, inducing a considerably greater assurance of specificity. In fact, one can increase the column length to 8 or 12 ft and in 30-40 min obtain chromatograms with an efficiency of 5000-9000 theoretical plates.

Finally, one must point to the much more involved highly specific and extremely sensitive types of procedures such as the plasma testosterone, estradiol and progesterone determinations so ably presented here previously. As Dr. Wilson already explained, one does not know that there is only one peak before the final analysis is carried through. Despite the necessarily more involved purification prior to GLC such procedures are still less time consuming than the more conventional techniques and I daresay, are now or will shortly be more sensitive as the latter. (Addition in proof - we are presently capable of measuring as little as $10^{-11}$ g estradiol standard as the heptafluorobutyrate and can measure $5 \times 10^{-11}$ g of this substance in human plasma.)

# GAS CHROMATOGRAPHY AND ITS ROLE IN THE VERSATILE ANALYSIS OF URINARY ESTROGENS

Herbert H. Wotiz and Sati C. Chattoraj, Department of Biochemistry, Boston University School of Medicine, Boston University Medical Center, Boston, Mass.

This study has been supported by a Research Career Development Award (3-K3-GM-15,369) and a Research Grant (CA 03135) from the U.S. Public Health Service, Bethesda, Md.

The difficulty in assaying estrogens is due largely to the fact that, except in pregnancy and in certain endocrinopathies, they are present in extremely low quantities in biological media which may contain large amounts of other substances with characteristics similar to those of estrogens. The discovery of a large number of new metabolites in recent years has further complicated the problems. As a result all of the methods of proven merit have suffered from a lack of sensitivity or practicability. The simultaneous reports on the feasibility of steroid gas chromatography (1, 2) and its application in quantitative analysis of estrogens (3) have brought a new technique to the field of estrogen analysis. As in some other fields, GLC has a great advantage over conventional chromatography through its high resolution, speed and separation, simultaneous quantification and high sensitivity through the use of ionization detectors. This communication deals with the different parameters involved in the development of a versatile and, sensitive method of determination of estrogens in biological media.

## GENERAL CONSIDERATIONS

Before coming to the detailed procedure for the estimation of various estrogens, it is important to consider some of the prerequisites necessary for the separation, detection and quantitative evaluation of estrogens, particularly as applicable to their analysis by GLC. I would like to now discuss some of the problems concerned with the formation and use of derivatives.

*Formation of derivatives.* Steroids with poly-functional groups show strong irreversible adsorption on the active site of the column (4). Formation of derivatives of a relatively non-polar nature with adequate volatility, not only overcomes these effects but generally enhances separations. In the present investigation two types of derivatives have been examined, namely, the acetates and TMS ethers. The first description of separation and analysis of steroid acetates was presented by Wotiz and Martin (5). The use of trifluoroacetates and TMS ethers in steroid separation was reported by VandenHeuvel et al. (6) and by Luukkainen et al. (7) respectively.

A comparison of a number of important characteristics of such derivatives is shown in Table 1, along with data for the mixed derivatives obtained by prior methylation at position 3, and the free parent compounds. The major objections to the use of acetates raised by some investigators are the longer retention times requiring higher column temperatures; asymmetry of the peak; and poor resolution. Examination of Table 1 and some of the chromatographic reproductions in the main body of the paper shows that with a properly prepared column excellent symmetry can be obtained. With respect to the speed of elution, identical retention times for any single compound may well be achieved simply by increasing the temperature. Most of the stationary phases presently in use can easily tolerate temperatures of 250 C and the silicone elastomer SE-30 has been used successfully with temperatures up to 300 C. Stability of a number of steroids during GLC at these higher temperatures has already been demonstrated (8). In fact, the only disadvantage of the acetates is their adsorption to the column, requiring separation prior to analyses. Furthermore, acetylation of all unhindered hydroxyl functions of the known estrogens can be achieved quickly and in excellent yield (95-100%). Their stability in solution, during GLC, and during further purification after derivative formation has been adequately demonstrated.

TABLE 1. Properties of different derivatives.

| | ACETATES | TMS | 3-METHYL ETHERS (as acetate or TMS) | FREE STEROID |
|---|---|---|---|---|
| Peak symmetry | Excellent | Excellent | Excellent | Poor |
| Completeness of derivative formation | 95 - 100% | Presumed quantitative | Presumed quantitative | -- |
| Reactivity | -OH of all estrogens | -OH of all estrogens | α-ketols decompose epiestriols-multiple peaks | -- |
| Stability in GLC | Proven for 10 estrogens | Presumed | Proven | Proven for $E_1$, $E_2$, $E_3$ |
| Stability in solution | Excellent | Tendency to hydrolyze (moisture or acid) | Excellent (as acetate) | Excellent |
| Stability for further procedures | Excellent | Poor | Excellent (as acetate) | Excellent |
| Ease of preparation | Good | Fair | Poor | -- |
| Comparative temperature of elution of $E_1$* for equal $R_t$ (5.4 min) on a 3% SE-30 column | 245 C | 239 C | Unknown | 232 C |
| Non-adsorptivity on column | Fair | Very good | Unknown | Poor |

*$E_1$-estrone, $E_2$ - estradiol, $E_3$ - estriol

3-0-1

Similar studies on the stability of TMS ethers are still lacking in the literature. The TMS ethers are presumed to be stable during GLC and quantitativeness of reaction has been established only on the basis of GLC examination alone. Symmetry of these derivatives is excellent. Similar to the acetates, all of the unhindered hydroxyls of known estrogens undergo reaction. Probably due to their greater volatility and bond shielding, adsorption phenomena on the column are reduced to the TMS ethers. On the other hand, stability in solution is more questionable especially where any traces of moisture or acid are likely to be present. The tendency to hydrolyze spontaneously is considerable. Once these derivatives are formed it has been our experience that they cannot be further purified without causing at least partial hydrolysis.

Use of the mixed derivatives, with the methyl group at position 3, while allowing the important advantage of purification through phase change prior to GLC, has not been employed by us. The process of methylation would cause destruction of the $\alpha$-ketols and was shown to produce more than one peak for both 16- and 17-epiestriols.

With respect to the suggestion that acetates do not allow adequate resolution of closely related estrogens, it must be stated that none of the known derivatives will permit the resolution of even the major known estrogen metabolites from pregnancy urine.

Table 2 shows the actual retention times of the free steroids, acetates, and TMS ethers on SE-30 and QF-1. The solid lines connect compounds which cannot be resolved despite the differences in retention time; the dotted lines link compounds which are only partially resolved with $R < 1.5$*. As can be seen, the best separations can be obtained for the acetyl derivatives on a QF-1 column, where only a single pair of compounds is not resolved. In view of the number of other elegant techniques, such as TLC, for preliminary separation, this inability to achieve complete resolution of all estrogen metabolites is of relatively little consequence. It is entirely feasible to separate the estrogens into groups, each sub-group then being resolved by GLC. This is not only necessary for separation of different metabolites, but also for removal of non-estrogenic material prior to GLC and is highly desirable as a means of enhancing the specificity of the various methods to be described. After evaluating the various qualities of these two types of derivatives, we have concluded that the use of estrogen acetates is preferable particularly when applied in a routine analytical laboratory. We do not advocate the use of free estrogens for direct analysis since resolution is decreased, serious adsorption on many types of columns occur, and peak skewing is generally obtained, resulting in poor specificity and accuracy.

## MATERIALS AND METHODS

All reagents used were of analytical grade unless otherwise stated. The distillation of solvents was carried out in an all glass apparatus.

__Ether__ is shaken with 13% (w/v) $AgNO_3$ (60 ml per l), followed by washing with 1N NaOH (100 ml per l) and water. The ether is freshly distilled before use.

__Petroleum ether__ (30-60 C), __n-hexane__ and __benzene__ are washed with concentrated $H_2SO_4$ (3 x 100 ml per l), water (3 x 180 ml per l), 8% M $KMnO_4$ in 4N $H_2SO_4$ (3 x 100 ml per l), water (150 ml per l) to neutral. The solvents are dried over $Na_2SO_4$ and then fractionally distilled.

__Dichloromethane__ is shaken with anhydrous potassium carbonate (10 g per l), filtered and passed through a column of active silica gel followed by fractional distillation.

__Ethyl acetate__ is refluxed with acetic anhydride (100 ml per l) and concentrated $H_2SO_4$ (5 ml per l) for four hr and distilled fractionally.

---

\* $R = \dfrac{V_{R_2} - V_{R_1}}{w_1 + w_2}$   Where:-
  $R$ = resolution
  $V_R$ = retention volume of each peak
  $w$ = peak width

TABLE 2. Comparison of retention times of different derivatives on two columns.

|  | SE-30 | | | QF-1 | | |
|---|---|---|---|---|---|---|
|  | Free[1] | TMS[2] | TMS[3] | Acetate[4] | Free[5] | TMS[6] | Acetate[7] |
| Estrone | 4.9 | 5.5 | 6.1 | 7.4 | 11.0 | 7.7 | 7.0 |
| Estradiol | 5.1 | 7.2 | 7.8 | 10.8 | 6.8 | 3.7 | 7.0 |
| 2-Methoxyestrone | 6.8 | 8.3 | 8.7 | 11.0 | 15.8 | 12.1 | 11.3 |
| 16α-Hydroxyestrone | 7.2 | 10.4 | 10.5 | 15.4 | 15.4 | 8.5 | 16.7 |
| 16-Methoxyestradiol | 6.7 | 10.6 | 11.5 | 16.0 | 12.4 | 12.4 | 20.0 |
| Estriol | 10.5 | 14.6 | 14.2 | 20.0 | 18.1 | 6.6 | 15.8 |
| 16-epiestriol | 10.7 | 15.9 | 15.1 | 23.3 | 18.6 | 7.7 | 23.5 |

(1) 0.75% coating, column at 205 C
(2) Same as above
(3) 3% coating at 239 C
(4) Same as above

(5) 1% QF-1 at 195 C
(6) Same as above
(7) 4% QF-1 at 208 C

Columns 1, 2, 5 and 6 taken from

Luukkainen et al, Biochim Biophys Acta 62: 153, 1962.

**Ethanol** and **methanol** are refluxed over solid NaOH and twice distilled under anhydrous conditions.

**Pyridine** is refluxed over pellets of NaOH for 4 hr and fractionally distilled after addition of fresh pellets under anhydrous conditions.

**Acetic anhydride** is fractionally distilled under anhydrous conditions.

**Aluminum oxide**, neutral, activity grade 1 (M. Woelm, Eschwege, Germany) is deactivated by the addition of 6 ml of water per 100 g, shaken thoroughly to break up any lumps and left overnight for equilibration in a tightly stoppered bottle. The activity is checked as follows:

The chromatographic column is partly filled with petroleum ether (according to the specifications given by Brown (9) and 3 g of deactivated alumina is added in a thin stream so that it is freed from air as it settles. The surface of the alumina is leveled by tapping and is covered with a few glass beads. The rate of flow is adjusted to approximately 50 drops per min. Five µg of three estrogen acetates (estrone, estradiol and estriol) in 25 ml petroleum ether is applied.

## ASSAY PROCEDURES

I. Pregnancy Urine Assays.

A. <u>Rapid determination of estrone and estriol by GLC of crude, late pregnancy extracts</u>. A detailed description of this procedure has been published (3, 10). Figure 1 shows a schematic outline of the method.

B. <u>Determination of estrone, estradiol and estriol throughout pregnancy using alumina chromatography and GLC</u>. An outline of the technique is shown in Fig. 2, and the specific details are described below under section II.

C. <u>Determination of estrone, 2-methoxyestrone, estradiol, 16α-hydroxyestrone, 16-ketoestradiol, estriol and 16-epiestriol</u>. A complete description of this procedure has been published (11) and a brief outline may be found in Fig. 3. One modification has been introduced. In order to quantify the ring D α-ketols separately, the zone containing them and estradiol in TLC System II, is chromatographed on a 4 ft 1/8 in i d column prepared by first coating the Diatoport S or Gas Chrom Q with 0.5% EGA (stabilized) and after evaporating the acetone a second coating corresponding to 2.5% SE-30 (in dichloromethane) is added. Elution is carried out with the following solvent mixtures and collected in 5 ml fractions.

1. 25% benzene in petroleum ether - 10 ml
2. 50%      "      "      "      "    - 15 ml
3. 75%      "      "      "      "    - 20 ml
4. Benzene                            - 20 ml

Estradiol diacetate should be eluted completely between second and third 5 ml fraction of mixture No. 3. The acetates of estrone and estriol are eluted quantitatively by the first 15 ml of benzene. Analysis is carried out by GLC using a 3% SE-30 column which allows the complete separation of estrone and estradiol acetates. If the alumina is properly standardized, there should not be any mixture of these two estrogens. Once a batch of alumina is standardized, it can be stored in an air tight container for a long time.

**Gas chromatography.** Columns were made from commercially prepared coated material (QF-1, lot #58835 on 80-100 mesh Gas Chrom P, Applied Science Laboratories, Inc.) and for SE-30 + EGA and NGS by filtration or evaporation of an alcohol washed commercially 80-100 mesh diatomaceous earth (Gas Chrom Z, Applied Science Laboratories, Inc. and Diatoport S, F&M Scientific Corp.). Spiral columns were packed under vacuum with limited vibration. U-tubes were packed by gravity and limited fibration.

SE-30 columns were cured at 300 C for 2 hr without gas flow followed by 250 C with 5-10 psi $N_2$.

1. Acid hydrolysis.
2. Ether extraction.
3. $Na_2CO_3$ wash (Brown, 1955).
4. Evaporation of ether.
5. Partition between benzene-pet ether and 1N NaOH.
6. Reextraction of aqueous solution at pH 9-10 with ether.
7. Evaporation of ether.
8. Acetylation with .1 ml pyridine and .5 ml acetic anhydride for one hr at 68 C.
9. GLC on 6 ft x 1/8 in 3% SE-30 column (2000 plates or better) at 250 C.

FIG. 1. Rapid analysis of estrone and estriol.

1. Acid hydrolysis.
2. Ether extraction.
3. $Na_2CO_3$ wash (Brown, 1955) (18).
4. Evaporate ether to dryness.
5. Separation of phenolic and neutral fractions by partition between benzene-petroleum ether and 1N NaOH.
6. Reextraction of the aqueous solution at pH 9.5-10 with ether.
7. Evaporation.
8. Acetylation with 0.1 ml pyridine and 0.5 ml acetic anhydride for one hr at 68 C.
9. Solvent partition.
10. Alumina column: (a) 25% benzene in pet ether, 10 ml - discarded
    (b) 50%  "    "   "   "  , 15 ml - discarded
    (c) 75%  "    "   "   "  , 20 ml - first 5 ml - discarded
                                       next 10 ml - $E_2A$
                                       last few ml - discarded
    (d) Benzene              20 ml - first 15 ml - $E_1A + E_3A$
11. GLC on 6 ft x 2 mm 3% QF-1 column.

FIG. 2. Measurement of classical estrogens in non-pregnancy urine.

1. Enzyme hydrolysis.
2. Ether extraction.
3. Wash with 8% $NaHCO_3$, then water.
4. Evaporate ether to dryness.
5. TLC in System I (benzene:ethyl acetate 1:1 incompletely saturated).

    Fraction A - Estrone, 2-methoxyestrone

    B - Estradiol, ring D-$\alpha$-ketols
        (16$\alpha$-hydroxyestrone, 16-ketoestradiol)

    C - 16-epiestriol

    D - Estriol

6. Elution of individual fractions with ethanol.
7. TLC in System II of fraction B (remove interfering androgens).
8. Elution with ethanol.
9. All fractions to dryness.
10. Acetylation with .1 ml pyridine and .5 ml of acetic anhydride for one hr at 68 C.
11. Addition of 5 ml of water and extraction with petroleum ether.
12. Evaporation of petroleum ether, redissolved in 50 or 100 µl of acetone.
13. GLC 3% SE-30 on 80-100 mesh alcohol washed Diatoport S

    4 ft x 1/8 in stainless steel column

    Fractions A and B - column temperature 228 C, 20 psi $N_2$

    "         C and D -    "           "        238 C, 20 psi $N_2$

FIG. 3. Measurement of classical estrogens and newer metabolites (40).

All other columns were cured at 240 C, with 5-10 psi $N_2$ for 48 hr. Three types of instruments have been used for the work reported here. An F&M Model 400 gas chromatograph, a Model 28-700 Jarrell-Ash gas chromatograph, and a Perkin-Elmer Model 801 gas chromatograph, all equipped with flame ionization detectors.

Injection of either 1 or 2 µl of the extracts are usually made after saturation of the column with standards. Quantitative measurement is carried out by comparison of the peak height of the unknown to that of a known standard at a concentration approximating the former.

Column conditions are variable and are given in the legends to the figures.

II. <u>Analysis of estrone, estradiol and estriol in low-titer urine.</u>

One-fifth of the 24 hr collection of urine is diluted with an equal volume of distilled water. Acid hydrolysis extraction and separation into the phenolic and non-phenolic fractions is carried out according to the method of Brown (9), the only exception being the use of 1N NaOH for the extraction of three classic estrogens in the same fraction. The pH of the alkaline solution is brought to between 9.5-10 by the addition of solid $NaHCO_3$. The mixture is then extracted, once with an equal volume of ether and twice with half the volume. The combined ether layers are washed with 8% $NaHCO_3$ solution (5 ml per 100 ml) followed by water (2.5 ml per 100 ml) until the discard is neutral. After evaporation to dryness in a rotating still under reduced pressure, the dried phenolic residue is transferred to a test tube with a little dichloromethane.

<u>Acetylation</u>. The residue is acetylated by dissolving it in a mixture of five parts of acetic anhydride and one part of pyridine and keeping the tube at 68 C for one hr. To the acetylated

mixture 10 ml of distilled water is added, while stirring thoroughly with a glass rod. The sample is then transferred to a small separatory funnel and extracted twice with 10 ml and once with 5 ml of light petroleum ether. The vessel used for acetylation is also rinsed with petroleum ether and the washings added to the separatory funnel. The combined petroleum ether fractions are washed with 8% NaHCO$_3$ (5 ml) followed by 2 ml portions of water until the discard is neutral. The solution is allowed to stand for 10 min and any water collected at the bottom is removed as completely as possible.

*Alumina column chromatography*. The preparation of the column and the elution of the estrogens are followed as described previously. The proper eluates containing the respective estrogens are evaporated to dryness in a 50 ml round bottom flask on a rotating still. The residues are transferred to 2 ml centrifuge tubes with petroleum ether or n-hexane. The solvent is dried and the residue is dissolved in 20 μl of hexane.

Two μl of this solution is then injected into a 3% QF-1 on 80-100 mesh support column and measurements taken by comparison to standards as described previously.

## RESULTS AND DISCUSSION

I. *Determination of Estrogen in Pregnancy Urine*.

A. *Rapid determination of estrone and estriol in crude extracts of late pregnancy urine*. The first practical application of GLC to the analysis of hormones in biological media (12) described the determination of the classical estrogens in pregnancy urine. This method was shown to be quite useful particularly for the rapid determination of estriol in the second half of pregnancy. Its great advantage was that it is possible to carry through such an analysis in two to three hours.

Evidence for the accuracy of the procedure was obtained through analysis of the estrogens following their addition to urine samples. Mean recoveries of estrone, estradiol and estriol of 75 $\pm$ 6, 80 $\pm$ 4 and 92 $\pm$ 6% respectively were found. For quadruplicate determination of estriol in urines containing 500 to 5000 μg a reproducibility of $\pm$ 4.3% was observed. In utilizing such a crude extract, serious doubts regarding the specificity of the procedure tend to occur. In order to ascertain that certain structures assigned to certain peaks contain primarily those designated, a large volume of urine was extracted and a phenolic extract prepared with minimum exposure to alkali. Following multiple injections into a gas chromatograph with a β-ionization detector and operating on argon gas, the respective peaks (Fig. 4) were trapped in a test tube submerged in liquid nitrogen. Rechromatography of individual peaks with further trapping resulted in an overall recovery of 68% of the original material. Table 3 shows the data obtained to verify the nature of the compounds designated as estrone, estradiol, estriol and 16-epiestriol. Adequate evidence for specificity has been obtained by means of paper chromatography, counter-current distribution and infrared data and the peaks designated as estrone, estradiol, estriol and 16-epiestriol are largely those compounds. Since much of our early work was done on urines obtained from only a few individuals, it was not noted until sometime later that a number of specimens contain an apparently phenolic impurity with a retention time very nearly that of estradiol. Figure 5 shows a chromatogram on an SE-30 column with approximately 3300 theoretical plates yielding a clean estrone and estriol peak but showing the presence of the unknown impurity. Estimation of estradiol under these circumstances is of course impossible.

It would appear therefore that use of crude extracts even in pregnancy urine should be relied on only for the rapid estimation of estriol. No evidence has been obtained by us as yet that interfering substances derived from drug metabolism or dietary factors produce false results.

B. *Determination of estrone, estradiol and estriol throughout pregnancy*. Because of the difficulties mentioned in the above pragraph and our previous inability to develop liquid-liquid partition systems which would remove such impurities and allow direct GLC, preliminary clean-up of the phenolic fraction on an alumina column has been found to yield eminently satisfactory results. The details of this procedure are discussed somewhat later in this paper under the low-level assay. Figure 6 shows the chromatogram of a phenolic pregnancy urine extract, following acetylation, alumina chromatography and GLC on a QF-1 column. It is noteworthy to point to the excellent purification that can be obtained by this simple and fast column prior to GLC. Other than estrone and estriol, no significant peaks are discernible, suggesting that the impurities present, are small enough compared to the estrogens to remain virtually undetectable at an attenuation of 200X for estrone and estriol, and an attenuation of 20X for estradiol. Quantitative measurement of such peaks may be achieved with considerable accuracy even for estriol which is superimposed on a des-

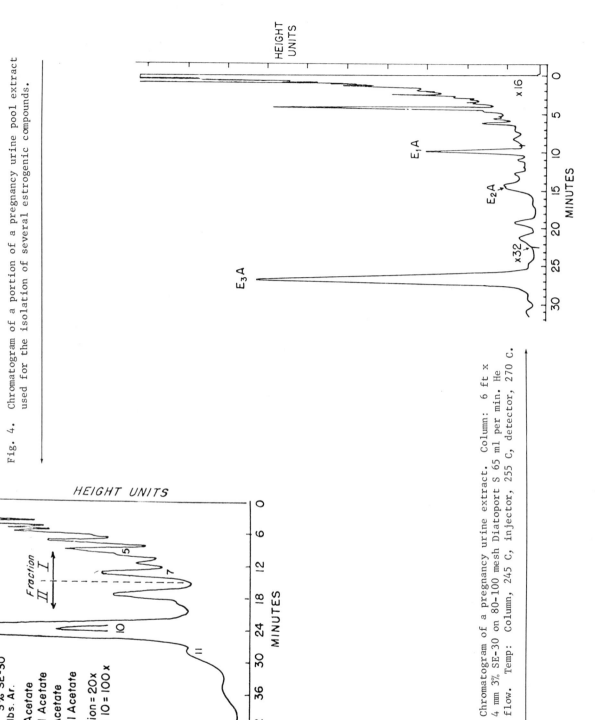

Fig. 4. Chromatogram of a portion of a pregnancy urine pool extract used for the isolation of several estrogenic compounds.

Fig. 5. Chromatogram of a pregnancy urine extract. Column: 6 ft x 4 mm 3% SE-30 on 80-100 mesh Diatoport S 65 ml per min. He flow. Temp: Column, 245 C, injector, 255 C, detector, 270 C.

TABLE 3. Characterization of pregnancy estrogens after GLC.

| Peak No. | Suggested Compound | Counter current* distribution (part. coefficient) | Paper chromatography $R_f$ | Infrared spectroscopy | Analytical GLPC** $R_T$ (min) Authentic | Unknown |
|---|---|---|---|---|---|---|
| 1, 2, 3 | Non investigated | ---probably non-steroidal --- | | | | |
| 4 | | | | Phenolic - no other oxygen | | |
| 5 | Estrone acetate | R = 1.3 (reported 1.3) | .77† (reported .78) | Positive identification | 9.7 | 9.7 |
| 7 | | | .42† (reported .45) | Positive identification | 13.7 | 13.7 |
| 8 | Mixutre of ketols | | | Steroidal phenol ketone present | | |
| 9 | Not investigated | | | | | |
| 10 | Estriol triacetate | | Runs with authentic standard toluene-prop. glycol | Positive identification | 24.6 | 24.6 |
| 11 | Epiestriol triacetate | | Runs with authentic standard toluene-prop. glycol | Positive indentification | 29.0 | 29.0 |

\* System $\dfrac{70\% \text{ MeOH}}{\text{C Cl}_4}$

\*\* Gas-liquid preparative chromatography - SE-30, 3%, 6 ft 1/8 in.

† System Isooctane 25, toluene 75, methanol 80, water 20.

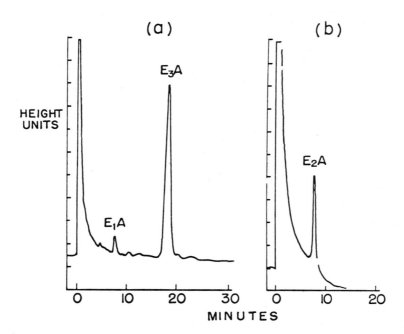

Fig. 6. Chromatograms of a pregnancy urine extract following acetylation and alumina chromatography. A. Analysis of estrone and estriol acetate on a 3% QF-1 column on 80-100 mesh Gas Chrom P, attenuation 200X, column temp: 222 C, injector temp: 270, flow rate: 60 ml per min. B. Estradiol acetate fraction at attenuation 20X; temp and flow rate as above.

cending baseline.

Evidence for the specificity of this particular assay as applied to pregnancy urine is obtained through collection of the paper aliquots from the alumina column, followed by GLC on both polar (NGS), non-polar (SE-30) and ketone-selective (QF-1) columns. Identical retention times for standards and the peaks from the alumina column eluates were observed. The excitation and emission spectra obtained with phosphoric acid were the same for standard estradiol and a sample collected from pregnancy urine and similar curves were obtained for an authentic mixture of estrone and estriol as compared to an aliquot of the respective eluates from the alumina column. Although this procedure is somewhat more involved than the crude assay of estrone and estriol, it has been our experience that approximately 25 assays a week can be carried out by one technician.

<u>Determination of seven estrogen metabolites in pregnancy urine</u>. The discovery of a number of estrogen metabolites beyond the classical three during the last decade has considerably complicated the analytical aspects of estrogen determination not only with respect to the separation of these compounds from each other and from interfering materials but, also because of the known lability and relative alkali insolubility of some of these compounds.

The ability to separate neutral substances from the urinary phenols has always been considered to be a great advantage in the prepurification steps, markedly reducing the number of interfering chromogens. The low solubility of 2-methoxy estrone in alkali and the well established lability of the ring D-$\alpha$-ketols, not only make enzyme hydrolysis mandatory for such assay procedures, but also prevent the partitioning of the extracts between alkali and organic solvent. As a consequence it has been found necessary to employ other preliminary steps in the purification prior to GLC analysis for the measurement of several of these compounds.

In an earlier paper (11) we have described the details of a procedure employing two different TLC systems following enzymatic hydrolysis and extraction of the urinary estrogens. The purpose of the first system was twofold. First, separation of a large number of known urinary steroidal components from unknown colored substances can be achieved. Second, discrete zones for 16-epiestriol, estriol, estrone and 2-methoxyestrone, and for estradiol and the ring D-$\alpha$-ketols may be obtained. As pointed out earlier, on a QF-1 column the major difficulty in the separation of estrogen acetates lies in the inability to resolve estrone from estradiol. Using TLC System #1 (Table 4) separation of estrone and estradiol is readily achieved. Furthermore, the fractions containing estriol and 16-epiestriol appear sufficiently clean for direct GLC analysis. Similarly, estrone and 2-methoxyestrone may be separated from each other by GLC as well as from a number of unknown peaks eluting faster than both of these estrogens. Elution and reapplication of Zone II containing estradiol and the ring D-$\alpha$-ketols, followed by development in a petroleum ether:dichloromethane:ethanol (10:9:1) system (Table 5) results in the separation of these three estrogens from a number of interfering neutral steroids. Separation of 16$\alpha$-hydroxyestrone from 16-ketoestriol finally is achieved by GLC on a 0.5% EGA (stabilized) and 2.5% SE-30 column.

The recovery of various estrogens after the addition of 1 µg of each to 50 ml of urine, varied from 72 to 91% with the two lowest values (72 and 78% respectively) occurring for 16-epiestriol and 2-methoxyestrone. The major losses of these two compounds appear to occur during TLC where only 18 and 82% respectively are recovered after chromatography alone, as compared to 90 to 97% for the other estrogens under discussion. The linear response of the flame detector to 0.02 µg to 2 µg amounts of six of these estrogens has been demonstrated (11).

II. <u>Determination of Estrone, Estradiol and Estriol During the Normal Menstrual Cycle</u>.

The great sensitivity of ionization detectors and the unique high resolution of the GLC column has raised the hope that this procedure could be applied directly to urinary fractions containing very low levels of estrogens following simple preliminary separations. We have not, as yet, determined the lower limit of sensitivity of the procedure described above, but a preliminary survey on a few urines suggested that the method would periodically run into difficulty if more than 5-10% of the urine was used for analysis. The major difficulty here appeared to be a serious overloading of the thin-layer plate. As a consequence we have reapproached this problem, attempting to develop a much more sensitive procedure which may be utilized in a more routine manner even at the sacrifice, at the moment, of analyzing only the three classical estrogens.

Using the method described in detail in the earlier part of this communication the recovery of the three estrogens was studied following column chromatography as well as after addition of steroid to a urine sample and carrying through the whole procedure as outlined in Table 6.

TABLE 4. $R_f$ and $R_{E_1}$ values of free steroids in system I*
(ethyl acetate:benzene 1:1).

| Compound** | Analyses | Mean $R_f$ | SD | Mean $R_{E_1}$*** | SD |
|---|---|---|---|---|---|
| Cortisol | 7 | .092 | ± .003 | .100 | ± .015 |
| Estriol | 10 | .117 | ± .010 | .119 | ± .005 |
| 16-Epiestriol | 11 | .289 | ± .015 | .321 | ± .032 |
| 11-Ketoestiocholanolone | 6 | .294 | ± .028 | .313 | ± .020 |
| Pregnanediol | 6 | .347 | ± .018 | .370 | ± .011 |
| Etiocholanolone and testosterone | 6 | .477 | ± .023 | .480 | ± .014 |
| 16-Ketoestradiol-17β | 12 | .633 | ± .019 | .675 | ± .017 |
| Dehydroepiandrosterone, androstenedione and androsterone | 6 | .643 | ± .023 | .610 | ± .023 |
| 16α-Hydroxyestrone | 6 | .658 | ± .017 | .707 | ± .004 |
| Estradiol | 11 | .679 | ± .035 | .760 | ± .017 |
| Cholesterol | 6 | .839 | ± .031 | .904 | ± .027 |
| 2-Methoxyestrone | 7 | .842 | ± .023 | .968 | ± .004 |
| 16-Ketoestrone | 6 | .881 | ± .039 | .953 | ± .019 |
| Estrone | 11 | .897 | ± .056 | 1.000 | |

\* The system was left incompletely saturated.

\*\*\* $R_{E_1}$ = Distance travelled relative to estrone.

TABLE 5. $R_f$ and $R_{E_1}$ values of free steroids in system II*

(petroleum ether:dichloromethane:ethanol 10:9:1).

| Compound | Analyses | Mean $R_f$ | SD | Mean $R_{E_1}$** | SD |
|---|---|---|---|---|---|
| Estriol | 8 | .050 | ± .007 | .012 | ± .001 |
| Cortisol | 6 | .052 | ± .010 | .013 | ± .002 |
| 16-Epiestriol | 7 | .079 | ± .007 | .202 | ± .005 |
| Pregnanediol | 7 | .189 | ± .040 | .468 | ± .025 |
| Estradiol-17β | 9 | .215 | ± .003 | .539 | ± .028 |
| 16α-OH-estrone | 8 | .216 | ± .003 | .538 | ± .029 |
| 16-Ketoestradiol-17β | 6 | .235 | ± .013 | .581 | ± .018 |
| 11-Ketoetiocholanolone | 8 | .236 | ± .021 | .592 | ± .013 |
| 16-Ketoestrone | 5 | .297 | ± .028 | .737 | ± .019 |
| Etiocholanolone and testosterone | 7 | .313 | ± .027 | .798 | ± .026 |
| Dehydroepiandrosterone and androstenedione | 8 | .355 | ± .034 | .901 | ± .036 |
| Estrone | 11 | .396 | ± .028 | 1.000 | |
| Cholesterol | 6 | .544 | ± .030 | 1.36 | ± .01 |
| 2-Methoxyestrone | 9 | .611 | ± .025 | 1.54 | ± .08 |

\* The system was fully saturated by lining the wall with filter paper.

\*\* Distance traveled relative to estrone.

TABLE 6. Recovery experiments in percentages.

| Experiment Number | Estrone | Estradiol | Estriol | Remark |
|---|---|---|---|---|
| 1 | 98.00 | 101.00 | 93.80 | 0.2 µg each of estrogens were added |
| 2 | 90.00 | 96.40 | 90.70 | |
| 3 | 93.00 | 88.50 | 95.00 | |
| 4 | 90.00 | 94.30 | 98.00 | |
| 5 | 95.00 | 89.00 | 85.00 | |
|   | 93.2 ± 3.5 | 93.8 ± 5 | 92.5 ± 5 | |

Overall recovery

| Experiment Number | Estrone | Estradiol | Estriol | Remark |
|---|---|---|---|---|
| 1 | 68.50 | 72.00 | 80.60 | 0.5 µg $E_1$ and $E_2$ and 1 µg $E_3$ was added to 1/20th of the total volume of urine (male) |
| 2 | 75.00 | 70.50 | 84.00 | |
| 3 | 72.00 | 81.40 | 86.00 | |
| 4 | 70.00 | 75.00 | 84.00 | |
| 5 | 76.00 | 79.00 | 79.00 | |
|   | 72.4 ± 3.2 | 75.6 ± 4.4 | 82.8 ± 3.2 | |

Although sufficient data for determination of the precision of the method are not yet available, in Table 7 are shown a series of nine duplicate analyses of a non-pregnancy urine. A variance of 0.26, 0.14 and 3.53 for estrone, estradiol and estriol respectively was found.

TABLE 7. Precision data on duplicate analysis.

| Experiment No. | | Estrone | Estradiol | Estriol |
|---|---|---|---|---|
| 1. | A | 7.00 | 1.10 | 17.50 |
|    | B | 6.00 | 0.90 | 17.00 |
| 2. | A | 10.00 | 1.20 | 20.50 |
|    | B | 10.90 | 1.20 | 21.80 |
| 3. | A | 9.30 | 2.25 | 26.80 |
|    | B | 9.30 | 2.35 | 23.00 |
| 4. | A | 7.00 | 3.52 | 33.00 |
|    | B | 8.25 | 4.80 | 32.40 |
| 5. | A | 15.66 | 4.50 | 43.00 |
|    | B | 15.00 | 4.30 | 37.70 |
| 6. | A | 5.30 | 1.50 | 46.40 |
|    | B | 5.30 | - | 52.00 |
| 7. | A | 3.00 | 0.90 | 22.00 |
|    | B | 2.70 | 1.30 | 21.60 |
| 8. | A | 5.20 | 2.30 | 26.30 |
|    | B | 5.00 | 1.80 | 25.80 |
| 9. | A | 8.00 | 2.40 | 27.00 |
|    | B | 8.90 | 2.00 | 28.20 |
| Variance of duplicates* | | 0.26 | 0.14 | 3.53 |

\* Calculated according to the formula $\dfrac{\Sigma (x - \bar{x})^2}{n-1}$

In Fig. 7, the excretion of the three classical estrogens and pregnanediol throughout the menstrual cycle has been plotted. The values obtained by the method described here appear to be in keeping with those reported in the past years by application of the elegant procedure of Brown (9). Pregnanediol was determined gas chromatographically by a method previously reported (14).

Typical chromatograms following GLC on SE-30 and QF-1 columns are shown in Figs. 8-11. It can readily be seen that much better separation from interfering substances is obtained during chromatography of the estradiol fraction (Figs. 9-11) on the fluoroalkyl silicone polymer. Although no specific experiments have been carried out to determine the sensitivity of the procedure, it seems unlikely that concentrations of less than 0.5 µg per 24 hr can be measured with the present instrumentation.

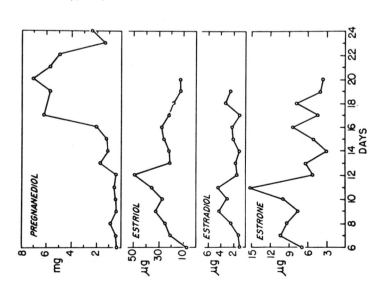

Fig. 7. Excretion of estrogens and pregnanediol during a normal menstrual cycle in a 27-year-old female.

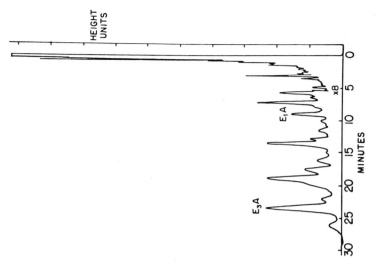

Fig. 8. Chromatogram of the $E_1A-E_3A$ eluate from a proliferative phase urine extract following alumina chromatography. Column: 3.8% SE-30 on Diatoport S. Temp: 250 C, flow rate: 65 ml per min.

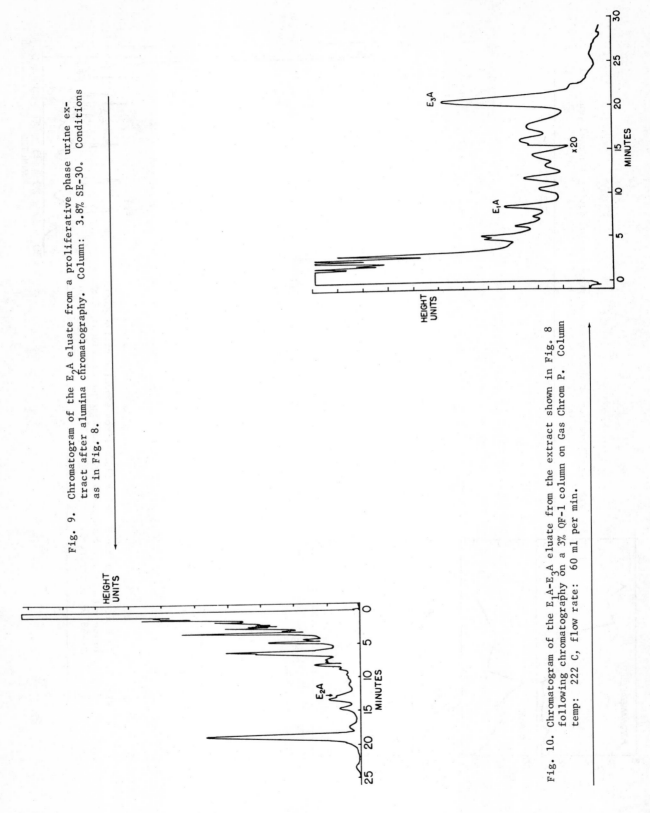

Fig. 9. Chromatogram of the $E_2A$ eluate from a proliferative phase urine extract after alumina chromatography. Column: 3.8% SE-30. Conditions as in Fig. 8.

Fig. 10. Chromatogram of the $E_1A-E_3A$ eluate from the extract shown in Fig. 8 following chromatography on a 3% QF-1 column on Gas Chrom P. Column temp: 222 C, flow rate: 60 ml per min.

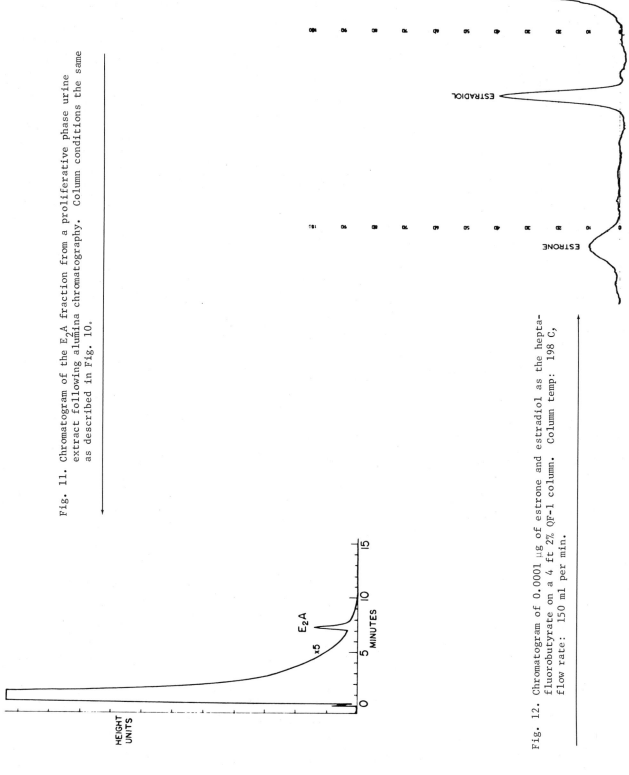

Fig. 11. Chromatogram of the $E_2A$ fraction from a proliferative phase urine extract following alumina chromatography. Column conditions the same as described in Fig. 10.

Fig. 12. Chromatogram of 0.0001 μg of estrone and estradiol as the heptafluorobutyrate on a 4 ft 2% QF-1 column. Column temp: 198 C, flow rate: 150 ml per min.

Evidence for the specificity is inherent in the procedure and depends to a large extent on the elution pattern from the alumina column and on multiple gas chromatographic determinations on polar (NGS), non-polar (SE-30), ketone-retentive (QF-1) and mixed phase (NGS/SE-30) columns. Identical retention times for the peaks from the extract and pure estrone, estradiol and estriol were observed. Further, saponification of the fractions of the acetylated extract and subsequent formation of the TMS ethers resulted in identical $R_t$ values for the respective estrogens and authentic estrogen derivatives on a QF-1 column. Fluorescence spectra of the eluate corresponding to estradiol produced a peak similar to that of an authentic standard. The aliquot corresponding to the mixture of estrone and estradiol following a single attempt at fluorescence spectroscopy yielded equivocal results.

Clark and Wotiz (13) described the formation of a new type of derivative formed by the reaction of hydroxyl groups with heptafluorobutyric anhydride. These compounds show not only excellent electron capturing properties, but also allow good resolution of various estrogens. They have the further advantage of being extremely volatile, allowing elution temperatures below 200 C, a condition necessary for the safe maintenance of the tritium-foil in the electron capture detector.

Figure 12 shows the separation and measurement of estrone and estradiol as the heptafluorobutyrates in concentrations of 0.1 mµg each. As expected, estrone produces a considerably smaller peak since it reacts with only one heptafluorobutyryl group. Estradiol appears much faster in the chromatogram because of the presence of two such groups thereby enhancing the volatility of the compound.

## REFERENCES

1. VandenHeuvel, W. J. A., C. C. Sweeley, and E. C. Horning, J Am Chem Soc 82: 3481, 1960.

2. Wotiz, H. H. and H. M. Martin, Am Chem Soc, New York, Spet. 1960.

3. Wotiz, H. H. and H. F. Martin, Anal Biochem 3: 97, 1961.

4. Horning, E. C., W. J. A. VandenHeuvel, and B. G. Creech, Methods of Biochemical Analysis, Vol. XI, p. 84, 1963.

5. Wotiz, H. H. and H. F. Martin, J Biol Chem 236: 1712, 1961.

6. Luukkainen, T., W. J. A. VandenHeuvel, E. O. A. Haahti, and E. C. Horning, Biochim Biophys Acta 52: 599, 1961.

7. Luukkainen, T., W. J. A. VandenHeuvel, and E. C. Horning, Biochim Biophys Acta 62: 153, 1963.

8. Wotiz, H. H. and H. F. Martin, Biochim Biophys Acta 60: 25, 1962.

9. Brown, J. B., Biochem J 60: 185, 1955.

10. Wotiz, H. H., Biochim Biophys Acta 74: 122, 1963.

11. Wotiz, H. H. and S. C. Chattoraj, Anal Chem 36: 1466, 1964.

12. Wotiz, H. H. and H. F. Martin, Fed Proc 20, March 1961.

13. Clark, S. J. and H. H. Wotiz, Steroids 2: 535, 1963.

14. Wotiz, H. H., Biochim Biophys Acta 69: 415, 1963.

# DETERMINATION OF URINARY ESTROGENS BY GAS CHROMATOGRAPHY

H. Adlercreutz and T. Luukkainen, Steroid Research Laboratory, Department of Medical Chemistry, University of Helsinki, Helsinki, Finland

The chemical methods available for the determination of urinary estrogens have been recently reviewed by Preedy (1). The limitations of these methods are due mainly to the difficulty of separating the closely related compounds from each other and from impurities without rendering the method impractical.

The reference standards of estrogens can well be separated and estimated by GLC as TMS ethers of non-methylated or methylated compounds (2, 3). Therefore, if the urinary extracts could be purified to such a degree that the sensitive but nonspecific detection systems of GLC could be used, estrogen measurement should be facilitated.

It was realized that an ideal method for the simultaneous determination of the whole steroid spectrum in urine is not feasible, similarly, the determination of all phenolic steroids in a single analysis is impossible. Therefore for different purposes, different methods were developed.

## EXPERIMENTAL

All reagents were of analytical grade. The solvents were redistilled before use through all-glass fractionating columns. The ethyl ether was shaken with $FeSO_4$ solution and distilled water, and redistilled with NaOH pellets. Pyridine was shaken with KOH pellets, redistilled and stored over KOH pellets.

*Gel filtration.* This was carried out as described by Beling (4) on Sephadex $G_{25}$ medium. The peak I and peak II estrogen conjugates described by Beling (4) were collected separately. The column sizes were varied from 1 x 50 cm to 5 x 50 cm according to the volume of the urine to be processed.

*Enzyme hydrolysis.* This was carried out as suggested by Adlercreutz (5) and Beling (4). Careful investigations of the inhibitors of $\beta$-glucuronidase in urine and the effect of gel filtration on the elimination of these compounds (4, 6, and Adlercreutz, to be published) show that the peak I and peak II conjugates must be hydrolyzed separately in order to obtain optimal hydrolysis. In both fractions the enzymic hydrolysis was carried out with 600 Fishman units of $\beta$-glucuronidase and 6000 phenol sulfatase units of a Helix pomatia extract per ml of 0.15 M acetate buffer, pH 4.1 to 4.2 at 37 C for 16 hr (overnight).

*Preparation of derivatives.* Methylation was carried out as described by Brown (7). TMS ethers were prepared according to Luukkainen et al. (8) and Adlercreutz and Luukkainen (9). The estrone methoxime was prepared according to Fales and Luukkainen (10).

*Quantitation.* The sample to be investigated by GLC was evaporated to the desired volume in a special calibrated glass tube that ends in a 50 μl capillary graduated in 5 μl. The calibration curves have been made by the technique which involves the measurement of the area of the peak by multiplying the height by the width at the half-height of the peak. An internal standard has been used only if the steroid concentration to be determined could be roughly estimated. The great variations in the concentrations of the different estrogens make the use of an internal standard possible for only one of the steroids in a given sample. With and without an internal standard the results have been the same in this laboratory.

Gas-liquid chromatography. Two types of gas chromatographs were used. A Chromalab model 110 (Glowall Corp., Penn.) and a Barber-Colman model 15 (Barber-Colman Co., Wheelco Industrial Instruments, Rockford, Ill.). The Chromalab was equipped with an argon ionization detector containing a radium foil source. The Barber-Colman apparatus has a modified hydrogen flame ionization detector of Carlo Erba (Milan, Italy). In both instruments only all-glass columns were used.

The following stationary phases were used: 1% SE-60 (cyanoethyl silicone), 1% Z (copolymer made from ethylene glycol, succinic acid and a methyl siloxane monomer), 3% F-60 (methyl-p-chlorophenylsiloxane polymer), 2% QF-1 (fluoroalkyl silicone) and 1% SE-30 (dimethylsilicone polymer). All phases were on 100-140 mesh Gas Chrom P and the columns were prepared as described by Vanden-Heuvel and Horning (11).

All the steps of this procedure can be carried out in test tubes, which greatly simplifies the procedure (Fig. 1). A high degree of purity is achieved, as shown in Figs. 2 and 3.

The difficulties of gas chromatographic work with urinary estrogens are illustrated in Fig. 4. The many compounds and the great differences in their concentrations are clearly indicated and these facts make it difficult to obtain reliable results even with regard to estrone and estradiol.

Acid hydrolysis may be used, which makes the method more rapid, if this is found necessary. However, in our opinion such treatment of urine samples should be avoided unless urgently needed for other reasons, since some compounds are destroyed in this way (Fig. 5).

The recovery of conjugated estrogens in this procedure has been tested with estrone sulfate and estriol-16 (17?)-glucosiduronate. The first mentioned compound was added to the urine samples after gel filtration and the last mentioned before gel filtration. The recovery of estrone sulfate was 87.2% (8 determinations) and of estriol glucosiduronate 85.5% (5 estimations).

The precision of the method in duplicate estimations can be seen from Table 1. The method is sensitive enough to be used throughout pregnancy; the range of estimations has been 0.2 to 60 mg per 24 hr urine specimens, which means that the method may be used from about the 6th week of pregnancy. Regarding estriol the specificity of the method is good, but for estrone and estradiol some other compounds interfere with the estimation and therefore the amount of these compounds can only be roughly determined, especially in the first trimester of pregnancy. Values for estrone higher than 0.5 mg per 24 hr can usually be estimated with reliability. The methylation of the estriol, including the phase-change purification step, makes the estimation of this compound highly specific, the specificity being further enhanced by GLC investigation, which can be carried out on several liquid phases, if necessary.

Method for the estimation of estrone, estradiol-17β and estriol in pregnancy urine, including screening of some other estrogens. For the estimation of the three "classical" estrogens in pregnancy urine, the methods of Brown (7, 12) in combination with the GLC estimation of the TMS derivatives of the methylated compounds may be used. An example of the GLC investigation of two estradiol fractions on the F-60 liquid phase is shown in Fig. 5.

The saponification procedure included in the method of Brown et al. (12) seems not to purify the extracts further if the final estimations are carried out by GLC. Enzymic hydrolysis is preferable, since many compounds are destroyed during acid hydrolysis. This is especially the case with 11-dehydroestradiol-17α, which has recently been isolated and identified in the estradiol fraction (Fig. 5) (Luukkainen and Adlercreutz, to be published).

Another procedure has been developed in this laboratory (Fig. 6). A high degree of purification is first obtained by gel filtration of the urine samples followed by enzymatic hydrolysis. The estrone-estradiol fraction is separated from the polar estrogens and is then methylated in order to obtain higher purification and specificity of estimation. The polar estrogen fraction is not methylated, since it is known that the methylation procedure cannot be carried out in the same way for all estrogens and the epimeric estriols and ketols especially seem to be converted to several different compounds during the methylation process (13). The fraction is pure enough, however, to make possible a reliable estimate of estriol without methylation of the compounds (Fig. 7).

The recovery of estrone sulfate and estradiol-3-sulfate added to gel-filtered urine samples was 86.5% (5 determinations) and 78.0% (5 determinations), respectively. The corresponding value for estriol-16 (17?)-glucosiduronate was 96.5% (5 determinations). At present, no values for the precision of the method can be presented but it has been found that the Barber-Colman gas chroma-

> Gel filtration of 10 ml urine on Sephadex $G_{25}$ medium. Peak I and peak II of conjugated estrogens collected separately.
>
> Peak I and peak II conjugates hydrolyzed separately for 16 hr at 37 C in 0.15 M acetate buffer, pH 4.1-4.2, with 600 Fishman units of β-glucuronidase and 6000 phenol sulfatase units per ml of a _Helix pomatia_ extract.
>
> Peak I and peak II combined.
>
> The same amount of Brown's carbonate buffer (7), pH 10.5 added. Extracted with 3 x 1/1 vol of ethyl ether.
>
> Ether evaporated to 4 ml under a stream of nitrogen. Extracted with 3 x 1/2 vol of 1 N NaOH.
>
> NaOH extract diluted to 15 ml with distilled water. 0.3 g boric acid added.
>
> Methylation.
>
> Methylated estrogens extracted with 8 ml of benzene. Benzene washed with 1 ml of distilled water. Benzene evaporated to dryness.
>
> Formation of TMS ether derivatives of the methylated compounds.
>
> Gas chromatography on Z and/or SE-30 liquid phases.

Fig. 1. Method for the estimation of estriol in pregnancy urine, including screening of estrone and estradiol.

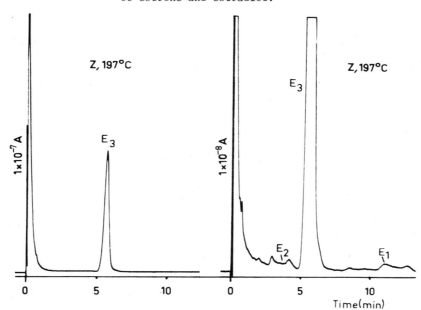

Fig. 2. GLC of the TMS derivative of a methylated "total" estrogen fraction obtained from late pregnancy urine with the method for estriol determination. The two figures show the same sample, but different sensitivities of detection. $E_1$ = estradiol-17β, $E_3$ = estriol. Chromalab model 110, argon inlet pressure 2 kg per $cm^2$. Phase on 100-140 mesh Gas Chrom P, 6 ft x 4 mm coiled glass column.

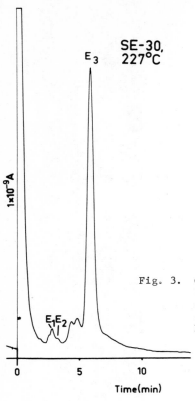

Fig. 3. GLC of a similar fraction as shown in Fig. 2 on another stationary liquid phase. Barber-Colman model 15 with a hydrogen flame ionization detector. Nitrogen inlet pressure 1.5 kg per $cm^2$. Phase on 100-140 mesh Gas Chrom P, 6 ft x 6 mm glass U-tube. $E_1$ = estrone, $E_2$ = estradiol-17β, $E_3$ = estriol.

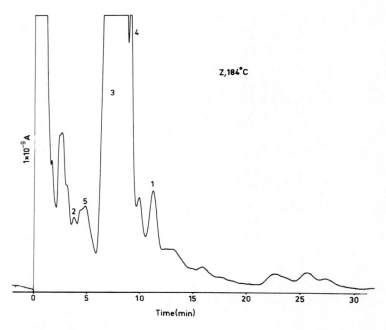

Fig. 4. GLC of a similar fraction as shown in Fig. 2 and 3. Conditions as in Fig. 3 with the exception that the column was a 6 ft x 4 mm glass U-tube. 1 = estrone, 2 = estradiol-17β, 3 = estriol, 4 = 16-epiestriol and 5 = ketoestradiol-17β.

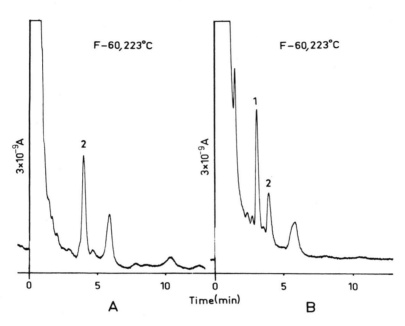

Fig. 5. GLC of the TMS derivatives of two methylated estradiol fractions obtained from late pregnancy urine with the method of Brown et al. (12). A = acid hydrolysis. B = enzymic hydrolysis. 1 = 11-dehydro-estradiol-17α, 2 = estradiol-17β. Conditions as in Fig. 2.

TABLE 1. Estriol excretion during pregnancy. Results of estriol determinations in duplicate samples of 10 ml from 24 hr urine specimens.

| Patient | Gestation week | Estriol mg per 24 hr | Patient | Gestation week | Estriol mg per 24 hr | Remarks |
|---|---|---|---|---|---|---|
| L.V. | 6 | 0.22<br>0.33 | H.N. | 16 | 4.96<br>4.85 | |
| B.M. | 6 | 0.24<br>0.24 | S.A. | 18 | 4.94<br>4.85 | |
| P.R. | 9 | 0.62<br>0.84 | E.R. | 23 | 8.12<br>7.88 | Hypertension |
| H.S. | 10 | 0.72<br>0.54 | E.R. | 26 | 4.96<br>4.85 | Hypertension |
| K.I. | 10 | 0.82<br>0.79 | E.P. | 28 | 5.78<br>5.52 | |
| E.S. | 13 | 3.84<br>4.03 | P.R. | 29 | 2.24<br>2.38 | Toxemia |
| L.E. | 14 | 2.20<br>2.14 | A.L. | 32 | 13.0<br>12.7 | |
| S.L. | 14 | 4.96<br>4.73 | P.R. | 35 | 10.3<br>10.2 | Toxemia |
| K.I. | 15 | 3.22<br>3.22 | K.A. | 36 | 13.7<br>13.5 | |
| R.K. | 15 | 6.20<br>5.85 | R.K. | 40 | 56.8<br>54.7 | |

Gel filtration of 10-20 ml urine on Sephadex $G_{25}$ medium (4). Peak I and peak II of conjugated estrogens collected separately and hydrolyzed for 16 hr at 37 C in 0.15 M acetate buffer pH 4.1 with 600 Fishman units of β-glucuronidase and 6000 phenol sulfatase units per ml of a Helix pomatia extract.

Peak I and peak II combined and the same amount of Brown's (12) carbonate buffer pH 10.5 added.

Extracted with 3 x 1/1 vol of ethyl ether. Ether evaporated to dryness.

Dry residue dissolved in 0.1 ml of ethyl alcohol. 3 ml of benzene and 3 ml of petroleum ether added.

| | |
|---|---|
| Benzene-pet ether extracted with 2 x 1/2 vol distilled water | Benzene-pet ether extracted with 2 x 1/2 vol of 0.4 N NaOH |
| Water extracted with 3 x 1/1 vol of ether Ether washed with 1/20 vol of 8% $NaHCO_3$-solution Ether washed with 1/40 vol of distilled water | NaOH extract diluted to 15 ml with 0.4 N NaOH, 0.3 g boric acid added Methylation according to Brown (12) Methylated estrogens extracted with 8 ml of n-hexane n-hexane washed with 1 ml distilled water |
| Ether evaporated to dryness | n-hexane evaporated to dryness |
| Formation of TMS derivatives | Formation of TMS derivatives of the methylated compounds |
| GLC on the XE-60; Z and SE-30 liquid phases | GLC on the XE-60 and Z liquid phases |

Fig. 6. Method for the estimation of estrone, estradiol-17β, and estriol in pregnancy urine including screening of some other polar estrogens.

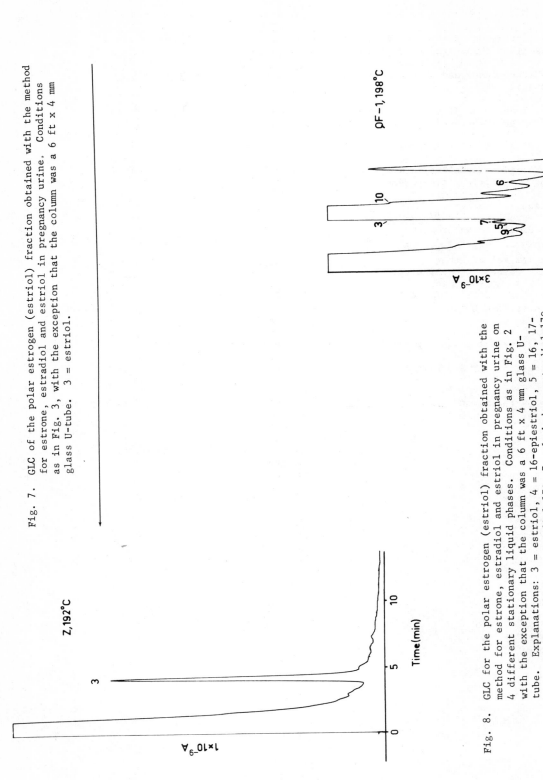

Fig. 7. GLC of the polar estrogen (estriol) fraction obtained with the method for estrone, estradiol and estriol in pregnancy urine. Conditions as in Fig. 3, with the exception that the column was a 6 ft × 4 mm glass U-tube. 3 = estriol.

Fig. 8. GLC for the polar estrogen (estriol) fraction obtained with the method for estrone, estradiol and estriol in pregnancy urine on 4 different stationary liquid phases. Conditions as in Fig. 2 with the exception that the column was a 6 ft × 4 mm glass U-tube. Explanations: 3 = estriol, 4 = 16-epiestriol, 5 = 16, 17-epiestriol, 6 = 16-ketoestradiol-17β, 7 = 6α-hydroxyestradiol-17β, 8 = 2 methoxyestriol, 9 = 6β-hydroxyestradiol, 10 = 16α-hydroxyestrone. The large peak in all gas chromatograms having a rather long retention time is an estetrol (Luukkainen, Adlercreutz and Fales, to be published).

tograph equipped with a hydrogen flame ionization detector can detect 0.02-0.03 μg of estrogen per sample injected. This means that an amount of about 20 μg estrone and estradiol per 24 hr can be estimated quantitatively. For estriol, the sensitivity limit at present is about 200 μg per 24 hr. Work is in progress to increase these sensitivity limits, which are mainly dependent on the technical features of the GLC apparatus. However, when 200 ml of nonpregnancy urine samples were processed it was found to be possible to estimate 2 μg of estrone and estradiol-17β per 24 hr urine specimen. As is seen in Fig. 8, 9, 10 and 11, it is possible to obtain information with regard to the following other estrogens: 6α-hydroxyestradiol-17β, 6β-hydroxyestradiol-17β, 16-epiestriol, 16α-hydroxyestrone, 16-ketoestradiol-17β and 2-methoxyestriol.

11-Dehydroestradiol-17α in the estrone-estradiol fraction may be estimated if the F-60 stationary liquid phase is used. An example of the gas chromatographic investigation of the estrone-estradiol fraction on the Z liquid phase is shown in Fig. 12.

Estimation of estrogens in non-pregnancy urine samples. The combination of gel filtration, enzymic hydrolysis and the method of Brown (7), including the final GLC analysis of the TMS derivatives of the methylated compounds, offers a possibility to estimate the three "classical" estrogens in non-pregnancy urine. However, such a procedure would not at present be of any advantage as compared with the procedure of Brown et al. (12) in clinical assays. Preliminary investigations with another procedure seem to be more promising. This method can be used for the estimation of estrone in both pregnancy and non-pregnancy urine.

Short method for the estimation of estrone in both pregnancy and non-pregnancy urine. The method is based on the short procedure of Brown (personal communication) for the estimation of estrone in non-pregnancy urine and involves acid hydrolysis. The estrone is estimated by GLC, using for instance the Z stationary liquid phase, as the 17-methoxime of the 3-methyl ether. An example of such a determination is shown in Fig. 13.

The sensitivity of the procedure is dependent mainly on the GLC equipment, since the final fraction is very pure.

The specificity of the methods presented here has been confirmed extensively by GLC, TLC and direct mass spectrography of gas chromatographic effluents (14).

## DISCUSSION

The concentrations and polarities of the different estrogens present in human urine vary widely. Therefore it is not possible to develop an ideal method for the simultaneous determination of all of them in a single GLC analysis. For clinical purposes, a specific and sensitive determination of estriol in pregnancy urine and estrone in non-pregnancy urine is usually sufficient. A method allowing the estimation of other estrogens besides the classical ones is necessary to provide information about their possible physiological and pathological significance.

The methods presented here make it possible to at least roughly estimate the amount of some other estrogens but is by no means a final solution to the problem. Recovery studies on the rare estrogens are at present not possible for lack of suitable reference compounds of their conjugates. It seems to be essential to use steroid conjugates in studying the recovery possible by the methods, in order to estimate the efficiency of the whole procedure, including the hydrolysis (5). The methods presented here give recovery for conjugated estrogens which are almost ideal and cannot be expected to be enhanced in any way. These results are almost exclusively dependent on the elimination of many enzyme inhibitors by gel filtration (4) and Adlercreutz (to be published), and on the completeness of the TMS ether formation.

The second very important point is the specificity of the procedures. When the specificity of a GLC method is under investigation, mass spectrography seems to be almost ideal, since less than 1 μg of estrogen is needed for their identification (Luukkainen and Adlercreutz, to be published). The use of several different stationary liquid phases enhances the specificity of a GLC procedure considerably. The mass spectrographic investigations indicate that if the TMS ether derivative of an unknown steroid shows identical retention times with the TMS ether derivative of an authentic reference steroid on at least four different stationary liquid phases, and in addition the area of the peak in every gas chromatogram is the same, the compound under investigation is identical with the reference compound. Until now we have not found any exceptions to this rule, a result which suggests that GLC using many different stationary liquid phases and if possible different derivatives of the compounds, is one of the best methods for the identification of microgram amounts of steroids.

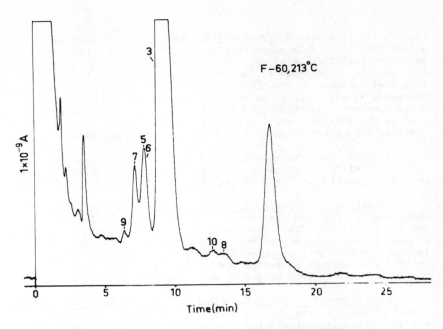

Fig. 9. GLC of the polar estrogen (estriol) fraction obtained with the method for estrone, estradiol and estriol in pregnancy urine on 4 different stationary liquid phases. Conditions as in Fig. 2, with same exception as in Fig. 8. The large peak in Fig. 8, 9, 10 and 11 having a rather long retention time is an estetrol (Luukkainen, Adlercreutz and Fales, to be published).

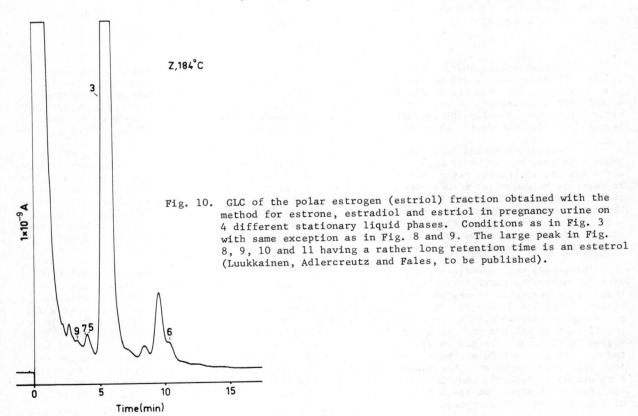

Fig. 10. GLC of the polar estrogen (estriol) fraction obtained with the method for estrone, estradiol and estriol in pregnancy urine on 4 different stationary liquid phases. Conditions as in Fig. 3 with same exception as in Fig. 8 and 9. The large peak in Fig. 8, 9, 10 and 11 having a rather long retention time is an estetrol (Luukkainen, Adlercreutz and Fales, to be published).

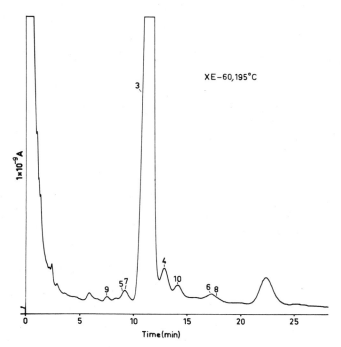

Fig. 11. GLC of the polar estrogen (estriol) fraction obtained with the method for estrone, estradiol and estriol in pregnancy urine on 4 different stationary liquid phases. Conditions as in Fig. 3 with same exception as in Fig. 8, 9 and 10. The large peak in all figures having a rather long retention time is an estetrol (Luukkainen, Adlercreutz and Fales, to be published).

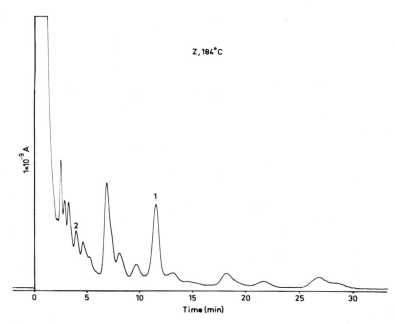

Fig. 12. GLC of the estrone-estradiol fraction obtained with the method for estrone, estradiol and estriol in pregnancy urine. Conditions as in Fig. 3, with the exception that the column was a 6 ft x 4 mm glass U-tube. 1 = estrone, 2 = estradiol.

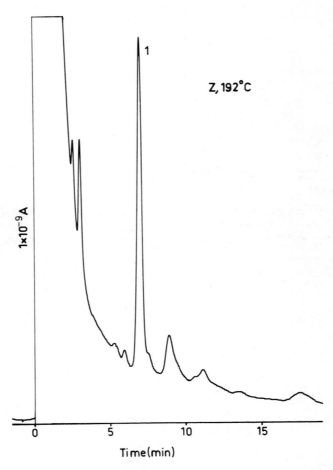

Fig. 13. GLC of the 17-methoxime of estrone 3-methyl ether obtained from pregnancy urine. Conditions as in Fig. 3 with the exception that the column was a 6 ft x 4 mm glass U-tube.

A method has recently been presented in which the overall technique is sensitive to a range of 0.1-0.2 μg of individual estrogens per 24 hr urine specimen, with a lower limit of detection of 0.02 μg (15). If the volume to be injected is 1/10 of the sample, then the whole 24 hr urine specimen has to be processed. For routine determinations this is most inconvenient. It seems that the minute amounts of estrogens present in a small sample of non-pregnancy urine can only be estimated with the electron capture detector. We have at present no experience of quantitation with this technique.

For scientific work the argon ionization and hydrogen flame ionization detectors are valuable because many different derivatives of steroids can be used and good separation of the epimeric steroids thus achieved. In this work TMS ether derivatives of estrogens have been found to be more suitable than the acetates because of: 1) better separation of epimeric estrogens and thus enhanced specificity, 2) short retention times at moderate temperatures, 3) completeness of the reaction, 4) ease with which symmetrical peaks are quantitated, 5) the fact that they emerge unchanged from the column, as judged by mass spectrography.

It was hoped that GLC could be used to analyze urinary extracts purified to a lesser degree than by previously used chemical methods. However, chemical methods include more specific final reactions for estrogens than the blind detection system used in GLC. Therefore an even higher purification is needed in GLC methods. The high separation capacity of GLC technique will compensate for the lack of specificity of the detector. Moreover, the chromogens that disturb the Kober reaction or the fluorometric estimations do not affect the GLC analysis and by using at least two different liquid phases, sources of error can be detected which is not the case with chemical methods. So far, we have not yet found any nonestrogenic drug which interferes with the determinations of urinary estrogens by the GLC methods presented here. This is, in our opinion, one of the most important advantages with these GLC procedures.

## ACKNOWLEDGEMENTS

For the generous gift of reference compounds for these investigations we would like to thank Prof. H. Breuer, Dr. R. Knuppen, Prof. W. Klyne and Schering AG, Berlin. We are also very indebted to Drs. Horning, VandenHeuvel and Fales, who supplied us with excellent liquid phases and placed their extensive experience in this field at our disposal. We would also like to express our thanks to Mrs. Maj-Britt Lofgren, Mrs. Sirkka Tiainen and Miss Anja Riihimäki for expert technical assistance. This work has been supported by grants from the Sigrid Juselius Foundation, the State Medical Commission in Finland and the Nordisk Insulinfond.

## REFERENCES

1. Preedy, J. R. K., in Methods in Hormone Research, R. I. Dorfman, ed., Academic Press, New York and London, Vol. 1, 1-50, 1962.

2. Luukkainen, T., W. J. A. VandenHeuvel, and E. C. Horning, Biochim Biophys Acta 62: 153, 1962.

3. Luukkainen, T. and H. Adlercreutz, Biochim Biophys Acta 70: 700, 1963.

4. Beling, C. G., Acta Endocr 43: Suppl. 79, 1, 1963.

5. Adlercreutz, H., Acta Endocr 42: Suppl. 72, 1, 1962.

6. Adlercreutz, H. and T. Luukkainen, Internationales Round Table-Gesprach uber Analytik von oestrogenen Hormonen und Testosteron, Jena 1964.

7. Brown, J. B., Biochem J 60: 185, 1955.

8. Luukkainen, T., W. J. A. VandenHeuvel, E. O. A. Haahti, and E. C. Horning, Biochim Biophys Acta 52: 599, 1961.

9. Adlercreutz, H. and T. Luukkainen, Biochim Biophys Acta 97: 134, 1965.

10. Fales, H. M. and T. Luukkainen, Anal Chem (in press).

11. VandenHeuvel, W. J. A. and E. C. Horning, Biochim Biophys Acta 64: 416, 1962.

12. Brown, J. B., R. D. Bulbrook and F. C. Greenwood, J Endocr 16: 49, 1957.

13. Nocke, W., *Biochim J* 78: 593, 1961.

14. Ryhage, R., *Anal Chem* 36: 759, 1964.

15. Wotiz, H. H. and S. C. Chattoraj, *Anal Chem* 36: 1466, 1964.

# GAS CHROMATOGRAPHY OF ESTRIOL IN PREGNANCY URINE

Joseph C. Touchstone, Department of Obstetrics and Gynecology, School of Medicine, University of Pennsylvania, Philadelphia, Penn.

Many methods have been reported for the determination of urinary estriol as an index of placental function. Frandsen (1) has summarized the status of this determination up to 1963 in a mongraph. Most chemical methods are modifications of the method of Brown (2). This laboratory has used a chemical procedure involving alumina column chromatography followed by colorimetry (3). The present report describes how the method was modified for use with a gas chromatograph.

*Apparatus*. The apparatus used in the present work is the single oven instrument described in a previous report (4) from this laboratory. Both the detector and column are contained in the same oven with the injection heater controlled separately. A dial faced integrator (Acromag) was used for numerical expression of the detector signal. A Lovelock detector using radium foil was employed. A timer was set between the detector and integrator to control the start of the integrator only to determine the desired component of the fractionation.

*Operating conditions*. The column (6 ft x 4 mm glass coil) was packed under vacuum with a combination of 10% QF-1 (Fluoroalkylsilicone polymer, 1000CS, Dow Corning) and 5% F-50 (chlorophenylmethyl silicone polymer, General Electric) on Gas Chrom Z of 80-100 mesh (Applied Science). Conditioning was carried out for two days at 270 C with argon passing through at 5 psi. The results reported here were obtained with the following operating conditions: column temperature 255 C, flash temperature 280 C; inlet pressure 30 psi; and detector voltage 1000. Prior to quantitative work continual injections of estriol were carried out until the column was equilibrated.

For calibration of the recorder response quadruplicate injections of 1, 2, 3, 4 and 5 µg of estriol dissolved in acetone (1 µg per µl) were made and a standard curve set up. At the beginning of each series of determinations a trial was made to determine whether the amount injected gave an integrator result which fell within the range of the standard curve.

By observation of the chart recordings it was determined the interval at which the timer could be set to activate the integrator just prior to the appearance of the estriol peak. The time for the peak of estriol to appear was usually 7 to 8 min in this work.

*Preparation of urine*. Routinely, 5 ml aliquots of pregnancy urine were diluted to 20 ml and conjugates hydrolyzed by reflux, after addition of 15 vol percent hydrochloric acid, for 15 min. This time was chosen as a result of a comparison of estriol determinations after 15 min, 30 min, and 1 hr reflux times. Table 1 shows that for practical purposes there was little difference among the results.

Following hydrolysis the urine was extracted once with 40 ml of ether. This was washed with 5 ml of 5% sodium bicarbonate. Phenols were extracted by partitioning the ether with 10 ml of N NaOH. After neutralization with 1 ml of conc HCl the estriol was re-extracted with 40 ml of ether. The ether was washed with 5 ml of 5% $NaHCO_3$ and 2 ml of water and evaporated in micro conical glass stoppered test tubes. To the residue was added 50 µl of acetone or in more recent work t-butanol. Two or 5 µl was injected into the inlet of the column using a 10 µl Hamilton syringe and integrator counts observed after the prescribed time. Recovery of estriol added to water and carried through the entire procedure averaged 67.3% in 8 determinations.

TABLE 1. Effect of reflux time on liberation of urinary estriol.

Results as mg per 24 hr

| Specimen | 15 min | 30 min | 60 min |
|----------|--------|--------|--------|
| 1  | 4.9  | 6.4  | 5.9  |
| 2  | 3.4  | 4.1  | 4.9  |
| 3  | 10.7 | 11.5 | 12.6 |
| 4  | 8.5  | 8.8  | 9.7  |
| 5  | 12.0 | 12.6 | 11.3 |
| 6  | 9.3  | 9.3  | 9.0  |
| 7  | 9.7  | 9.4  | 9.7  |
| 8  | 8.0  | 8.4  | 8.5  |
| 9  | 6.9  | 7.8  | 7.1  |
| 10 | 12.6 | 13.6 | 14.2 |
| 11 | 10.7 | 11.4 | 11.1 |

## RESULTS

Figure 1 shows the chart recording of a typical separation by GLC of a urinary estriol prepared by the method described. The dotted line shows the point at which the fine bucking voltage control was used to bring the curve back to base line just prior to emergence of the estriol peak. This is necessary because all urines did not give a tracing that returned to the base line before the estriol began to appear. This was done because the integrator was calibrated to operate at 1000 counts per min from the zero base line to 100% of the chart paper. However, if the integrator operated faster, denoting too high an estriol concentration, a repeat injection must be made.

The calibration curve obtained for integration of the estriol curves was the same as that shown in the previous work using a different column. The slope of the curve changes little over a two month period and was linear between 1 and 4 μg of estriol. Thus, for accurate work it was necessary to work within these limits.

Table 2 gives the results of 12 quadruplicate determinations of different pregnancy urines. The standard deviation of this series of determinations was 0.52 mg per 24 hr with a percentage derivation 4.8%. Thus, one could expect 95% of the results to be within two standard deviations. At present a comparison is being made between the colorimetric and GLC methods for determination of estriol in extracts prepared by the method just described.

## DISCUSSION

Other investigators have reported methods to determine urinary estrogens by GLC (5, 6). Wotiz and Martin (5) and Luukkainen et al. (7) resorted to the preparation of derivatives to separate the estrogens. Using the QF-1 and F-50 combination columns described here it was possible to quantitate estriol in pregnancy urines without the extra manipulation involved in preparing derivatives. The use of GLC on a routine basis for determinations of this type depend to a great extent on the ability to prepare extracts which will give well defined peaks for the substance to be measured. The technique of injection of the samples and operation of the instruments must be mastered.

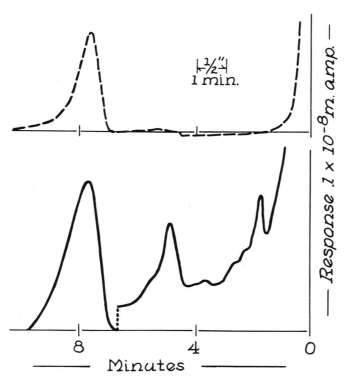

Fig. 1. GLC of estriol in pregnancy urine. Dotted line - reference estriol 2 μg.

TABLE 2. Results of quadruplicate determination of urinary estriol by GLC.

Results as mg per 24 hr

| Specimen | Result | | | |
|---|---|---|---|---|
| 1  | 6.8  | 6.4  | 6.9  | 7.0  |
| 2  | 9.7  | 10.0 | 9.8  | 9.7  |
| 3  | 8.1  | 8.1  | 8.4  | 8.5  |
| 4  | 30.5 | 28.4 | 28.4 | 32.0 |
| 5  | 9.6  | 7.9  | 7.7  | 8.21 |
| 6  | 2.6  | 2.3  | 2.5  | 2.7  |
| 7  | 9.3  | 10.4 | 10.1 | 9.5  |
| 8  | 12.1 | 9.8  | 11.8 | 10.9 |
| 9  | 2.3  | 2.5  | 3.0  | 2.2  |
| 10 | 11.3 | 11.1 | 11.5 | 11.3 |
| 11 | 6.3  | 6.8  | 7.2  | 6.8  |
| 12 | 3.0  | 3.0  | 3.0  | 2.4  |

## ACKNOWLEDGEMENTS

This work was supported in part by U.S. Public Health Grant HD-01199 and Research Career Development Award AM-K-14,013. The cooperation of the various members of the Department of Obstetrics and Gynecology is greatly appreciated.

## REFERENCES

1. Frandsen, V. A., *The Excretion of Oestriol in Normal Human Pregnancy*, Munksgaard, Copenhagen, 1963.

2. Brown, J. B., *Biochem J* 60: 185, 1955.

3. Greene, J. W., Jr., and J. C. Touchstone, *Am J Obst & Gynec* 85: 1, 1963.

4. Touchstone, J. C., *J Gas Chromatog* 3: 170, 1964.

5. Wotiz, H. H., and H. F. Martin, *Anal Biochem* 3: 97, 1962.

6. Yousem, H. L., and D. Strummer, *Am J Obst & Gynec* 88: 375, 1965.

7. Luukkainen, T., W. J. A. VandenHeuvel, E. O. A. Haahti, and E. C. Horning, *Biochim Biophys Acta* 52: 599, 1961.

# ANALYSIS OF ESTROGENS IN THE URINE OF NON-PREGNANT WOMEN BY GAS LIQUID CHROMATOGRAPHY

E. Menini, Medical Research Council, Clinical Endocrinology Research Unit, Edinburgh, Scotland

Mixtures of pure estrogens and many of their derivatives, such as the 3-methyl ethers, the acetates, the trifluoracetates and the TMS ethers, have been successfully analyzed by GLC using a variety of stationary phases (1-5). However, analysis of estrogens in the urine of non-pregnant women by this technique presents particular difficulties due to the fact that these compounds are excreted in a concentration which is very low in relation to the amounts of interfering substances. The need of extensive purification of urinary extracts before analysis of estrogens by GLC has been emphasized by Wotiz and Chattoraj (6).

During the past months several purification procedures have been critically examined in an attempt to prepare estrogen extracts, suitable for GLC analysis, from the urine of non-pregnant women. Since for reliable quantitative analysis by GLC, it is desirable that the compounds to be estimated should be the main components on the recording chart, the criterion for assessing the efficiency of the different purification procedures has been the elimination from the gas chromatograms of peaks due to impurities, without any major loss of estrogenic material.

A procedure described recently (7) affords extracts, the gas chromatograms of which show that the derivatives of the estrogens are among the major components and that in most instances they produce discrete and readily measurable peaks. Briefly the procedure is as follows: urine (usually 50-100 ml) is first treated with potassium borohydride (to reduce estrone to estradiol) then hydrolyzed with conc. HCl and the products of the hydrolysis are extracted into ether. The ether extract is purified, evaporated to dryness and saponified (8, 9). After this purification step, estradiol and estriol are separated from the neutral steroids by partition between a mixture of benzene, light petroleum and ethanol and aqueous sodium hydroxide and the phenolic fraction is methylated (8). The methylated products are chromatographed on alumina and the fractions containing estradiol-3-methyl ether and estriol-3-methyl ether are collected together. After evaporation of the solvent the residue is acetylated with acetic anhydride in pyridine and the acetates of the estrogen-3-methyl ethers are again chromatographed on alumina and eluted together. The final fraction contains the acetate of estradiol-3-methyl ether, derived from the urinary estrone and estradiol and the diacetate of estriol-3-methyl ether derived from the urinary estriol. Cholesterol is added to the total or a fraction of this purified extract as an internal standard, the extract is evaporated to dryness, and transferred onto a Dixon ring by means of a teflon spotting tile (10) and analyzed by GLC.

In urines with high or relatively high concentrations of estrogens, this procedure can be substantially shortened by omitting the saponification and in some cases also the chromatography on alumina of the estrogen-3-methyl ethers.

The instrument used for the final analyses was the Pye Argon Chromatograph (Pye Scientific Instruments, Cambridge, England), equipped with a Lovelock type argon ionization detector fitted with a $^{90}$Sr source, operating at 1000 v. Glass column, 4 ft long, were packed with silanized Gas Chrom P (100-120 mesh) coated with 1% SE-30. Other operating conditions were: temperature of the column, 210 C; flow of argon, 30 ml per min; temperature of the evaporation chamber (Flash heater), 250-260 C. Under these conditions estradiol-3-methyl ether acetate and estriol-3-methyl ether diacetate had retention times of 13 and 26 min respectively.

As in steroid metabolism interconversion of carbonyls and secondary carbinols appears to be freely reversible, reduction of estrone to estradiol and estimation of the latter compound will probably afford the same information about the secretory activity of the estrogen-producing tissue as the separate determination of the two compounds.

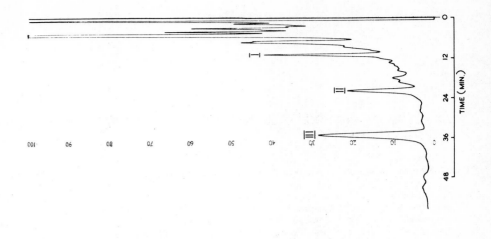

Fig. 1. Gas-liquid chromatogram of a purified extract containing a mixture of derivatives of urinary estrogens. The fraction chromatographed is equivalent to 1/25 of a 24 hr specimen of urine from a normal female. This urine contained 5.6 μg of estradiol, 12.2 μg of estrone and 28.8 μg of estriol. Compounds II and IV have the same retention times as the acetate of estradiol-3-methyl ether and the diacetate of estriol-3-methyl ether, respectively. Compound V (cholesterol) was added as internal standard.

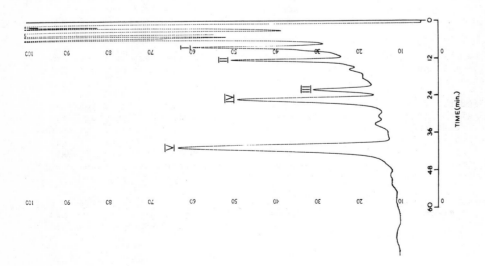

Fig. 2. Gas-liquid chromatogram of a purified extract containing a mixture of derivatives of urinary estrogens. The urine was hydrolyzed enzymatically (Helix pomatia). The fraction chromatographed is equivalent to 1/25 of a 24 hr specimen of urine from a normal female. This urine contained 5.2 μg of estradiol, 18.8 μg of estrone and 31.0 μg of estriol. Compounds I and II have the same retention times as the acetate of estradiol-3-methyl ether and the diacetate of estriol-3-methyl ether, respectively. Compound III (cholesterol) was added as internal standard.

TABLE 1. Analysis of estrogens by the method of Brown (1955, 1957) and by GLC. Comparison of results. (Values in μg per 24 hr).

| Urine No. | Condition | Method (hydrolysis) | Estrone + estradiol | Estriol |
|---|---|---|---|---|
| 3 | Normal | Brown | 26.5 | 23.5 |
|  |  | Present (H+) | 16.8 | 19.3 |
| 19 | Stein-Leventhal | Brown | 27.0 | 68.0 |
|  |  | Present (H+) | 35.6 | 59.0 |
| 21 | Virilizing adrenal adenoma | Brown | 98.4 | 45.0 |
|  |  | Present (H+) (*) | 75.0 | 51.2 |
| 29 | Normal | Brown | 24.0 | 31.0 |
|  |  | Present (H+) | 21.6 | 38.0 |
|  |  | Present (Hel.pom.) | 19.0 | 26.0 |
| 30 | Normal | Brown | 7.5 | 16.6 |
|  |  | Present (H+) | 6.0 | 21.0 |
|  |  | Present (Hel.pom.) | 6.0 | 10.5 |

(*) Saponification was omitted.

From the methodological point of view, reduction of estrone to estradiol has some obvious advantages:

1. It reduces the complexity of the analytical sample by reducing the number of compounds;

2. It brings the reduction products within a narrower range of polarity, with the result that the compounds can be purified as a group.

3. The accuracy of the determination of the sum of estrone and estradiol will be greater than in the determination of the two compounds separately.

Pure estradiol and estriol added to water and submitted to the operations and reactions described gave only the expected GLC peaks and were recovered in yields of 68-77%. Pure estriol added to hydrolyzed urine (two experiments) was recovered in yields of 72 and 79%.

Figures 1 and 2 show two typical gas-liquid chromatograms of urinary extracts obtained by the procedure described above.

Quantitative evaluation of the gas chromatograms was based on the proportionality between the amounts of estrogen derivatives and the ratio of the heights of their chromatographic peaks and of the height of the peak corresponding to a known amount of the internal standard.

In a few instances, urines have been analyzed by the method of Brown (8), Brown, Bulbrook and Greenwood (9) and by the present method. A comparison of the results obtained is presented in Table 1, which includes also the results of two experiments in which urine was hydrolyzed enzymatically using a preparation from the snail Helix pomatia.

Although much additional information needs to be accumulated on the reliability of the present method, before any values can be accepted with confidence, the results obtained up till now are promising and confirm that GLC preceded by extensive purification of the starting material, is a valuable technique for the analysis of estrogens in the urine of non-pregnant women.

## REFERENCES

1. VandenHeuvel, W. J. A., C. C. Sweeley, and E. C. Horning, Biochem Biophys Res Comm 3: 33, 1960.

2. Wotiz, H. H. and H. F. Martin, J Biol Chem 236: 1312, 1961.

3. VandenHeuvel, W. J. A., J. Sjövall, and E. C. Horning, Biochim Biophys Acta 48: 596, 1961.

4. Luukkainen, T., W. J. A. VandenHeuvel, E. O. A. Haahti, and E. C. Horning, Biochim Biophys Acta 52: 599, 1961.

5. Kroman, H. S. and S. R. Bender, J Chromatog 10: 111, 1963.

6. Wotiz, H. H. and S. C. Chattoraj, Anal Chem 36: 1466, 1964.

7. Menini, E., Biochem J 94: 15, 1965.

8. Brown, J. B., Biochem J 60: 185, 1955.

9. Brown, J. B., R. D. Bulbrook, and F. C. Greenwood, J Endocr 16: 49, 1957.

10. Menini, E. and J. K. Norymberski, Biochem J (in press) 1965.

# THE MEASUREMENT OF ESTRIOL IN URINE AND AMNIOTIC FLUID BY GAS-LIQUID CHROMATOGRAPHY

Adolf E. Schindler, Marjorie C. Lindberg and Walter L. Herrmann, Department of Obstetrics and Gynecology, University of Washington School of Medicine, Seattle, Washington

This research was supported in part by U. S. Public Health Service Grants AM-06055-03 and T1 AM-5428-01 from the National Institutes of Arthritis and Metabolic Diseases.

The excretion pattern of estriol is of special interest to the clinician. During pregnancy, urinary levels of this compound can be directly correlated with several parameters applying directly to the fetus, ranging from weight to fetal embarrassment and death. The reason for this is the active participation of the fetus in the metabolism of estriol.

In the nonpregnant individual, it appears that estriol, in all likelihood, is uniquely derived from estradiol-17$\beta$, and that its excretion pattern parallels those of estrone and estradiol-17$\beta$. Since $E_3$ is excreted in larger amounts than $E_2$ and $E_1$, it is reasonable to assume that the measurement of estriol will be easier to carry out with some degree of reliability. In addition, its greater polarity facilitates separation from estrone and estradiol. With this in mind, it has been proposed that, for clinical purposes, the measurement of estriol in urine might well serve as an indicator for estrogenic activity.

Since rapidity of such measurements, next to the other criteria (reproducibility, specificity, sensitivity, accuracy), will be of utmost importance, particularly in pregnancy when clinical management may depend upon levels of urinary estriol, GLC was found to be particularly useful as part of the method. The need for immediate information concerning levels of estrogen exists also under other circumstances, i.e., in patients being treated with Pergonal in an attempt to induce ovulation. In these patients, the dosage of the potent human menopausal gonadotropin is still very problematic, and insufficient stimulation as well as overstimulation have led to failure of therapy or excessive ovarian enlargement respectively. Here again, the possibility of following the results of such therapy with day by day information concerning estrogen excretion appears to be very promising.

Another part of this study concerns the estriol content of amniotic fluid. It has been observed that estriol, both in its free form and as glucuronide, is cleared very slowly from this compartment, as compared to maternal plasma whence the clearance mechanism is very efficient. This rapid disappearance of estriol from maternal plasma must be considered as one of the factors responsible for the relatively low levels (5-15 µg per 100 ml) found even in late pregnancy. These considerations would give the amniotic compartments special significance and the $E_3$ concentration in amniotic fluid might yield information concerning the fetus. In addition, the opportunity to obtain the necessary samples by a single puncture rather than collection of 24 hr urine specimens seems to be an additional factor.

## METHODS

*Urine.* The method for determination of estriol in urine includes the preparation of a phenolic extract, following the addition of tritium-labeled estriol as internal standard, the formation of the TMS ether of estriol, followed by isolation and quantitation of the compound by GLC (Table 1).

TABLE 1.

1. 50 ml of urine plus tritium-labeled estriol as internal standard.

2. Acid hydrolysis (15 volume percent HCl, reflux 1 hr).

3. Ether extraction (3 x 50 ml).

4. Washing with saturated sodium bicarbonate pH 8 (1 x 50 ml).

5. Extraction with sodium carbonate 1N pH 10 (2 x 25 ml).

6. Sodium hydroxyde 2N pH 10 (2 x 25 ml).

7. Acidification with concentrated HCl to pH 8.

8. Extraction with ether (3 x 50 ml).

9. Dried over anhydrous sodium sulfate and evaporation of the ether.

10. Aliquot is removed for counting (Packard Tri-Carb Liquid Scintillation Spectrometer).

11. A measured amount (100 μg) of cholestane is added as internal standard for the rest of the procedure.

The samples (1 to 5 μl) are introduced with a Hamilton microsyringe. As final control of the chromatographic system, and for purposes of quantification, TMS ethers of authentic estrone, estradiol and estriol, either individually or as a mixture, are injected in varying amounts. The standard curve obtained in this fashion shows linearity up to 4 μg per single injection, and is illustrated in Fig. 1. Area values are calculated from peak heights and width at half heights. The retention time is calculated relative to androsterone or cholestane. The GLC apparatus was a Barber-Coleman Model 10 equipped with a $^{90}$Sr argon ionization detector.

Figure 2 represents the chromatogram of the injection of 1 μg of the derivatives of the three forementioned estrogens.

The accuracy of the method was evaluated by analyzing several samples with a known amount of estriol (water or male urine with estriol added), at different time intervals, and was found to be 78 ± 3%.

The precision of replicate estimations performed at different times on pregnancy urine near term was ± 1.5 mg.

The specificity of the method was judged by the chromatographic behavior of the tritium-labeled estriol (purified by celite column and paper chromatography) and the identical retention time on the gas chromatogram of standard and unknown compounds.

A single injection of .2 μg of estriol gave a satisfactory deflection of the recording system. This places the sensitivity of the procedure in the range of 1 mg per total sample. Figure 3 illustrates a chromatogram of pregnant urine obtained in the early part of the third trimester.

TMS ethers are prepared by treating the residue with 0.2 ml of pyridine, 0.2 ml of hexamethyldizilazane, and 0.1 ml of trimethylchlorosilane. The conical glass-stoppered tubes are heated for 1 1/2 hr in a 55 C water bath or allowed to stand overnight at room temperature. The solvents are then evaporated by dry nitrogen stream. The residue is dissolved in 50-200 μl of normal hexane, depending on the week of gestation, in order to keep the area of these peaks within the linearity of the gas chromatograph.

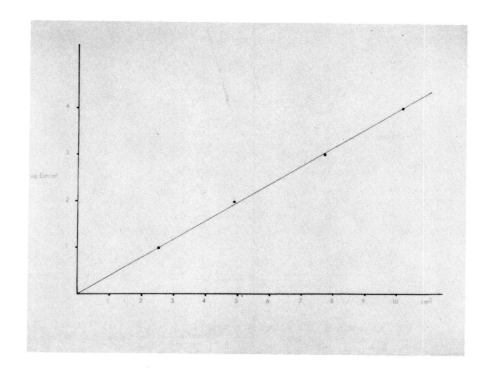

Fig. 1. Linear response of estriol TMS ethers.

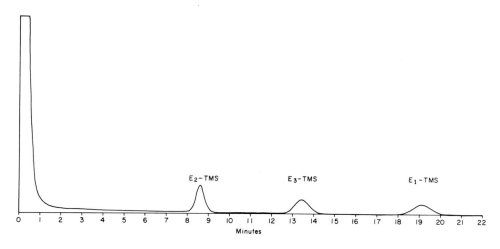

Fig. 2. GLC separation of the TMS derivatives of estrone ($E_1$), 17β-estradiol ($E_2$) and estriol ($E_3$). The chromatogram represents 1 μg of each compound. Column conditions: 6 ft x 4 mm glass U column, 2% XE-60 on 80-100 mesh Gas Chrom P, Argon 16 psi, column temp 224 C.

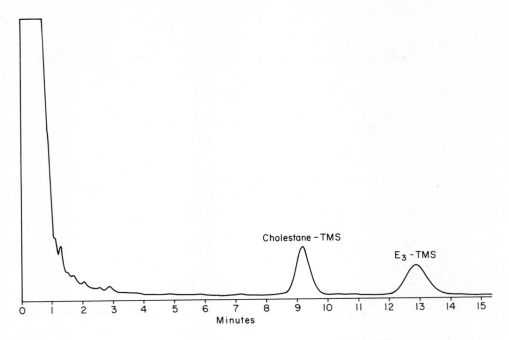

Fig. 3. GLC of estriol TMS ether obtained from human third trimester urine. Column conditions: 6 ft x 4 mm glass U column, 2% XE-60 on 80-100 mesh Gas Chrom P, Argon 16 psi, column temp 224 C.

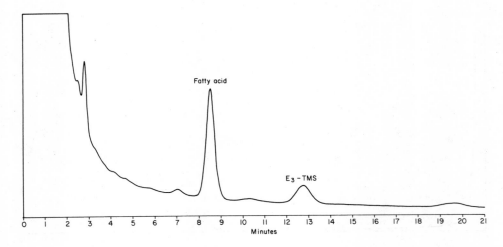

Fig. 4. GLC representing 1/25 of an extract obtained from 75 ml amniotic fluid (39th week of gestation). Column conditions: 6 ft x 4 mm glass U column, 2% XE-60 on 80-100 mesh Gas Chrom P, Argon 16 psi, column temp 224 C.

Amniotic fluid. Amniotic fluid was obtained by transabdominal puncture and processed as described above. Additional steps include centrifugation at 2,000 rpm for 15 min to eliminate cellular debris, and washing of the combined alkaline extracts with 50 ml of normal hexane.

Figure 4 represents a chromatogram of amniotic fluid obtained by transabdominal amniocentesis prior to delivery. The major additional peak which was observed consistently in all samples of amniotic fluid was identified as a mixture of fatty acids (palmitic acid, oleic acid, and stearic acid).

# GAS CHROMATOGRAPHY OF FREE AND CONJUGATED ESTROGENS IN BLOOD PLASMA

Joseph C. Touchstone and Tara Murawec, Department of Obstetrics and Gynecology, School of Medicine, University of Pennsylvania, Philadelphia, Penn.

Several studies describing estrogen levels in blood have been reported (1, 2, 3). These studies have evaluated the total concentration rather than differentiating between free and conjugated estrogens. A preliminary report (4) from this laboratory showed the concentrations of estrone, estradiol-17β and estriol in the free, sulfate, and glucosiduronate fractions of extracts of blood plasma. The present report describes the methods used for determination of estrogens from these fractions by GLC.

*Apparatus and column conditions*. The gas chromatograph used in this work was a Chromalab Model A110 (Glowall Corp.) equipped with a Lovelock detector using radium foil. The column was a glass coil (6 ft x 4 mm) packed with a combination of 12% QF-1 (fluorosilicone polymer, Dow Corning) and 6% F-50 (chlorophenylmethylsilicone polymer, General Electric) on Gas Chrom Z, 80-100 mesh (Applied Science). The substrates were coated on the support in one application by dissolving them in methylene chloride (5-6 vol X the weight of substrate) and evaporating the solvent with constant stirring after addition of the support. The packing was dried in a vacuum oven before addition to the glass coil under a vacuum. The columns were conditioned at 270 C for 2 days with argon at an inlet pressure of 10 psi. Operating conditions were: column and detector temperature, 250 C; flash temperature, 270 C; inlet pressure 30 psi; and voltage 1000. Samples were dissolved in 25 μl of acetone and 2-4 μl aliquots injected.

*Preparation of blood plasma extracts*. Pooled blood samples were moved from the antecubital vein, in syringes containing heparin, from women in the third trimester. Cord blood was milked from the umbilical cord at the time of separation from the fetus. Blood plasma was extracted within half an hr after removal or frozen until extraction was undertaken. Proteins were precipitated by addition of 4 vol of acetone and removed by filtration. The acetone was evaporated *in vacuo* in a warm water bath. The aqueous residue was then extracted in three sequential steps, as follows:

1. *Free estrogens*. The aqueous residue was extracted three times with equal volumes of ether and separated. The aqueous phase was carried to the next step.

2. The aqueous phase was adjusted to pH 6.8 by addition of maleate buffer. For each 100 ml of deproteinized plasma, 5 ml of 0.2 M Monosodium maleate, and then pH was adjusted with 0.2M sodium hydroxide addition using a pH meter. Bacterial glucuronidase (Sigma) (1000 units per ml) was added and incubation carried out at 37 C for 24 hr. The freed steroids were extracted by partition three times with equal vol of ether. The aqueous phase was carried to the next step.

3. The aqueous portion was adjusted to pH 6.0 by addition of maleate using a pH meter. Mylase P (Nutritional Biochemicals) was added (10 mg per ml) and incubation carried out at 50 C for 24 hr. Freed steroids were extracted as above.

The ether obtained in the three instances was washed once with 5% sodium bicarbonate and once with water (10% of ether vol). The ether was evaporated. The residue was dissolved in 30 ml of toluene and shaken with 20 ml of 1N NaOH three times. Back wash of the alkaline solution with toluene was performed. The alkali was neutralized by addition of concentrated HCl to pH 6-7. The phenols were then extracted three times with equal vol of ether. Combined ether extracts were then washed with bicarbonate and water as above. The ether was evaporated and the residue subjected to alumina column chromatography.

*Preparation of alumina and chromatography of phenols*. Activated alumina (Harshaw, catalyst grade) was washed copiously with absolute ethanol and then dried in a vacuum oven at 120 C for 18 hr. It should be protected from moisture. One g of this alumina was poured into a 10 ml syringe fitted with a 22 gauge needle and filled with benzene (5). The extract was added to the column in benzene followed by the eluting solvents. These were benzene-30 ml, 1% ethanol in benzene-20 ml and 20 ml of 10% ethanol in benzene and finally 20 ml of 50% ethanol in benzene. The fractions were evaporated in a vacuum oven and transferred with 1-1:5 ml of acetone into micro conical test tubes and evaporated in a vacuum oven. Acetone, 25 μl, was added to each tube and aliquots taken for GLC or other identification procedures. Estrone and estradiol-17β are found in the 10% ethanol in benzene fraction while estriol was found in the 50% ethanol in benzene fraction.

## RESULTS

Figure 1 shows the results of GLC of known amounts of estrone, estradiol-17β and estiol after chromatography on alumina columns prepared as described. Recoveries of estrone, estradiol-17β and estriol from this alumina column have been quantitative.

Figure 2 shows the results of GLC of extracts of maternal blood plasma prepared in the manner described. These results are qualitatively typical of a series of determinations done with both cord and maternal plasma.

Table 1 gives the results of quantitation of estrone, estradiol-17β and estriol by comparison of the areas of the peaks on chart recordings with the areas of known amounts of reference steroids.

In a previous paper (6) it was shown that the recovery of estriol by this method average 80% in six determinations when quantitation was performed with the Bachman reagent. The recovery of estrone and estradiol-17β in preliminary experiments was 70%. More recoveries must be performed to obtain the accuracy for estrone and estradiol-17β.

Table 1 shows the concentrations of free and conjugated estrogens found in both cord and maternal blood by the present method. Estriol sulfate was the major conjugate found in both the maternal and cord bloods being tenfold higher in the cord blood. Estrone sulfate was present in higher concentration in the maternal blood.

## DISCUSSION

Early in this work, columns used in the gas chromatograph contained only QF-1 as substrate. This silicone gave the best separation of estrone and estradiol-17β. However, it had the disadvantage that cholesterol and other steroids had the same retention time as estrone. In spite of phenol separation and column chromatography prior to GLC the estrone quantitation was misleading. The use of combination columns has given a more specific separation and more confidence can be had in the results.

It should be noted that the present method using a phenol partition is not suitable for the determination of some of the 16, 17-oxygenated steroids since they are not stable under the alkaline conditions used. The presence of the sulfo-glucosiduronate diconjugate of estriol reported previously (6) is not reflected in the results since cross hydrolysis with two different aliquots of the same sample was not performed, consequently the values for estriol sulfate concentration may be high.

It is apparent that the success of GLC as an analytical tool for determinations of this type depends to a great extent upon the method used to prepare the samples. In some instances it is advantagenous to prepare derivatives for more efficient separation on the column. The present method for determination of estrogens may be limited to pregnancy plasma since the sensitivity at present is not of the order to permit determination of the lower levels of estrogens in non-pregnancy blood unless larger quantities of blood are extracted.

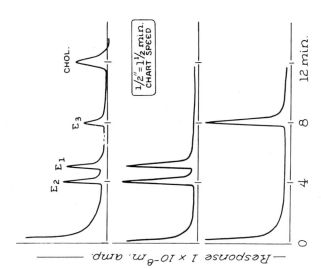

Fig. 1. GLC of estrone ($E_1$), estradiol-17β ($E_2$), and estriol ($E_3$) after alumina column chromatography. Upper figure reference steroids. Chol = cholesterol. Middle figure 10% methanol in benzene fraction. Bottom figure 50% ethanol in benzene. Reference steroids $E_1$ and $E_2$ = 1 μg; $E_3$ = 2 μg.

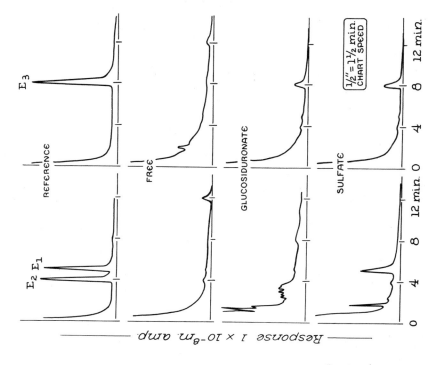

Fig. 2. GLC of estrone, estradiol-17β, and estriol in maternal plasma. Left panels = 10% methanol in benzene fraction; right panels = 50% methanol in benzene fractions from the alumina column. $E_1$ and $E_2$ = 1 μg; $E_3$ = 2 μg.

TABLE 1. Free and conjugated estrogens in plasma (μg per 100 ml).

| Plasma | Free | | | Sulfate | | | Glucosiduronate | | |
|---|---|---|---|---|---|---|---|---|---|
| Maternal | $E_1$* | $E_2$ | $E_3$ | $E_1$ | $E_2$ | $E_3$ | $E_1$ | $E_2$ | $E_3$ |
| 1 | 1.0 | ** | 0.1 | 3.0 | 0.2 | 4.0 | 0.2 | 6.2 | 1.5 |
| 2 | --- | --- | --- | 3.0 | --- | 10.0 | 0.1 | --- | 4.0 |
| 3 | 1.0 | 1.0 | --- | 2.0 | 0.2 | 4.0 | 6.5 | 0.1 | 2.0 |
| 4 | 1.0 | 1.0 | --- | 4.0 | --- | 5.0 | 1.5 | 1.0 | 6.0 |
| Cord | | | | | | | | | |
| 1 | 1.0 | --- | 2.0 | 1.2 | --- | 56.0 | --- | --- | 12.0 |
| 2 | 2.0 | 0.6 | 10.0 | 1.0 | 0.1 | 70.0 | 0.1 | 0.1 | 25.0 |
| 3 | 2.0 | --- | --- | 1.5 | 0.1 | 86.0 | 6.1 | --- | 10.0 |

\* $E_1$ = estrone, $E_2$ = estradiol-17β, $E_3$ = estriol.

\*\* --- = Not detected at level of sensitivity (0.1 μg per 100 ml).

## ACKNOWLEDGEMENTS

This work was supported in part by U.S. Public Health Grant HD-01199 and Research Career Development Award AM-K-14,013. The cooperation of the various members of the Department of Obstetrics and Gynecology is greatly appreciated.

## REFERENCES

1. Maner, F. D., B. D. Saffan, R. A. Wiggins, J. D. Thompson, and J. R. K. Preedy, J Clin Endocr 23: 445, 1963.

2. Roy, E. J., J Obst Gynec Brit Comm 69: 196, 1962.

3. Oertel, G. W., C. D. West, and K. B. Eik-Nes, J Clin Endocr 9: 1619, 1959.

4. Touchstone, J. C. and T. Murawec, Fed Proc 22: 469, 1963.

5. Eberlein, W. R., A. M. Bongiovanni, and C. M. Frances, J Clin Endocr 18: 1274, 1958.

6. Touchstone, J. C., J. W. Greene, Jr., R. C. McElroy, and T. Murawec, Biochem 2: 653, 1963.

# ESTIMATION OF ESTRADIOL-17β BY GAS-LIQUID CHROMATOGRAPHY WITH ELECTRON CAPTURE DETECTION

Kristen B. Eik-Nes, Asbjorn Aakvaag, and Lee J. Grota, Department of Biochemistry, University of Utah, College of Medicine, Salt Lake City, Utah.

This work was in part supported by training grant #T4-CA 500 and grant #AM 06651-03 from the U.S. Public Health Service, Bethesda, Md.

## INTRODUCTION

Over the past three years our laboratory has been interested in steroid biosynthesis in vivo by the canine ovary (1-4). Information on pathways of steroid biotransformation in this organ has been obtained by infusing labeled steroids via the ovarian artery and determining radioactive steroids in venous blood of the infused organ. Much knowledge could be gained from such experiments if the specific activity of the radioactive metabolites could be determined rather than only the amounts of isotope accumulated in isolated compounds (5). Methods are available for the estimation of small amounts of steroids in blood and urine. Most micro techniques, however, use isotopes of different energy spectra -- one isotope being used for correcting losses of steroids occurring during processing of the sample, the other one serving as a reagent to quantitate the steroid to be measured (6). Such techniques cannot be applied to the estimation of steroids in biological specimens already containing these isotopes (7).

Landowne and Lipsky reported on electron capture spectrometry of haloacetates of sterols (8), and recently our laboratory has applied this observation to the measurement of testosterone in normal human plasma (9). In this work it was also observed that small amounts of the chloroacetate of 20α-hydroxyprogesterone could be determined by GLC with electron capture detection, thus constituting a possible basis for the estimation of progesterone in human blood. Details of both methods will be discussed during this symposium (Brownie, A. C., and van der Molen, H. J., this book).

Since estradiol ($E_2$) can be acetylated (10), attempts were made to measure estradiol as its monochloroacetate by electron capture detection.

## MATERIALS AND METHODS

All solvents were purified as described (9) and assayed for impurities by GLC with electron capture detection. It was observed early that the purity of the benzene was critical for adequate assay. Commercially available benzene was washed with concentrated sulphuric acid until the acid became free of visible color. The benzene was then washed with water until water washes were neutral and the washed benzene was shaken with sodium hydroxide pellets and then distilled adding fresh sodium hydroxide pellets to the benzene before distillation.

Evaporation of solvents was done under a stream of nitrogen at 45 C. Preparations of thin-layer plates using silica gel as support medium, as well as the preparation of columns for GLC, have been described by our laboratory (9, 11). When working with biological samples containing tritium, counting of radioactivity in the purified samples was done as published in previous work (1-4). The average efficiency of tritium counting was 19.4%.

A standard Barber-Colman Model 10 gas chromatograph was used throughout this investigation. This instrument was equipped with an Aerograph electron capture detector obtained from Wilkins Instrument and Research, Inc. Pulsating voltage was applied to the electron capture cell using a Model 214a pulse generator (Hewlett-Packard Company). Fifty v, a pulse width of 10 μsec, a pulse position of 100 μsec and an internal repetition rate of 10 KC were applied to the cell. The source impedance was 50 Ohms. The output of the electron capture detector was fed into the electrometer of the chromatograph.

## RESULTS

Commercially available $E_2$-methylether (Steraloids, Inc.) was purified by TLC in several solvent systems (12, 13) or by recrystallization. One hundred mg of the purified product was thoroughly dried and reacted in a desiccator overnight and in the dark with 5 parts of monochloroacetic anhydride in dry tetrahydrofurane (100 mg per 100 ml) and 1 part dry pyridine. The reaction was stopped by the addition of water and the solution was extracted with benzene. The extract was washed once with 1/4 vol 6 N hydrochloric acid and then twice with 1/4 vol water. The washed benzene was then evaporated to dryness and the residue was recrystallized several times from acetone-water to give crystals with a mp of 121.5 C (corrected). When this material was subjected to TLC in benzene only one band reacting with iodine (14) was found ($R_f$ 0.4). Furthermore, when the material was chromatographed on a 1% XE-60 column only one peak could be observed.

In Fig. 1 are recorded the actual gas chromatographic recordings of small amounts of $E_2$ (3-methylether-17-chloroacetate) chromatographed on a 3 ft 1% XE-60 column. The detector system used appears to be able to distinguish between 0.005 µg (about 0.004 µg $E_2$) and 0.01 µg (about 0.008 µg $E_2$) of the estrogen derivative.

Calibration curves for the quantitation of $E_2$ (3-methylether-17-chloroacetate) were made and when concentrations from 0.005 to 0.05 were plotted against the size of the peak (cm$^2$) on the gas chromatographic tracing a straight line was obtained. Table 1 demonstrates that from 0.005 to 0.080 µg steroid can be determined with adequate accuracy.

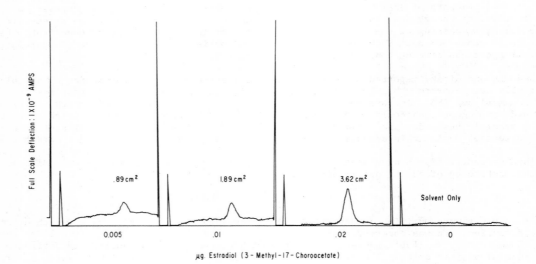

Fig. 1. GLC tracings of different amounts of estradiol (3-methylether-17-chloroacetate). The samples were applied to the column in toluene. The column temperature was kept at 207 C with the detector at 212 C and the flash heater at 240 C. High purity nitrogen was used as gas phase and was lead through a tube filled with molecular sieve, the gas pressure was 25 psi. Chromatography was done on a 3 ft glass column containing 1% XE-60. Quantitative evaluation of the areas of the recorded peaks was done by triangulation.

TABLE 1. Mean peak area (cm$^2$) for different amounts of $E_2$ (3-methylether-17-chloroacetate) when exposed to GLC with electron capture detection. Ten estimations were done on each concentration of the steroid.

| Amount (μg) | Peak area (cm$^2$) |
|---|---|
| .005 | 0.92 ± .17* |
| .010 | 1.80 ± .14* |
| .020 | 3.98 ± .31* |
| .040 | 7.96 ± .43* |
| .080 | 15.58 ± .53* |

* One standard deviation.

Ovarian vein blood of animals infused with tritiated dehydroepiandrosterone via the ovarian artery was divided into equal parts and processed as described by our laboratory (1-4). After a toluene-sodium hydroxide partition, the layer containing the estrogens was neutralized, extracted with methylene dichloride and the extract evaporated to dryness. The residue was chromatographed on paper in the solvent system benzene/formamide. Radioactive material on the chromatograms of the plasma extracts (2, 4) with running rate of authentic $E_2$ (chromatographed on a separate strip of paper in the same chromatography tank and localized as described (1) was eluted and chromatographed on paper in the solvent system chloroform/formamide. Material in the plasma behaving chromatographically like authentic $E_2$ was eluted, methylated (15) and extracted from the reagent mixture with hexane. The hexane was washed with water until the washes became neutral and the hexane evaporated to dryness. The residue was chromatographed by TLC in the solvent system benzene:ethyl acetate (4:1 v/v) and material in the plasma extract chromatographing like authentic $E_2$-methylether ($R_f$ 0.47) was scrapped off the plate, dissolved in water and extracted with benzene (recovery: methylation, TLC, elution and extraction: about 89%). The benzene was evaporated to dryness, chloroacetylated and extracted as described (recovery: chloroacetylation, TLC, elution, extraction and purification with 6 N hydrochloric acid and water: about 80%). The hexane was evaporated to dryness and the residue chromatographed by TLC in benzene. Material in the plasma extract with $R_f$ of 0.4 was eluted, the eluate was evaporated to dryness and the residue was dissolved in 1 ml benzene containing 0.4 μg cholesterol chloroacetate (9). One tenth ml of the solution was removed for estimation of radioactivity, the rest of the sample was evaporated, dissolved in 15 μl toluene and chromatographed on a 3 ft glass column using XE-60 1% as column substrate. The conditions of chromatography can be found in the legend of Fig. 1. The specific activity of $E_2$-$^3$H was calculated in the following fashion:

$$\text{DPM per μg} = \frac{10 \times N}{\dfrac{E_x}{E_s} \times \dfrac{C_s}{C_x} \times \dfrac{272}{362.5} \times 0.01}$$

N: dpm in aliquot removed

$E_x$: area (cm$^2$) of $E_2$ (3-methylether-17-chloroacetate) in sample

$E_s$: area (cm$^2$) of 0.01 μg of authentic $E_2$ (3-methylether-17-chloroacetate)

$C_s$: area (cm$^2$) of 0.4 μg of the cholesterol chloroacetate standard

$C_x$: area (cm$^2$) of the cholesterol chloroacetate in sample

272: mol weight of $E_2$

362.5: mol weight of $E_2$ (3-methylether-17-chloroacetate).

As can be seen from Table 2 the precision of assay is good. Plasma samples processed as outlined were in such state of purity that the small concentrations of $E_2$ in the plasma appeared as a distinct peak on the tracings of the elution pattern from the GLC column (Fig. 2).

TABLE 2. Specific activity of estradiol in duplicate samples of ovarian vein blood. The animals were infused with tritiated dehydroepiandrosterone via the ovarian artery. All data are given in dpm per µg estradiol.

| Dog Index | Sample A | Sample B |
|---|---|---|
| 5264 | 16,000 | 21,200 |
| 5281 | 7,600 | 6,100 |
| 5284 | 4,100 | 5,700 |
| 5291 | 37,400 | 41,600 |

Furthermore, dog ovaries were homogenized and incubated in phosphate buffer containing tritiated dehydroepiandrosterone and a NADPH-generating system (16). After a 2 hr incubation at 37.5 C in an atmosphere of 95% $O_2$-5% $CO_2$, the content of the incubation flasks was extracted with ethyl acetate and the extract from each flask divided into two equal parts and evaporated to dryness. The residues were processed as already described for ovarian vein plasma. Also, in this experiment the duplicate samples agreed fairly well with respect to the specific activity of the steroid under investigation (Table 3). An exception, however, is sample #VIII. Finally, whole ovaries from normal dogs were weighed, homogenized in 0.9% sodium chloride, and the homogenate extracted and purified as outlined for ovarian vein plasma. Duplicate samples of the ovary of one dog gave concentrations of estradiol of 0.036 and 0.034 µg per g of wet weight, and in another dog these values were 0.076 and 0.068 µg. These samples were also in a high state of purity as judged from their GLC tracings (Fig. 3). In these experiments no correction was made for loss of $E_2$ through the method. When amounts of $E_2$ ranging from 0.03-0.07 µg are processed by the method, the recovery of the steroid has varied from 27-46%. With experience this recovery seems to improve.

TABLE 3. Specific activity of estradiol in duplicate samples. In all experiments homogenate of dog ovary was incubated with tritiated dehydroepiandrosterone. The data are given in dpm per µg estradiol.

| Incubation No. | Sample A | Sample B |
|---|---|---|
| I | 350,400 | 305,000 |
| II | 340,200 | 347,300 |
| III | 346,000 | 368,000 |
| IV | 131,100 | 114,000 |
| V | 404,900 | 420,400 |
| VI | 147,200 | 166,800 |
| VII | 270,800 | 301,700 |
| VIII | 14,200 | 59,400 |

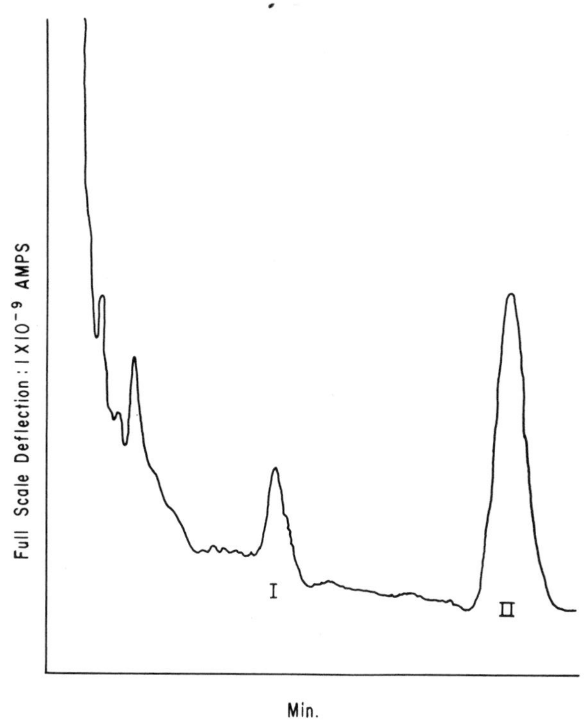

Fig. 2. GLC tracing of estradiol (3-methylether-17-chloroacetate) extracted from ovarian vein blood of a dog. The purification of the steroids has been described in the text of this paper and the conditions of GLC can be found in the legend of Fig. 1.

I: Compound with retention time of authentic estradiol (3-methyl-ether-17-chloroacetate).

II: Cholesterol chloroacetate.

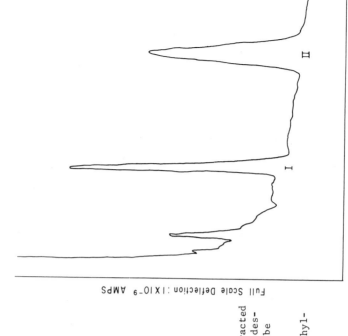

Fig. 3. GLC tracings of estradiol (3-methylether-17-chloroacetate) extracted from ovarian tissue. The purification of the steroid has been described in the text of this paper and the conditions of GLC can be found in the legend of Fig. 1.

I: Compound with retention time of authentic estradiol (3-methyl-ether-17-chloroacetate).

II: Cholesterol chloroacetate.

## DISCUSSION

The data demonstrate that relatively small amounts of $E_2$ can be determined in biological samples following extensive purification and GLC with electron capture detection. Admittedly, this method does not have the sensitivity of the bromine-82 method of Slaunwhite and Neely (17), but as stated by the authors, their method "is too involved to be used as a routine clinical procedure" (17). The method proposed by us seems to be applicable to the estimation of $E_2$ in ovarian vein blood of the dog, of $E_2$ in the canine ovary and also to the quantitation of estradiol produced in vitro by this tissue. This method can furthermore be employed to the determination of specific activity of $E_2$ biosynthesized from tritium and/or carbon 14 labeled substrate(s) (7). The method can probably be improved upon using other column substrates than the one presented employed.

Systems of purification less elaborate than those used by us should also be explored. Working with estradiol in the canine it is, however, felt that the step of methylation cannot be omitted. We have considerable trouble crystallizing ovarian vein blood estradiol to constant specific activity in animals receiving radioactive precursors via the ovarian artery (1), unless the biotransformed estradiol (containing tritium and/or carbon 14) is methylated before crystallization (18). Currently we are trying to apply this method to the measurement of estradiol in peripheral plasma of the dog. Since the recovery of estradiol is important in such experiments, tritiated estradiol of high specific activity is added to the plasma prior to extraction and the toluene-sodium hydroxide partition.

## SUMMARY

About 0.005 µg of the 3-methylether-17-chloroacetate of estradiol can be measured by GLC with electron capture detection. This method has been applied to the estimation of the specific activity of estradiol biosynthesized from tritiated dehydroepiandrosterone. Extensive purification of the biological sample is needed before adequate assay is obtained.

## ACKNOWLEDGEMENT

The authors would like to express their deepest gratitude to Mrs. Minalee Woodward for technical assistance.

## REFERENCES

1. DePaoli, J. and K. B. Eik-Nes, Biochim Biophys Acta 78: 457, 1963.

2. Aakvaag, A. and K. B. Eik-Nes, Biochim Biophys Acta 86: 380, 1964.

3. Aakvaag, A., A. A. Hagen, and K. B. Eik-Nes, Biochim Biophys Acta 86: 622, 1964.

4. Nishizawa, E. E. and K. B. Eik-Nes, Biochim Biophys Acta 86: 610, 1964.

5. Eik-Nes, K. B., and M. Kekre, Biochim Biophys Acta 78: 449, 1963.

6. Kliman, B. and R. E. Peterson, J Biol Chem 235: 1639, 1960.

7. Hagen, A. A. and K. B. Eik-Nes, Biochim Biophys Acta 90: 593, 1964.

8. Landowne, R. A. and S. R. Lipsky, Anal Chem 35: 532, 1963.

9. Brownie, A. C., H. J. van der Molen, E. E. Nishizawa, and K. B. Eik-Nes, J Clin Endocr 24: 1091, 1964.

10. Dominguez, O. V., J. R. Seely, and J. Gorski, Anal Chem 35: 1243, 1963.

11. DePaoli, J., E. E. Nishizawa, and K. B. Eik-Nes, J Clin Endocr 23: 81, 1963.

12. Lisboa, B. P. and E. Diczfalusy, Acta Endocr 40: 60, 1962.

13. Struck, H., Microchim Acta 29: 634, 1961.

14. Barrett, G. C., Nature 194: 1171, 1962.

15. Bush, I. E., *Microchemical Reactions for Steroids*. *In* The Chromatography of Steroids, New York, Pergamon Press, 1961, p. 367.

16. Hall, P. F., E. E. Nishizawa, and K. B. Eik-Nes, *Proc Soc Expr Biol Med* 114: 791, 1963.

17. Slaunwhite, W. R., Jr. and L. Neely, *Anal Biochem* 5: 133, 1963.

18. Aakvaag, A. A. and K. B. Eik-Nes, *Biochim Biophys Acta* 86: 380, 1964.

# GAS CHROMATOGRAPHY OF ESTROGENS AND OTHER STEROIDS FROM ENDOCRINE TISSUES

Kenneth W. McKerns and Egon Nordstrand, Department of Obstetrics and Gynecology, University of Florida College of Medicine, Gainesville, Florida

This investigation was supported by research grant AM 04930 from the National Institute of Arthritis and Metabolic Diseases, National Institutes of Health, Bethesda, Md.

We wish to thank Carleen Jernigan for technical assistance during part of this investigation.

A simple and satisfactory method has been developed for the extraction, partial purification and separation by GLC of many steroids synthesized from endocrine tissues. The method was developed to aid our studies on the mechanisms of the regulation of biosynthesis of estrogens, androgens and progesterones from human, rat, and cow ovaries, and from human placenta. Since steroids are synthesized in our in vitro incubation systems of endocrine tissues in microgram quantities, extensive purification of the steroid extracts before GLC was not found to be necessary. The methods have been described previously (1, 2). Additional modifications and further studies on the biosynthesis of steroids are described here.

## EXPERIMENTAL

A. <u>Extraction and Partial Purification of Steroids</u>.

Biosynthesis of steroids is studied in a homogenate system of ovarian or placental tissue consisting of 25-60 mg of tissue incubated for various time periods in 2 ml of Krebs-Ringer bicarbonate buffer at 37 C under $O_2$-$CO_2$(95:5). The homogenate is extracted with 60 ml of reagent grade of methylene chloride methanol (2:1 v/v) with occasional stirring for 30 min. This single phase system, which is very efficient in extracting steroids, is filtered through Whatman #44 filter paper and the precipitate washed through with 30 ml of the methylene chloride:methanol mixture. The combined filtrates are evaporated to dryness on a rotary evaporator at 60 C. The residue is dissolved and transferred to a separatory funnel with 60 ml of methylene chloride. The methylene chloride is washed by shaking with 10 ml of water and the water phase is back extracted with 30 ml of methylene chloride. Water is removed by the addition of 10 g of calcium chloride to the combined solvent fractions. This is filtered and the filtrate evaporated to dryness on the rotary evaporator. The residue is transferred to a 4 ml vial using methylene chloride:methanol (2:1) to a total vol of 4 ml. The solvent is evaporated at 60 C with a stream of nitrogen. The TMS ethers are prepared in the same 4 ml vial by the addition of the following reagents: 1 ml pyridine, 0.2 ml hexamethyldisilazane, and 0.05 ml trimethylchlorosilane. Reaction time is 1 hr at room temperature (26 C). The solvents are evaporated with a stream of nitrogen and the residue dissolved in 3 ml of ethyl acetate. Silica gel G (0.5 g) is then added and the mixture centrifuged. The silica gel is washed with 3 ml ethyl acetate. The combined solvent fractions are evaporated to dryness with nitrogen and dissolved in 50 μl of n-hexane. The TMS ether derivatives of many steroids have been separated by GLC by Luukkainen et al. (3).

This procedure gives efficient recovery of steroids and sufficiently clean solvent fronts for extracts of human or rat ovarian tissues and for human placental tissues. More pigmented and fatty tissues such as the cow corpus luteum require additional silica gel or other purification steps (2).

B. <u>Column Systems and GLC Techniques</u>.

A Research Specialties Co., Model 600 Gas Chromatograph was used with a $^{90}$Sr ionization detector at 1500 v with an attenuation setting in the range from 5-20 depending on the amount of steroid. A 4 ft or 6 ft x 4 mm U-shaped glass column was packed with a mixture of 3% XE-60 on 80-100 mesh Gas Chrom P (Applied Science Labs., St. College, Pa.) and 3% SE-30 on 100-120 mesh Gas Chrom P (Applied Science Labs.) in a 1:1 w/w ratio. This mixed polar and non-polar phase system is useful in the separation of a wide range of steroids. The phase ratio can be changed to change the retention times for the best separation of steroids in a particular tissue under study. The column temperature of this system was maintained at 220 C for the 4 ft column and 230-235 C for the 6 ft column. Argon was used as the carrier gas at an inlet pressure of 25 psi for the 4 ft column and 35 psi for the 6 ft column.

An F&M Scientific Corp. Model 400 Gas Chromatograph with a hydrogen flame ionization detector was also used. The range control was usually set at 1 with the attenuation at 32 or 64. Four ft x 4 mm glass columns were used with just the 3% SE-30 phase on 100-120 mesh Gas Chrom P support medium. The column temperature was maintained in the range 210-225 C and the helium flow at 60 ml per min. The XE-60 phase was also used with either GLC instruments.

Between 1 and 5 µl of the TMS ethers of the steroid samples in n-hexane was injected with a Hamilton microsyringe into the top of the column. If a second run was not made shortly, the solvent in the remainder of the 50 µl sample was evaporated under nitrogen. The dry solvent-free samples can be kep for several days in a dry atmosphere at room temperature without noticeable hydrolysis of the TMS ether derivatives. The steroid residue can be redissolved and re-run with or without the addition of added steroid standards. Cholestane, 1-androstenedione, 4-androstenedione, and progesterone were used as reference standards. These standards were also run each morning to check operating conditions and changes in retention times. Cholesterol has also been used as an internal standard and the relative retention times of many TMS steroids to the TMS ether of cholesterol has been calculated (2). Changes in retention times of steroids is probably due to stripping of the XE-60 coating, contamination of the column or fluctuations in the column temperature. Tentative identification of steroids has been by their relative retention times to standards in the several column systems and by the increase in peak height obtained with added standard. As well as the TMS ether derivatives, acetate derivatives have been run according to the methods of Wotiz (4). Carbon-14 labeled derivatives have also been added to the incubation mixtures and the radioactive steroid products have been trapped from the column effluent for scintillation counting and for additional identification procedures. Extracts from the incubation medium have also been chromatographed on TLC systems. Silica gel G with added phosphor (Adsorbosil-P 2, Applied Science Labs.) was used as a plate coating with benzene-ethyl acetate (3:2 v/v) as the developing medium according to Futterweit et al. (5). The steroids were located, outlined under UV light, extracted and run through the various GLC systems described above. The plates have also been sprayed with 3% $H_2SO_4$ in 50% aqueous ethanol and the color developed by heating in an oven at 100 C for 15 min. This procedure is similar to the method of Lisboa and Diczfalusy (6). A wide range of colors have been produced from the various steroids and the method is sensitive to detect steroids in the range of 1-3 µg. The adsorption spectra of a great number of steroids after $H_2SO_4$ treatment has been described (7).

## RESULTS

The retention times in min of a number of steroids as the TMS ethers is given in Table 1. From the retention times on the SE-30 and on the XE-60 systems, it is possible to calculate various proportions of the mixed phase system for more efficient separation of the steroids under study. Mixed phase systems have been described for non-steroid compounds (8). Hartman and Wotiz have described the relationship of retention time to the structure of steroids (9). Examples of the separation of mixtures of pure steroids as their TMS ethers on the three types of column systems are given in Fig. 1-3. Figure 1 shows the separation of various steroids on an XE-60 column. Figure 2 shows the pattern of separation of steroids on an SE-30 phase system. Figure 3 demonstrates the separation of steroids on a mixed phase system of SE-30/XE-60. The quantity of steroid can be most readily measured by the peak height method for steroids that appear relatively close to the solvent front. The relationship between peak height and amount of steroid standard, the regression lines and the correlation coefficients for a number of steroids on various column systems has been previously described (2). The total area under the peak is more satisfactory for steroids at some distance from the solvent front such as progesterone on the SE-30/XE-60 system. Figures 4 and 5 show the steroid pattern obtained with extracts of homogenate of ovaries (60 mg per 2 ml buffer). The lower trace of Fig. 4 shows the steroid pattern obtained from an extract of homogenate of ovaries from rats given two doses of 100 IU pregnant mare serum gonadotrophin over a two day period. The homogenate was incubated for 1 hr at 37 C in a Krebs-Ringer bicarbonate buffer under $O_2:CO_2$ (95:5). The upper trace of Fig. 4 shows the additional steroids synthesized when an NADPH generating system consisting of NADP + GLC-6-P was added to the incubation medium. Figure 5 shows additional steroids synthesized by the same homogenate containing added 5-pregnenolone + NADP and GLC-6-P.

TABLE 1. Retention times in min of a number of TMS ethers of steroids on three different phase systems. Gas chromatographic columns were 4 ft x 4 mm at 225 C.

|  | XE-60 | SE-30/XE-60 | SE-30 | Code |
|---|---|---|---|---|
| 17α-estradiol | 2.3 | 3.8 | 9.4 | $E_2$ |
| 17β-estradiol | 2.7 | 4.1 | 10.6 | $E_2$ |
| Cholestane | 3.0 | 5.8 | 18.4 | C |
| 17α-hydroxypregnenolone | 4.1 | 7.4 | 17.3 | 17-H-Preg |
| 16α-estriol | 4.3 | 7.4 | 19.6 | $E_3$ |
| Dehydroepiandrosterone | 4.3 | 4.7 | 7.4 | DEA |
| 5-pregnenolone | 5.9 | 6.7 | 12.1 | 5-Preg |
| Cholesterol | 6.4 | 12.7 | 37.4 | CH |
| 19-nortestosterone | 6.7 | 6.5 | 8.4 | 19-NT |
| Estrone | 6.8 | 6.2 | 8.7 | $E_1$ |
| 20α-hydroxycholesterol | 7.2 | 20.8 |  | 20α-CH |
| Testosterone | 7.7 | 7.4 | 9.8 | T |
| 19-hydroxytestosterone | 10.7 | 10.3 | 16.2 | 19-HT |
| 5-androstene-3β, 17β-diol | 10.7 | 9.6 | 9.9 | 5-A-diol |
| 4-pregnene-20β-ol-3-one | 13.5 | 12.8 | 17.4 | 20β-HP |
| 20α,22β-dihydroxycholesterol | 14.2 | 29.3 |  | 20α,22β CH |
| 4-pregnene-20α-ol-3-one | 15.6 | 14.1 | 19.1 | 20α-HP |
| 1-androstene, 3, 17-dione | 17.1 | 12.4 | 7.3 | 1-AD |
| 4-androstene-3, 17-dione | 23.5 | 16.2 | 8.0 | 4-AD |
| 16α-hydroxy-androstene 3, 7-dione | 25.8 | 18.5 | 13.5 | HAD |
| 17α-hydroxyprogesterone | 27.0 | 20.0 | 17.4 | 17α-HP |
| Progesterone | 31.5 | 22.4 | 13.0 | P |

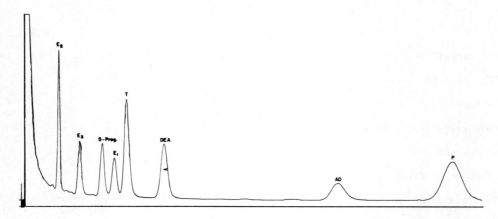

Fig. 1. Separation of a steroid mixture as TMS ethers on an XE-60 column system. The steroid code is given in Table 1.

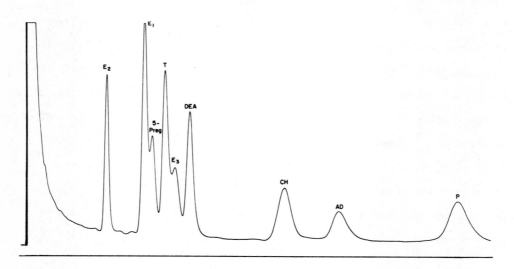

Fig. 2. Separation of a steroid mixture as TMS ethers on an SE-30 phase system.

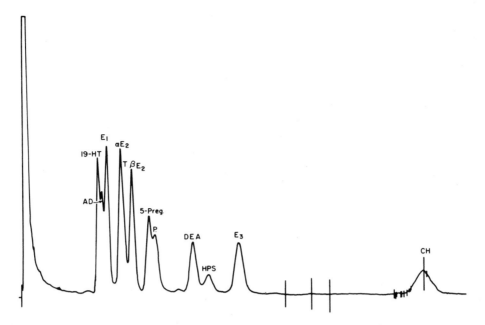

Fig. 3. Separation of a steroid mixture on a mixed phase system of SE-30/XE-60 (1:1).

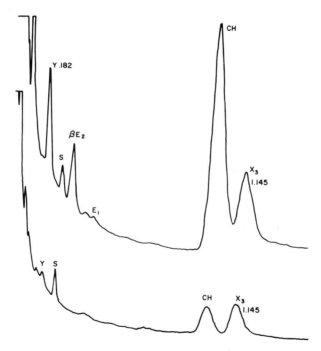

Fig. 4. GLC tracing obtained from an extract of rat ovaries. In the lower trace, ovaries were obtained from immature rats given two injections of 100 IU each of pregnant mare serum gonadotrophin. The upper trace shows additional steroid synthesized when NADP and GLC-6-P was added to the incubation medium. Reproduced from Biochim Biophys Acta, ref. 2, by permission of Elsevier Publishing Co.

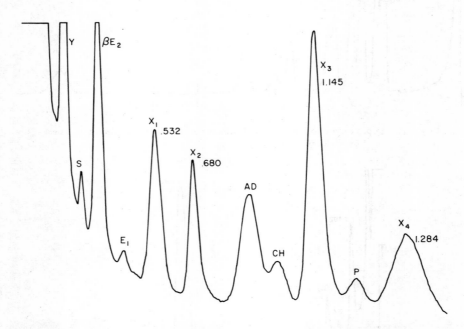

Fig. 5. Additional steroid synthesized from 5-pregnenolone by the same homogenate of Fig. 4, containing NADP and GLC-6-P. Several unidentified steroids are indicated. Reproduced from ref. 2.

## DISCUSSION

The simple method of extraction and purification of steroids that has been developed is suitable for the study of metabolic pathways of steroid synthesis in the ovary and placenta. The incubation of a small amount of rat ovarian tissue (25-60 mg) especially if previously stimulated with gonadotropins, leads to the synthesis of amounts of steroids (0.1 μg or more) that are easily measured with a hydrogen-flame ionization detector or $^{90}Sr$ ionization detector. Thus, the amount of steroid relative to interfering or contaminating substances is proportionally high. The small number of purification steps necessary to extract and measure the steroids under these conditions accounts for the high degree of efficiency of the extraction and reproducibility of the method. Furthermore, it was not necessary to separate and purify individual steroids. Thus, the total spectrum of steroids synthesized in the whole ovary, follicle, corpus luteum or placenta can be visualized on one chromatograph. The changing pattern of total steroid synthesis under various experimental conditions can be studied. Correlations have also been made between enzyme activities and steroids synthesized under various conditions of gonadotropin stimulation in the immature rat (2). Tentative identifications of the steroids synthesized can be made by various GLC procedures. Confirmation of structure depends of course on additional methods such as infrared analysis. The methods described may be applicable to the measurement of plasma steroids especially those whose levels are relatively high as with estrogens and progesterones in human pregnancy. The measurement of low levels of plasma steroids such as testosterone has required additional techniques, both in purification before GLC and in more sensitive detection devices (10).

## SUMMARY

A simple method has been developed for the extraction, separation, and tentative identification by GLC of estrogens, androgens, and progesterones from ovarian tissues and human placenta. The simple extraction and purification system gave good efficiency and reproduction of extraction. Polar, non-polar, and mixed phase column systems were used to separate a number of steroids as the TMS ethers. The method has been useful in studies on the regulation of steroid synthesis in the ovary and human placenta.

## REFERENCES

1. McKerns, K. W. and E. Nordstrand, *Biochim Biophys Acta* 82: 198, 1964.

2. McKerns, K. W. and E. Nordstrand, *Biochim Biophys Acta* (in press).

3. Luukkainen, T., W. J. A. VandenHeuvel, E. O. A. Haahti, and E. C. Horning, *Biochim Biophys Acta* 52: 599, 1961.

4. Wotiz, H. H. and H. F. Martin, *J Biol Chem* 236: 1213, 1961.

5. Futterweit, W., N. L. McNiven, and R. I. Dorfman, *Biochim Biophys Acta* 71: 474, 1963.

6. Lisboa, B. P. and E. Diczfalusy, *Acta Endocr* 40: 60, 1962.

7. Bernstein, S. and R. H. Lenhard, *J Org Chem* 25: 1405, 1960.

8. Hildebrand, G. P. and C. N. Reilley, *Anal Chem* 36: 47, 1964.

9. Hartman, I. S. and H. H. Wotiz, *Biochim Biophys Acta* 90: 334, 1964.

10. Brownie, A. C., H. F. van der Molen, E. E. Nishizawa, and K. B. Eik-Nes, *J Clin Endocr* 24: 1091, 1964.

# DISCUSSION ON ESTROGENS

DR. FINKELSTEIN: We have been working on the estimation of the urinary estrogens for about 20 years and the results which we have obtained do not always agree with the results presented here using gas chromatography.

I would like to mention a small personal point and that is the so-called Brown's partition which separates estrone and estradiol from estriol in the solvent system benzene:petroleum ether and water. This partition was published by us in 1951 and 1952 in every detail (Finkelstein, M., Nature 168: 830, 1951; Acta Endocr 10: 149, 1952). It was borrowed in 1955, I believe, by Brown (Brown, J. B., Lancet 268: 320, 1955).

With this partition method followed by further purification of the respective fractions, and estimation of the estrogens by fluorometry in phosphoric acid we obtained results similar to those of Brown. We felt however, that we overestimated the urinary estrogens when they were present at low concentrations.

By careful purification of the estrone, estradiol-17β and estriol fractions, by paper chromatography and TLC in several systems, reducing estrone to estradiol-17β, acetylation of the respective compounds and cleavage of the acetates we isolated each of the above compounds in nearly pure state (Ladany, S. and M. Finkelstein, Steroids 2: 297, 1963). The estrogens were identified on the paper or TLC by the fluorescence which they develop when excited with UV light at 365 mµ, following spraying the chromatograms with 70% phosphoric acid and subsequent heating at 85 C for 20 min.

The colors are yellow for estrone and estrone-acetate, yellow-greenish for estradiol-17β and its diacetate, and orange-pink for estriol and estriol-triacetate. The colors are clearly visible with 3 µg of each compound on the papergrams and 1 µg or less on the thin-layer plates.

The final chromatograms of the isolated estrogens from urinary extracts of menstrual cycle did not show fluorescent bands other than the three estrogens in question.

Quantitative estimation of the estrogens was performed by fluorometry using 85% phosphoric acid to develop the fluorescence. The readings were made with a Farrand spectrofluorometer. The fluorescence spectra of the isolated estrogens were identical with reference estrone, estradiol-17β and estriol. When submitted to bioassay, the activity of the isolated estrogens or their acetates (quantitated fluorometrically), was equal to crystalline standards.

For calculating the recoveries through the procedure labeled estrone and estradiol-17β were used as tracers. Losses for estriol were estimated by processing in parallel samples of urine with added crystalline estriol. The mean recoveries for all three compounds were in the range of 60%.

With this method, which admittedly is not easy to perform, we have estimated the urinary excretion of estrone, estradiol-17β and estriol in the menstrual cycle, in pregnancy and also in some pathological conditions. During the menstrual cycle the excretion of each of these estrogens rarely exceeds a few micrograms, with the exception of the "peak period". Here, the highest concentrations we have recorded were 15-20 µg per 24 hr, for estrone and estriol and about 5 µg per 24 hr for estradiol-17β. The lowest readings during the menstrual cycle were around 1 µg for estrone and estriol and considerably less than 1 µg for estradiol-17β.

Thus our figures and especially those of estriol are considerably lower than those reported using Brown's colorimetric procedure. I may also add that we could not detect any of the above estrogens in the urine of bilaterally ovariectomized patients. Similarly, in some cases of amenorrhea the concentration of estrone, estradiol-17β or estriol were too low to be detected by our method. This emphasizes the problems that must be solved when using GLC methods.

On the other hand we find a high increase of urinary estriol already about four weeks following conception. In such cases the usual concentration of estriol is 50-80 μg per 24 hr urine.

Finally, I wish to mention that in most of our cases the urine was hydrolyzed with "Glusulase" or "Ketodase". To exclude the possibility that our low figures, especially in the menstrual cycle, were due to glucuronidase inhibitors, controls were performed with aliquots of late pregnancy urine of known concentration of estriol added to the urine.

DR. ENGEL: May I comment that the fluorescent technique has a sensitivity which rivals GLC. Here is a fine opportunity for comparison of a nonspecific method of detection with a relatively specific one.

DR. FALES: I have the strong feeling that GLC is very useful for biochemical studies in following certain biochemical reactions. But after this morning's session it seems to me that people are really beginning to use it for clinical purposes.

I would like to ask the participants this morning, particularly Dr. Luukkainen, if he is actually using GLC data to make clinical decisions. Is this the case and, if so, how many other people are using it in this same sense?

DR. LUUKKAINEN: That's correct. We are using it in selected obstetrical patients. It is particularly important when the patients are receiving medication that interfere with chemical assay of estrogen. We determine estriol daily in certain patients and when the values begin to fall, a decision for operation is made.

DR. LIPSETT: Gas chromatography and fluorescence can complement each other. When we measure urinary testosterone for specific activity determinations we find it very useful to check mass measurements by GLC and fluorescence and are reassured when they agree closely.

I would like to continue Dr. Fales' discussion and ask if estriol measurements in late pregnancy cannot be made more simple by solvent extraction and either colorimetry or fluorescence.

DR. VAN DER MOLEN: I appreciate Dr. Fales' question because we have actually been trying to establish a set-up for the routine estimation of the conventional steroids in a department of obstetrics. Dr. Lipsett mentions in this connection the estimation of estriol in urine during pregnancy. Everybody who has been working in this field will realize that this estimation is not the most difficult steroid estimation. And I still believe that in a case like this you might prefer to use a simple specific fluorometric or colorimetric technique. This might well prove to be as efficient compared to the efforts that you might have to take in order to obtain a specific isolated estriol peak following GLC.

I actually would like to express a word of caution to those people expecting to find over here THE method for the estimation of estrogens in normal urine. Not because I don't appreciate the thorough studies, that have been presented, but I think we should stress that much attention has to be paid and actually is being paid, to the separation from several contaminating compounds of the relatively very small amounts of estrogens isolated from normal urines. Several different columns and derivatives have been presented for this purpose. Adequate specificity may be obtained using rather time consuming techniques and in the future more attention should be paid to the overall reliability. I do not see why GLC should not be able to compete even in the estrogen field with other methods that admittedly may carry a history of more than 20 or 30 years. I am convinced that as a result of the potential sensitivity of detection following GLC comparable to the sensitivity using the most sensitive fluorometric and colorimetric detection techniques, these other methods might eventually also be taken over by GLC, that has the added advantage of a recorded tracing indicating the specificity of the isolated compound.

DR. SOMMERVILLE: Just a quick comment. In our Institute we reserve GLC for problems where we do not already have some simple convenient method. For example, urinary estriol can be determined by a modified Brown method and the result obtained on the same day. In my opinion it would be premature to advocate the application of GLC methods for routine clinical work of this type in laboratories where expert advice is not available and especially where the equipment is not suitable for simultaneous determination of a labeled internal standard.

DR. LEVITZ: In connection with the question Dr. Engel asked earlier, although I haven't done any GLC, we have developed a fluorometric method to determine estriol in pregnancy plasma. The high school graduate or untrained medical resident can be tought the procedure in short time. The technician can easily do eight analyses in the working day. At the stage of purification where either gas phase chromatography or fluorescence analysis may be applied, we estimate one technician can do about 12 determinations in 1.5 to 2 hr. Can this be topped by gas phase chromatography?

DR. GOLDFIEN: Dr. Touchstone, do you have any data using this method so that we could have some estimates of the levels which you are measuring in plasma?

DR. TOUCHSTONE: Although this meeting was supposed to be on GLC methods only, I did bring some slides of plasma values with me. Table 1 shows the average estrogen levels obtained from eight different determinations of estrogen fractions in maternal blood and cord blood. The maternal blood contains considerably more estrone sulfate than the cord blood.

TABLE 1. Estrogens in plasma during pregnancy.

Estrone in blood plasma (µg per 100 ml)

|  | Cord | Maternal |
|---|---|---|
| Sulfate | 0.8 ± 0.2 | 3.7 ± 0.7 |
| Gluc. | 0.4 ± 0.1 | 0.4 ± 0.1 |
| Free | 1.1 ± 0.4 | 0.3 ± 0.1 |

Estradiol-17β in blood plasma (µg per 100 ml)

| Sulfate | 1.4 ± 0.3 | 2.9 ± 1.3 |
|---|---|---|
| Gluc. | 0.7 ± 0.3 | 1.0 ± 0.3 |
| Free | 0.8 ± .4 | 1.7 ± 0.4 |

Estriol in blood plasma (µg per 100 ml)

| Sulfate | 102.0 ± 9.8 | 3.6 ± 1.9 |
|---|---|---|
| Gluc. | 26.1 ± 5.3 | 3.2 ± 1.2 |
| Free | 7.5 ± 5.3 | 1.0 ± 0.2 |

We have wondered for a long time whether incubation of estradiol in maternal blood would convert estradiol to estrone. We have in the past months incubated estradiol in maternal plasma and there is evidence that some estradiol is converted to estrone. This is on the order of only one or 2%.

There is more free estradiol in the maternal blood than in cord blood. Estradiol sulfate is found in large quantities.

We are all aware that the major estrogen in cord blood is estriol sulfate. There are still considerable amounts of the free estriol present. The total conjugated estrogens which we have found agree amazingly well with the total estrogen determinations reported by three other investigators who used acid hydrolysis prior to extraction.

DR. WOTIZ: Dr. Touchstone, I am a little concerned about the statement you made that there is greater specificity with a mixed phase column. I think that the use of polar and non-polar and selective phases alternately rather than mixed would imply much greater specificity. This has been our experience in proving or disproving identity of steroids. Such evidence is coupled to the use of either free estrogens and their various derivatives on all of these columns.

There is a further point. I don't know whether you would take issue with me on it, but, considerable implied specificity can be gained by the use of a high efficiency column. This would require low stationary phase concentration resulting in three or four thousand theoretical plate columns, rather than mixed phase columns which, as a rough estimate from your slide had 100-200 theoretical plates. Do you have any evidence that the mixed phase column is in fact more specific?

DR. TOUCHSTONE: Actually, I am sure it all depends on the definition of the word "specificity". As we pointed out before, single phase columns give poor separations in many of the gas chromatograms. We have found as many as six to eight different steroids in a single peak of a gas chromatogram whereas when we use mixed phase columns we can cut it down to one or two. It has been pointed out that two different columns in series give the same results as the same columns mixed together.

Now, whether this is increasing specificity is a matter of argument. It would appear that more confidence can be had in the results when the mixed phases are used.

DR. FALES: I think there is really no great disagreement. The point simply is that if you are going to separate two compounds, you can get a certain separation with either one phase of another phase, and in this sense the separation of two particular compounds on a mixed phase by definition cannot separate them better than it could on either one phase or the other phase. It is pretty much the same as running two columns hooked in series without a detector in the middle. You may obscure the resolution in the second column that you gained in the first column.

On the other hand, if you have a mixture of two or four materials, the optimum separation may indeed be achieved with propitious choice of the two phases or three phases or four phases.

DR. KARMEN: I have two questions for Dr. Eik-Nes and his group.

In your description of the electron capture technique you stress the importance of pulsating voltage. Did you try the simpler approach using direct current, and was the pulsating voltage necessary? The second question is, how much cleanup have you avoided by using chloroacetates?

DR. EIK-NES: I will take care of Dr. Fales' second question and let Dr. van der Molen answer the first question since I was in Australia when Drs. Brownie, van der Molen and Nishizawa discovered the usefulness of pulsating voltage for electron capture detection.

In our method we have a very high degree of sensitivity and the only solution we have found in cutting down background noise is to apply the final sample extract onto the GLC column in a high state of purity. Over the past year we have observed that the purity of the solvents used in our methods is crucial, benzene has to be purified extensively; if not the residue of 10 ml benzene carried through our testosterone, progesterone, or estradiol method will give peaks interfering with the calculation of the sample peaks. I have thus little hope that steroid methods using electron capture detection in association with GLC will permit adequate assay of biological samples unless such samples are free of competing signals from non-steroidal compounds. I discussed the problem of the Atlantic Ocean yesterday and if the tracings of a sample purified by our methods is of that quality, I at least would not date to calculate a sample concentration but would throw the tracing in the wastebasket. If, in future work, we will be able to work with detector systems of higher sensitivity than those currently available and thus, with rationale, could try to measure plasma levels of estradiol in the dog and in the rat during the menstrual cycle, I am convinced that such investigations can only be done on highly purified plasma samples. At least up to now, electron capture detection of steroid-chloroacetates has not proven to be specific enough to ignore sample purity.

DR. VAN DER MOLEN: I am by no means an expert on the construction of electron capture detectors. As is well known, it has taken quite a lot of effort from the manufacturers to build some good detectors and even now I personally have the feeling that you might have trouble in finding a good electron capture detector and, if you had bad luck, might not find a good one at all. Maybe we should direct these complaints to the manufacturers.

When we started our investigations there were only one or two detectors available on the market and they were operating on a direct voltage basis. When we began measuring chloroacetylated compounds in this way, we obtained unstable and drifting baselines and it was impossible to obtain quantitatively reproducible results. It was at that time that Lovelock (<u>Anal Chem</u> 25: 474, 1963) advocated the use of a pulsed sampling detection technique. We therefore modified our detector to use the pulsed sampling technique and subsequently obtained quantitatively reproducible results with the compounds described earlier.

Over in Europe I am using a commercially available electron capture detector that has been built for operation on a pulsating basis and we are most happy with this one. We have no trouble with anomalous responses or drifting of baseline. But I am sure that other people may also want to comment on this.

DR. VANDEN HEUVEL: I want to say several things. You can obtain a great deal of structural information from the use of both non-selective phases such as SE-30 and selective phases such as QF-1. Different stationary phases yield different kinds of information. In this case I agree with Dr. Wotiz; use a wide variety of columns.

But in addition to obtaining structural information, you should also achieve your separation. This will be determined with selective phases mainly by the functional groups on the steroid but molecular size and shape are also important.

In general, the more efficient the column, the better. Let's not use three ft columns where we can use six and 12 columns.

In our experience in Houston and Bethesda, we were able to separate every pair of steroids that was given to us, either through the use of selective stationary phases or through the use of derivatives which accentuate functional group differences.

DR. EIK-NES: It is fine to be able to separate authentic steroids on GLC. Those of us who are left with the unattractive task of purifying plasma or urinary steroids cannot predict at what stage of purity our samples are ready for GLC. As already stated, sensitive detectors require biological samples with negligible contamination of non sample material. The fact that six synthetic steroids will separate on column so and so does not necessarily indicate that this column is good for the separation of three of these same steroids when extracted from a biological sample. An impurity, difficult to remove, may have the same retention time as one of the steroids, and you have either to get rid of this impurity or also forget about using GLC with a sensitive detector as analytical endpoint.

Some of us lived through the early time of paper chromatography of steroids and can now exchange experience on published systems for paper chromatography of authentic steroids which could never be applied to the separation of the same steroids when extracted from blood or urine. Maybe the time has come when authors of articles dealing with the separation and quantification of synthetic steroids by GLC also should try to investigate what separation and what quantification can be achieved when the same steroids are extracted from tissue, blood or urine. I am sorry to express this opinion, but I feel that this suggestion would greatly add to the usefulness of observations made on synthetic steroids and their retention times on a new column.

DR. CRANE: I would like to show one slide to help to stimulate further discussion on the difficulty of purifying the extracts and keeping them free of contaminants.

You will notice three chromatograms on this figure. The one on the left is that of 10 μg of etiocholanolone. The chromatograph in the center is that of the same sample of etiocholanolone after paper chromatography on Bush system A to get rid of that small peak just before the major peak of etiocholanolone.

The chromatogram on the right is that of a sample from a patient. The standard etiocholanolone had been chromatographed on the same paper at the same time as the patient's sample.

You will notice there is a new peak which has appeared in standard etiocholanolone at 4 min, and it is comparable to one in the patient's sample at the same time. By paper chromatography we were able to get rid of that small peak at 9 min, but we introduced some others. I would like to hear a discussion on how to purify chromatography paper and solvents along with the discussion of purification of samples by thin layer and GLC.

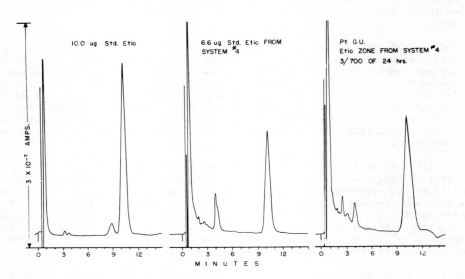

Fig. 1. Chromatography of standard etiocholanolone before and after paper chromatography on system No. 4 (See Table 1 of article by M. G. Crane and J. J. Harris). Column conditions: 1% XE-60 on Gas Chrom P in 1/4 in x 20 in stainless steel tubing; flash heater 35 C, detector 270 C, column 250 C; argon gas 110 ml per min at 15 psi.

MISS PATTI: Dr. Crane, we have had the same problem with extraneous peaks on chromatograms of eluates obtained from column and TLC. The same eluting system was used for each procedure.

Although purification and distillation of the reagent grade solvents removed most of the interfering contaminants, one extraneous peak remained. Unfortunately, the retention time of the impurity was similar to one of the steroids being studied. We decided, therefore, to analyze by GLC a "blank" corresponding to the steroid eluate obtained from TLC, column, and even paper chromatography. In this way, one can account for the interference of the solvents and, perhaps, even of the supports (absorbents, etc.) employed.

Dr. Karmen, I would like to ask you some questions about electron capture detection. I spent the past three years attempting to find phases suitable for our work. Finally, I found three which work well with hydrogen flame and ionization detectors. After observing the data presented here, I am a little reluctant to become involved with electron capture.

Recently we investigated the acetylation of testosterone employing both acetic anhydride and the mono chloro analogue. The acetates were analyzed on SE-52 with hydrogen flame detector. It was found that .002 μg could be detected. I would like to know what electron capture would offer me if I can determine nanogram amounts with hydrogen flame? Is it truly as specific as everyone says? I was under the impression that when one went to GLC that one left the radioactive isotope dilution technique with paper and TLC. If I have to use radioactive materials with GLC, what does GLC offer me?

DR. KARMEN: I can't answer all those questions. However, people who have worked with pesticides have used electron capture detection for two purposes. One, to detect a compound like DDT in the presence of a very murky biological mixture such as the extracts of foodstuffs without as much cleanup as would be necessary to detect the same amount of DDT using a hydrogen flame detector. For this reason the use of electron capture is very widespread and it has become a very useful tool.

The problem we would like to get the answer to is: when we take a biological extract serum or try to use electron capture for the detection of chloroacetates, is there a sufficient concentration of electro-negative compounds in the urine itself to interfere?

The second question is: are we making the electro-negative compound in significant concentration?

What is the benefit of using a double isotope technique when you can do this with paper? Hopefully, GLC offers higher resolution than many of the other kind of chromatography. As such, it should materially increase your confidence in the answer that you get, using a double isotope technique.

With respect to the electron capture detector and its relative sensitivity, I would hope Dr. Eik-Nes and his group would tell us a little bit more about the level of sensitivity that one might expect. For example, how would you compare the sensitivity you would get after the cleanup procedures you use if you substituted a hydrogen flame detector?

DR. EIK-NES: I would like to comment on Miss Patti's question about estimating specific radioactivity by the technique of GLC. Maybe I did not make this point clear in my presentation.

If you infuse the dog ovary in vivo with two labeled precursors like progesterone-$^{14}$C and 17α-hydroxypregnenolone-$^{3}$H or incubate ovarian homogenates with these very same radioactive steroids, you will find that both isotopes are present in the intermediates as well as in the final product, in our investigation: estradiol. Two ways are open for assessing biosynthetic capacity of the ovary, either by measuring the amount of both isotopes present in the biosynthesized compounds or also by measuring the specific radioactivity of the steroids produced. The latter investigation requires a technique for the estimation of steroid mass, the concentration of some of the intermediates is often very low. The most adequate technique for mass estimation of small amounts of steroids utilizes isotopes of different energy spectra, one isotope is added to the sample to correct for losses occurring during processing of the sample, and the other isotope serving as a reagent to quantitate the steroid to be measured. This technique of mass estimation cannot be applied to samples already containing $^{14}$C and $^{3}$H. I don't care how steroid mass is being estimated in such experiments but the principles of isotope dilution cannot be applied. We have found that the estimation of steroid mass in plasma samples containing $^{14}$C and $^{3}$H from infused labeled substrates, can be done by electron capture detection following separation of the compound on the GLC column. The concentration of steroid available for mass estimation is low in these samples, if you have detection systems with better sensitivity than the electron capture cell, go ahead and apply them. What I would like to point out is the fact that in certain areas of research the most sensitive methods for estimation of steroid mass cannot be used and for investigators interested in the problem of multiple pathways for steroid formation, the technique of GLC with sensitive detectors offers an excellent tool for obtaining additional information on the relative utilization by endocrine tissues of steroid substrates with different radioactive labeling. We have used this technique to estimate the amounts of $\Delta^5$-pregnenolone, progesterone, dehydroepiandrosterone, androstenedione, testosterone and estradiol biosynthesized by the canine ovary and testis from $^{14}C$ and $^{3}H$ containing steroid precursors.

DR. VAN DER MOLEN: We have compared for several compounds the sensitivity of electron capture detection with that of flame ionization detection (Fig. 2). This figure shows sensitivities obtained with testosterone, testosterone-acetate and testosterone-chloroacetate using the flame ionization and electron capture detector under identical conditions and sensitivity settings for all three compounds. Electron capture was clearly more sensitive. It may be difficult, however, to compare these results with the absolute sensitivity that Miss Patti is claiming for her flame ionization detector. If we should try to detect the small amounts of testosterone isolated from plasma in its free form using the electron capture detector, it is clear, that we would not obtain any signal at all under the conditions used here.

Though we have been talking mainly about urinary matters this morning, I may add that it is our impression, in using GLC for the final estimation, that it is a lot more difficult to obtain specific isolation of free testosterone from urine than it is for plasma using the chloroacetate derivative.

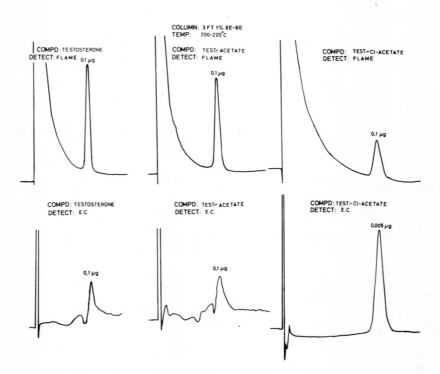

Fig. 2. Comparison of sensitivity of electron capture with flame ionization detector.

DR. LIEBERMAN: Twice this morning the question of the high resolving power of GLC has arisen and I wonder whether one of the experts could comment on the following question.

The property of GLC which we all believe is superior to the other forms of the chromatography is the property of the detector. Two or three months ago there appeared a paper in <u>Analytical Chemistry</u> in which a mathematical analysis was made of the resolving powers of the GLC and liquid-liquid chromatography, and the conclusion of that paper was that GLC was not as good a resolving tool as was liquid partition chromatography.

I heard Dr. Wotiz say that he thought one of the columns that he saw had a theoretical plate number of only 400 or so. Even if the number of theoretical plates was as high as 2,000, I think 2,000 theoretical plates is not at all beyond the possibility of liquid-liquid partition chromatography.

DR. VANDEN HEUVEL: I don't want to become involved in a discussion of the theory of chromatography. I am afraid most of us wouldn't understand that paper, anyway. However, we should remember that in GLC we do have a distribution or partition between the two phases. I suspect it may be easier to prepare a 2,000 plate gas chromatography column than to prepare a 2,000 plate liquid-liquid partition column. Furthermore, it is not terribly difficult to make five or six thousand-plate GLC columns, and with these one does observe excellent resolution coupled with a short retention times.

DR. BROWNIE: I would like to make a simple statement. In our particular method it is obvious that the most important factor that makes us use GLC is the sensitivity of the detection technique available.

I think Dr. Karmen was giving the impression that we were doing a tremendous amount of purification before we applied the material to the column. Although we really think we are doing efficient purification, after the chloroacetylation we do only one TLC. This seems to me to be a simple prepurification.

DR. WOTIZ: I just wanted to add a comment in response to Miss Patti's question. We have been doing some work with electron capture detection of plasma testosterone. Under the best circumstances we can detect .0002 to .0003 µg. With a flame ionization detector the lower limit of sensitivity would be between .01 and .03 µg, depending on the retention time, and the nature of the derivative. Thus there is a definite advantage to the use of the electron capture detector if you are inclined to measure such very low concentrations of steroid.

MISS PATTI: In reply to Dr. Wotiz's comment about the low levels that we try to measure, it is not that we desire to work at low level but, unfortunately we are working with small children. The biological specimens obtainable are small. We work with one or at the most three of four ml of plasma. In order to work up these specimens adequately for GLC analysis, it is necessary to employ a lengthy purification procedure. I am sure that anyone who has worked with plasma is fully aware that it is more difficult to process than urine. For the steroids of interest to us, it is necessary to include a series of purification steps. When one is working with small amounts of plasma, it is important to select suitable purification without losing too much of the material for subsequent quantitation with GLC. However, the preparation of urinary extracts can be readily accomplished by differential hydrolysis and extraction because of the larger amount of starting material.

Initially we use column purification which is usually followed by TLC. GLC analyses are performed on eluates from both procedures. If we do not detect steroid peaks which are sufficiently separated from the solvent front and adjacent steroid peaks, the extract remaining is subjected to paper chromatography. For the precise quantitation in which we are interested, it is necessary to get a peak that is well separated from adjacent peaks and from the solvent front. I am a little dubious about trying to quantitate a peak that is separated from another one by only one minute of more.

We do not use GLC for routine clinical work. We have in isolated instances used it clinically for routine analysis of DHEA and urinary cortisol.

DR. VAN DER MOLEN: I should like to make a final remark, at least as far as I am concerned, about the problem that Miss Patti brought up here. In our hands, and I don't know if we are the only ones, it definitely has been easier to purify plasma extracts for electron capture detection following chloroacetate formation of steroids, than it has been to purify urine extracts. This may of course be different with the problems that you are studying Miss Patti. If I may compare the estimation of plasma testosterone that Dr. Brownie has described to us and the estimation of urinary testosterone published by Dr. Lipsett's group, a series of 10 plasma estimations can be done in two or three days. For the urine samples we need at least seven to ten days, as a result of the larger volumes that have to be proceeded and the paper chromatographies that have to be used in order to obtain extracts that contain the pure compound.

DR. LURIE: I would like to describe a dry injector system which is very simple and highly accurate in operation. The instrument used was made from the wire stilette of a wide bore aspiration needle. The stilette shaft diameter is very slightly less than 1/8 in. The stilette shaft was filed down and grooved as shown in the diagram (Fig. 3). It is important that the trough or groove should start proximal to the carrier gas inlet and that the concave surface of the trough itself should face the carrier gas inlet when inserted.

The other components of the injector are a rubber septum with a hole pierced through it to take the stilette shaft; and a spring steel retainer bent to hold the needle tightly in place. The trough, which is shown in Section AA in Fig. 3, was made in order to accomodate a larger volume of organic solvent for evaporation. Evaporation of the organic solvent can be done under an ordinary light bulb or at room temperature.

In spite of a brief removal of the rubber injector septum when the dry applicator is inserted, carrier gas flows through the column since the Perkin-Elmer gas chromatograph presently in use has a proportional gas flow control which maintains sufficient pressure. When the device and sample are inserted, the device is left in position until the next sample is ready for chromatography. Two such devices used concurrently have proved convenient.

It is possible to achieve remarkable accuracy with this dry applicator. In 16 consecutive injections of 0.1 µg progesterone, a mean peak area of 5.41 sq cm was found and a standard deviation of 0.0306. Furthermore and in spite of removal of septum to insert the dry injector the retention time in all the 16 consecutive injections described above was exactly identical.

The advantages of the dry injector described are that it allows the application of the whole of the extracted sample quantitatively. It has far greater accuracy and reproducibility than fluid injection. There is no solvent front (Fig. 4) and it is quick (about 15 min). The injector is simple to make and maintain in good order. Finally, it can be adapted to any kind of injector port. After continuous use for a seven month trial period, it is felt that the instrument can be recommended for more general use.

DR. WOTIZ: I would like to go back for a moment to the subject of solid injection and show just one slide. Figure 5 is a chromatogram of cholesterol and pregnanediol. Cholesterol has been used by several investigators as an internal standard.

The injection block was kept at 300 C. The left side of the picture shows a chromatogram when the glass-liner was in position in the injection port. The right-hand side shows a chromatogram under the same conditions but with the glass sleeve from the injection port removed, allowing exposure of the steroid to the hot stainless steel injector. The resulting steroid breakdown is apparent. Perhaps Dr. McNiven's silver injection tip would be less deleterious.

MR. THOMAS: I would like to ask Dr. Touchstone a question about his dual phase column. He has 10% QF-1, a not very stable phase and 5% of another phase. I would like to ask, how much of these phases are left on the column after baking?

The column I described yesterday, of the mixture, of two substances which have much higher thermal stability but even these would not stand the temperature that you mentioned.

DR. TOUCHSTONE: According to the manufacturer, QF-1 is stable to 275 C. However, we have heated QF-1 columns to 300 C for as long as a day and we are still able to use them.

I believe Miss Patti said she uses a temperature of 300 C for 5 hr to condition a QF-1 column. Most of these other phases, SE-30 and other silicones are stable up to 300 C.

We have made a point to screen these substrates so we can add temperature stability. Most of our estrogen determinations are performed between 240 and 250 C. Therefore we can condition these columns between 270 and 300 C and use them at 250 C with very little bleeding over a period of as long as four months.

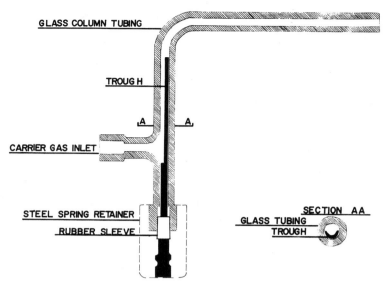

Fig. 3. The figure shows the dry injection device inserted in the inlet of the glass column. It is important to note that the trough begins proximal to the carrier gas inlet and that the concavity of the trough faces the carrier gas inlet. The steel spring retainer in this case is made to catch around the collar of the glass injection port and a gas tight fit can be maintained at pressures above 60 psi over long periods.

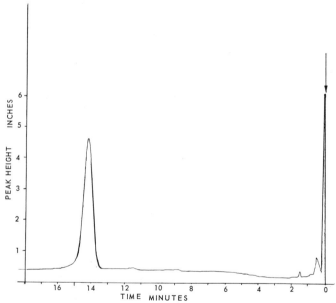

Fig. 4. The figure shows the tracing of a typical chromatogram when the dry injection device is used. Note the symmetry of the peak and the absence of the usual solvent front present in fluid injection. In this case the peak is free progesterone 0.3 µg on XE-60 3% Gas Chrom P, col. temp. 230 C. Injector temp. 280 C. Helium carrier gas 50 psi. Hydrogen flame detector, Perkin-Elmer model 801 gas chromatograph.

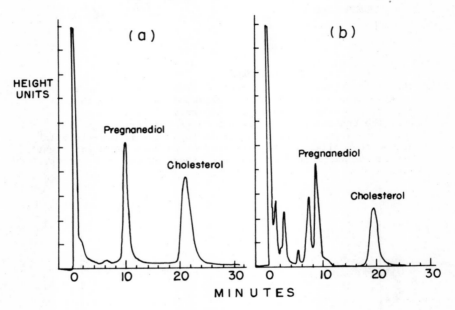

Fig. 5. Left hand panel - chromatogram using solid injector with glass sleeve. Right hand panel - chromatogram of the same steroid mixture with exposure of steroids to hot steel surface.

MISS PATTI: We buy QF-1 on Anachrom ABS from Analabs and the maximal temperature stipulated is 300 C. We condition our 3% QF-1 columns at 290 C for five hr with a gas flow of 10 psi. The column can be operated between 210-250 C with the carrier gas (nitrogen) between 203 psi.

The current QF-1 column has been in use for five months and we have observed no significant decomposition of the phase. Now to determine when a column loses its efficiency, it is necessary to analyze a selected reference standard daily and determine the area and the retention time under similar experimental conditions. If there is any change in the retention time or area and if it is necessary to change the working condition, one can conclude that the column is slowly decomposing.

About the question of measuring steroids in urine or plasma, I do not like to become embroiled in any controversy. In our experience with plasma, the estrogens have given us the least problems. We are not working with the conventional 17-ketosteroids. We have done some work with sulfates of androsterone and DHEA. We have been unable to detect etiocholanolone sulfate or the free steroid. It is necessary to remove many contaminants before we can separate these three conjugates and a variety of $\Delta^5$-compounds. We have had problems with plasma 17-ketosteroids and plasma pregnanediol whose isolation has not been adequately investigated.

DR. VAN DER MOLEN: I would completely agree with Miss Patti as far as the estrogens during pregnancy are concerned. I don't know about anybody who has successfully estimated estrogens in plasma of normal females and males using GLC. Even with double isotope labeling techniques it is quite a job and you have to use large volumes of plasma; if I recall properly, the group in Copenhagen (Svendsen and Sorensen, Acta Endocr (Kbh) 47: 245, 1964) recently obtained estimates of the of 0.01-0.1 μg per 100 ml plasma.

DR. EIK-NES: We have worked with methylated estradiol. The reason for this is the fact that when we tried to crystallize ovarian vein blood estradiol biosynthesized from infused 17α-hydroxy-progesterone-$^{14}$C and dehydroepiandrosterone-$^3$H to constant specific radioactivity, we lost 50% of the radioactivity during the first crystallization. The starting material, estradiol-$^{14}$C,$^3$H had been purified by extensive paper chromatography. The lack of radiochemical purity of estradiol was a major headache in our laboratory until Dr. Engel visited us and said: "Try to make the methylated derivative." We followed Dr. Engel's suggestion and had no more trouble in crystallizing estradiol biotransformed from labeled steroids by the canine ovary. I doubt that you can to too much with ovarian vein blood estradiol at least in the dog, unless you take the trouble for forming the methylated derivative of this compound before any attempt of steroid quantification is made.

DR. KARMEN: As far as the benefits of using solid versus liquid injection, a few years ago we went down the path with solid injectors. They are a lot less convenient to use than liquid injectors.

You have seen how it can be done and how it can have strong advocates. One of the things I would like to suggest is that carbon disulfide be used in conjunction with the hydrogen flame detector since this detector does not respond to carbon disulfide. If you can get the material to dissolve in it, which is probably the major problem, you can end up with no solvent peak whatsoever. This is one way of using a liquid injection system with all its convenience. Incidentally, carbon disulfide is toxic.

During these two days we have noted that a lot of people are bothered by large solvent fronts but nobody seems to have applied temperature programing. One of the ways to separate the steroid peak from the solvent peak is by temperature programing.

DR. ENGEL: I would like to ask a question and make a comment. I am both puzzled and disturbed and I am sure Dr. Gallagher shares my sentiments on this, that in all of the discussions of urinary estrogens, that no mention was made this morning of one of the major metabolites, 2-methoxyestrone. No one has mentioned this compound this morning. I would like to hear some comment to this.

DR. WOTIZ: Sometime ago we published a method for the determination of several estrogen metabolites, including 2-methoxyestrone. This method is useful for pregnancy urines including the early stages of pregnancy. We ran into occasional difficulties when the method was applied to non-pregnancy urines. The thin-layer chromatograms necessary for preliminary clean-up were overloaded because we were forced to use too large a urine volume.

DR. SOMMERVILLE: I entirely subscribe to Dr. Eik-Nes' comments about the importance of careful preliminary purification. At the same time this need not be extremely time-consuming and I should like to put in a word for two-dimensional TLC. In our radiochromatographic technique for progesterone (<u>Nature</u>, 1964) preliminary purification depends upon two-dimensional TLC and the entire procedure takes only 3 1/2 hr. I am sure that Dr. Lieberman will agree that this is a very different story from the time involved with the early partition columns which were used in the late 1940's. We also find two-dimensional TLC of great value for the purification of plasma extracts for estrogen determination. The steroids (with labeled internal standards) are located by autoradiography of the TLC plate and this is a safer procedure than extrapolation from side standards since the $R_f$ may be affected by the presence of impurities in these biological extracts. I should also like to emphasize the sensitivity of the Argon detectors which are available in England and the fact that sensitivity is maintained by regular cleaning of the detectors (this costs 15 dollars in England as against 100 dollars in the U.S.A.).

# SOME ASPECTS OF THE CHEMISTRY OF GAS-LIQUID CHROMATOGRAPHY

W. J. A. VandenHeuvel, Merck Institute for Medical Research, Rahway, New Jersey.

It is unfortunate that Dr. Horning could not be here because it is thanks to his efforts that this field exists and we are holding this meeting. Many of the speakers of the last two days have either worked in his laboratory for periods of weeks or more, or at least have drifted through for a few hours to pick up some of the information that could be found there.

One of the favorite words in his laboratory is "incredible." It is used to describe everything from cold weather to the antics of the Colt 45's. I must admit that after having seen some of the slides presented in the last two days I was tempted to say "incredible." But after having heard of the formidable obstacles that are faced by the many people here who are using GLC to analyze very small amounts of compounds in very complex mixtures - I can appreciate their problems. On the other hand, perhaps this morning I could spend a little time showing you what GLC can do.

I think we will at least see how GLC can give a great deal of structural information. As you can tell, I am fully committed to GLC as a method of analysis. It is essentially the chemistry of this method that makes it work or allows a failure to occur. Nearly everyone in the early days held that steroids could never come through a hot tube. This is because, as everyone knows, many steroids are labile, due mainly to the fact that heat plus light or oxygen cause decomposition. In a GLC column you have neither light nor air. Your carrier gas is argon or nitrogen, so all one has is heat. It is usually only when light and/or oxygen are present that you run into trouble.

Furthermore, when one is dealing with microgram or submicrogram amounts of steroids and injecting them into a column, one is working with an essentially infinitely dilute or ideally dilute solution. There are no strong interactions between the steroid molecules such as occur in melting point tube or in a distilling flask, but only interactions between the steroid molecules and the stationary phase. These interactions are much weaker than the former type, and therefore one observes greater volatility than might be expected, and indeed these compounds come through a column quite well.

Now that we have talked a bit about steroids in the gas phase, let us turn to their analysis by GLC. This method is a separation method. You will recognize much of the information in Table 1. You can effect separation by having an efficient column, by preparing derivatives, or by the choice of the stationary phase. What we are dealing with are the relative volatilities of the various components of mixtures. These volatilities are determined by the interactions between the steroid molecules and the stationary phase, and these are based on the chemical nature of the steroid and the chemical nature of the stationary phase employed. Separation patterns can be altered effectively by changing the stationary phase.

TABLE 1. Separation methods.

1. <u>Choice of phase</u>

    Non-selective:  molecular size
                    molecular shape

    Selective:      functional group retention

2. <u>Derivatives</u>

    Functional group transformations

3. <u>Column efficiency</u>

    Increase in theoretical plates

Basically, there are two kinds of stationary phases. The so-called non-polar or non-selective types, effect separations based on molecular size, weight, or gross molecular shape. The polar or selective stationary phases separate solute molecules on a functional group basis also.

A phase like SE-30 is very useful for separating certain types of steroids. It can separate A/B cis-trans isomers quite readily. You can separate pregnenolone from progesterone with SE-30 because the latter possesses a conjugated ketone. You can separate them both from cholesterol because of the difference in molecular weight. But I would say it is quite futile with an SE-30 type phase to try to distinguish between compounds that are similar in molecular weight and functional groups. It is difficult to separate testosterone from androstenedione or androsterone from dehydroisoandrosterone. You don't have significant differences in structure and hence there are no significant differences in volatility.

On the other hand, stationary phases are available which are selective. Thus androsterone and dehydroisoandrosterone can be separated on neopentylglycol succinate, and the fluoro-silicone QF-1 will readily separate testosterone from androstenedione.

In Table 2 we see some of the stationary phases that are in widespread use; many polysiloxanes, polyesters and modified phases are available. There are selective phases and non-selective phases, and often each phase exhibits a desirable separating quality. With so great a variety of stationary phases one can effect many separations and also obtain considerable amounts of structural information.

TABLE 2. Liquid phases for steroid separation.

1. **Non-selective**

    SE-30, JXR, F-60

2. **Selective**

    | QF-1 | NGS, NGS-PVP |
    | CNSi (XE-60) | CHDMS, CHDMS-PVP |
    | PhSi | EGSS-Z |

3. **Intermediate selectivity**

    JXR-CHDMS

One may also modify separation patterns by preparing derivatives of functional group-containing steroids. The derivativatization of keto and hydroxyl groups can result in large changes in volatility, often leading to improved separations. As can be seen in Table 3, many derivatives are in use.

Derivatives can also be used to impart specific detection properties to sterols (as exemplified by the work of Dr. Karmen and Dr. Eik-Nes) and to improve GLC properties (reduce irreversible adsorption). We shall return to this discussion of phases and derivatives in a minute.

I would like to turn to the components of the instrument itself. Now, what is the heart of the instrument? The heart is the column. The column contains the column packing. The column packing is a support, hopefully inert but in fact usually not, upon which is coated the stationary or liquid phase.

The next three figures will show chromatograms. In each case the stationary phase was F-60. If one coats the liquid phase onto a nondeactivated (non-acid washed) diatomaceous earth preparation, the resulting packing will be of poor quality, as can be observed in Fig. 1.

TABLE 3. Derivatives.

<u>Alcohols</u>

    Acetyl, propionyl, butyryl
    Trifluoroacetyl, pentafluoropropionyl, heptafluorobutyryl
    Chloroacetyl
    Methyl ether
    Trimethylsilyl ether
    Methyl carbonate, methyl thiocarbonate

<u>Acids</u>

    Methyl ether

<u>Ketones</u>

    N,N-dimethylhydrazone
    N-aminopiperidine, N-aminohomopiperidine
    Pentafluorophenylhydrazone
    Methoxime

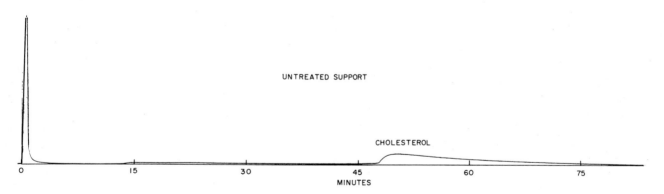

Fig. 1. GLC of cholesterol on non-deactivated support coated with F-60.

Here we have cholesterol. It looks like a whale surfacing. Note the great amount of trailing due to irreversible adsorption. Such a column cannot be used for qualitative or quantitative work with free sterols. In addition, notice the rise in baseline at 15 min. Some of this cholesterol is dehydrating to the olefin. On the other hand, if one chromatographs the TMS ether of cholesterol, a very different result is seen in Fig. 2.

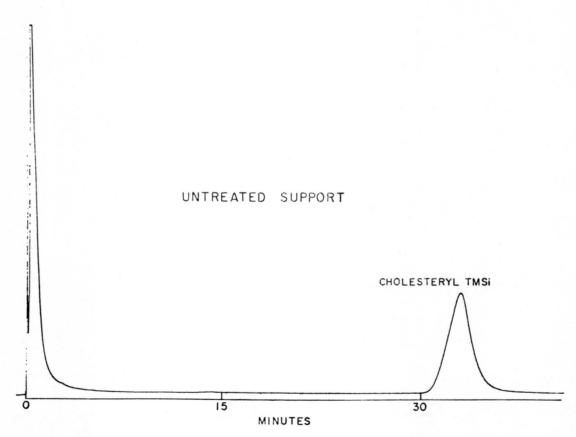

Fig. 2. GLC of cholesterol-TMS on non-deactivated support coated with F-60.

Under the same conditions, and with the same column, the peak resulting from the derivative, while not of completely theoretical shape, is a remarkable improvement over that for the parent sterol. We have gone from a polar sterol to a non-polar derivative, nearly hydrocarbon-like in its properties. It is clear that with respect to irreversible adsorption the chemical nature of the solute molecule is of great importance.

In Fig. 3 we have the results from the application of a sample of cholesterol to another column. This was a support that had not been acid washed, but had been silanized. Again, the peak is poor. Cholesterol should actually be eluted at 20 min. Complete dehydration has occurred and the peak is due to the dehydration products rather than the alcohol. The need for the use of properly deactivated supports is clear. Acid washing followed by silanizing is particularly satisfactory when plysiloxane liquid phases are to be applied to the support.

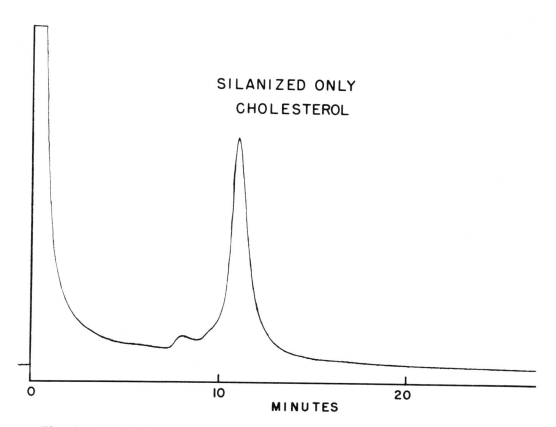

Fig. 3. GLC of cholesterol on non-deactivated silanized support coated with F-60.

Figure 4 shows the effect of the treatment of the support (neopentyl glycol succinate) upon the separation of the TMS ethers of four steroids, androsterone (5α), etiocholanolone (5β), dehydroisoandrosterone ($\Delta^5$), pregnanediol (Pd), with neopentyl glycol succinate. Both columns contained packing prepared from acid washed (con. hydrochloric acid) support; however, the packing used to give the chromatogram on the right was prepared from support that had also been silanized with dichlorodimethylsilane. The normally observed separation pattern for these four steroids is seen in the right panel. If one uses the NGS packing from the acid washed (only) support, the elution order for 5α and Pd is reversed. It is clear that not only the stationary phase but also the nature of the support can influence separation patterns.

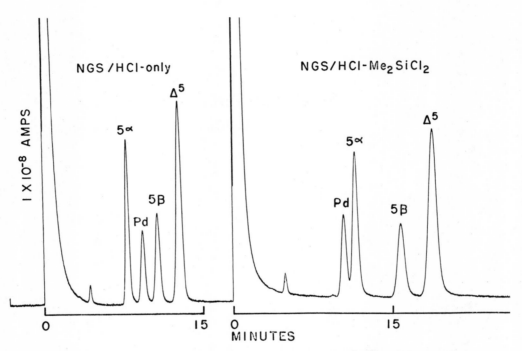

Fig. 4. Effect of treatment of support. Left-hand panel - acid washed support, F-60 phase. Right-hand panel - acid washed and silanized support.

5α = androsterone
5β = etiocholanolone
$\Delta^5$ = dehydroisoandrosterone
Pd = pregnanediol.

In Fig. 5 the upper chromatogram was obtained with a CHDMS stationary phase coated on acid washed and silanized support. The first peak is androstane-3,17-dione, and the peak just to its right is androstane-3β,17β-diol. The last peak out toward the right is testosterone. You don't observe much of a separation between the dione and the diol, and there is a great amount of tailing with the diol. If you take the acid washed support and treat it with polyvinylpyrrolidinone prior to the coating of the polyester one can obtain the result seen in the bottom chart.

The ketone is eluted at the same time on this column, but the diol is eluted late. Thus the PVP has altered the partition characteristics of the stationary phase. Furthermore, the amount of tailing has been sharply reduced. Notice that testosterone has shifted out and also has a peak shape which is improved over that found with the silanized support. The PVP treatment of the support not only imparts increased selectivity for hydroxyl groups, but also reduces irreversible adsorption.

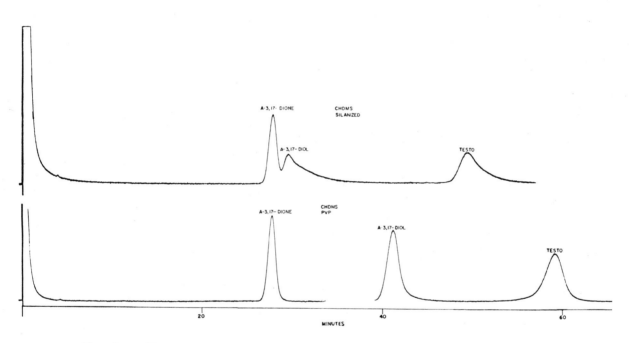

Fig. 5. Effect of treatment of support with polyvinylpyrrolidone.
Upper panel - acid-washed silanized support coated with CHDMS.
Lower panel - acid-washed support treated with PVP prior to coating.

Let's go back to derivatives for a moment. Figure 6 demonstrates an attempted separation of cholesterol and epicholestanol with a JXR column. Although there is a stereochemical difference between the two sterols (epicholestanol has a 3α or axial hydroxy group, cholestanol is 3β or equatorial), there is no separation. The volatilities are very similar with this non-selective phase. Therefore there is one peak. If you ran this mixture on QF-1 or on another silicone such as XE-60, you would get a nice separation with the equatorial isomer cholestanol, eluting later than the epicholestanol. One can also improve separation patterns by preparing derivatives. As is evident from the figure, the TMS ethers are very well separated with JXR. Notice that the TMS ether of cholestanol comes out considerably later than the alcohol. The derivative of the axial epimer possesses the same retention time as the parent alcohol. This separation can be explained by the change in molecular shape which occurs when you put on these bulky derivative groups.

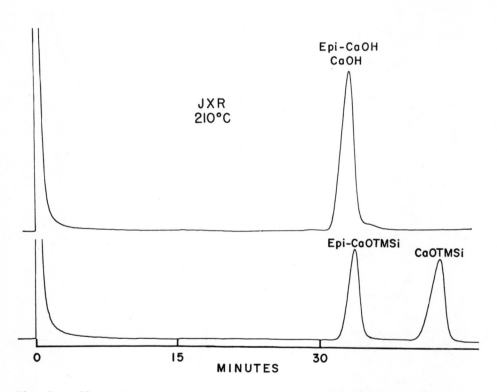

Fig. 6. Effect of derivative formation on resolution. Upper panel - GLC of epicholestanol and cholestanol. Lower panel - GLC of TMS ethers of epicholestanol and cholestanol.

Figure 7 illustrates the separation of the TMS ethers of pregnanediol (Pd), androsterone (5α), etiocholanolone (5β) and dehydroisoandrosterone ($\Delta^5$) with the selective stationary phase EGSS-Z. Under these conditions the di-TMS ether of pregnanediol, due to its rather non-polar nature, is eluted faster than the mono-derivatized 17-ketosteroids, which still possess the polar keto group. The fine separation of these compounds is made possible by the formation of the TMS ethers, which accentuates the existing stereochemical differences.

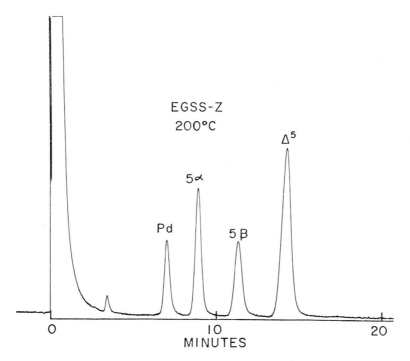

Fig. 7. Chromatogram of TMS ethers on selective phase EGSS-Z. Symbols as in Fig. 4.

Dr. Touchstone has been using mixed phases, and Fig. 8 illustrates the use of a two-component phase. They are often very useful. In the upper chromatogram we have the standard mixture of the TMS ethers of the three classical 17-ketosteroids and pregnanediol. With this phase, which is less selective than EGSS-Z, the pregnanediol derivative is eluted later than the 17-ketosteroids, a consequence of the greater molecular weight of the former. The lower chromatogram shows the results obtained with a sample of early pregnancy urine. The high level of pregnanediol is evident.

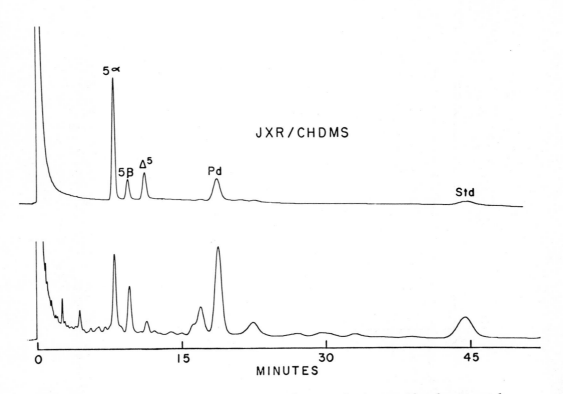

Fig. 8. Use of mixed phase. Upper panel - standard steroids; lower panel - pregnancy urine. Symbols as in Fig. 4.

In Fig. 9, there are two chromatograms obtained with JXR. The upper chart shows estrone, estradiol, and estriol; note the lack of satisfactory separation between the first two compounds. However, they exhibit relatively little trailing. Estriol on the other hand shows a fantastic amount of trailing. This was a good column packing. But when you apply to it a compound of such great polarity as estriol you observe this very poor result. Transformation of these compounds to the silyl ethers results in two rather dramatic changes. The amount of irreversible adsorption is markedly reduced. Furthermore, notice the great improvement in separation for the silyl ethers. The ethers are retained longer than the free steroids.

Here again you see how derivative formation can help you. The monohydroxyl compound has only shifted a little with derivative formation. Estradiol has formed the di-derivative, and has moved out even more. The kind of changes in retention time that one observes upon reaction with specific reagents for ketones or alcohols can often yield a great deal of information.

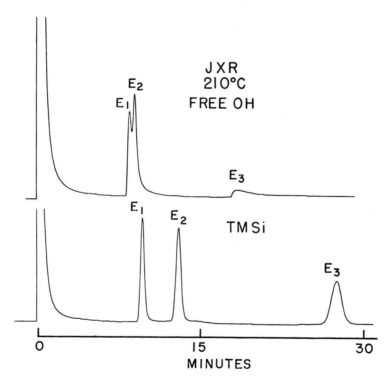

Fig. 9. Effect of derivative formation on column performance. Upper panel - free estrogens; lower panel - estrogen-TMS ethers. $E_1$ = estrone, $E_2$ = estradiol, $E_3$ = estriol.

Figure 10 shows what happened when a sample of cholesterol trifluoroacetate was applied to and SE-30 column in which the column packing extended up into the flash heater zone which was at about 290 C. Obviously decomposition has occurred. It turns out that under these conditions we are losing the trifluoroacetyl group. The early peaks, including the large one, are olefinic hydrocarbons. The last peak, at 37 min, is the authentic trifluoroacetate.

If you remove four or five inches of the packing from the top of the column, there is no longer packing at 290 C, but only at 220 C (the column temperature) and under these conditions one observes only the late peak, and none of this decomposition. You have to be careful even if you have good packing. There may be hot spots along the column, or the column itself may be too hot, and then one may observe decomposition with certain steroids.

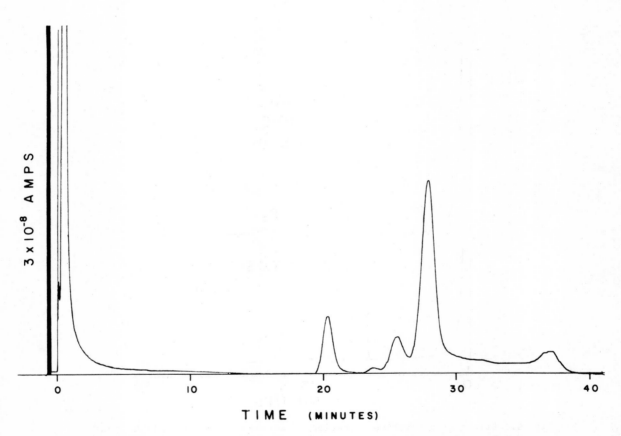

Fig. 10. Chromatogram of cholesterol trifluoroacetate with column packing in flash heater zone.

I am going to briefly discuss on what I consider to be a rather interesting phenomenon. If one prepares the methanesulfonate or p-toluenesulfonate of cholesterol and subjects the derivative to GLC an elimination reaction occurs. One of the hydrocarbon products observed is a so-called "i" steroid, possessing a three-membered ring. The trifluoroacetate evidently can behave in a similar manner. It is clear that there are certain rearrangements which do occur during GLC. Such transformations may also occur with the sulfates of sterols.

If you have a derivative and you find its retention time with SE-30 is one-half that of the parent sterol, I think you can be quite sure you are losing the functional group. One must be careful about such things. Sometimes the compounds are altered.

A few words about structure and retention time. Many of you are probably aware of the so-called "carbon number" concept of Woodford and Van Gent. If you plot the log of the retention times against the number of carbon atoms for a homologous series of fatty acid methyl esters, you get a straight line. You can observe the same phenomenon with the steroid hydrocarbons, androstane, allopregnane and cholestane. A plot of the log of the retention time of these compounds versus their "steroid numbers" (number of carbon atoms) gives a straight line (Fig. 11).

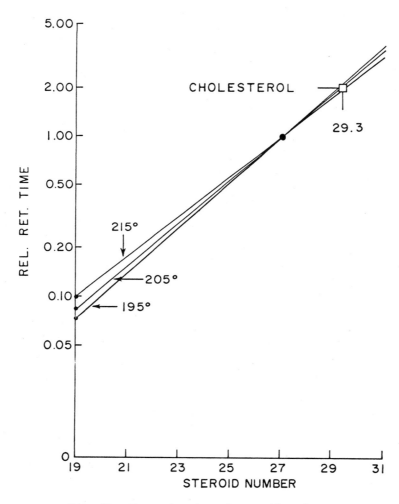

Fig. 11. Determination of steroid number.

Let us now turn our attention to another effect. It is known that if you change the temperature or pressure or the length of column or a variety of other things you are going to change the retention time of the compound. Most of us work with relative retention times, which are quite unaffected by most of the variables which are generally encountered. Unfortunately, these values are strongly temperature dependent; the retention time of androstane relative to cholestane is clearly temperature dependent, as is seen in Fig. 11.

We have found that if you use two reference standards (androstane and cholestane with SE-30), and this is the so-called steroid number approach, you can obtain temperature independent values. Steroid numbers are determined experimentally in the way seen in Fig. 11. Determinations of retention times are made for the two reference substances (steroid numbers of 19 and 27) and the compound under study. Relative retention times are plotted in graph form as log values with respect to steroid number values. The steroid number of the compound under study may be read from the graph as the intercept on the steroid number axis. Figure 11 shows the determination of the steroid number of cholesterol. Steroid numbers are defined in terms of the equation seen in Table 4.

What can you do with these values? Their temperature-independent nature makes them highly useful for inter-laboratory comparisons of retention behavior. The F values for the various steroid functional groups may be determined through the use of the equation. Once these values are known, the equation can then be used to predict behavior of known compounds and to indicate possible structures for unknown peaks.

Other structure-retention time work has been done by Dr. Knights, Dr. Thomas, Dr. Chamberlain, Dr. Brooks, Dr. Wotiz, and Dr. Clayton.

TABLE 4. Steroid numbers.

$$SN = S + F_1 + F_2 + \ldots + F_n$$

SN = Steroid number

S = Steroid skeleton carbon content

$F_n$ = Functional group contribution

Reference compounds

Androstane, 19.0

Cholestane, 27.0

Table 5 presents the steroid number data for pregnane-3α,20α-diol. The $C_{21}$ steroid nucleus gets a value of 21. The two hydroxyl groups contribute, with SE-30, 4.6 as F values, and if you add these values up you get 25.6. However, as you see, the steroid is A/B *cis*. You have to subtract out 0.3 because *cis* compounds are eluted somewhat faster than A/B *trans*. You calculate 25.3 as your value and observe 25.3. One can carry out the same calculations with selective phases. It is possible to predict accurately what the steroid number should be.

In Table 6 the steroid number data with SE-30 as shown for a compound given as an unknown to a first year medical student who was doing a project in our laboratory. He found the steroid number to be 24.1. Upon treatment with acetic anhydride the value increased to 25.4.

The compound also condensed with dimethylhydrazine without acetic acid catalyst, indicating the presence of a ketone at the 3-position. This reagent is like Dr. Fales' methoxime.

TABLE 5. Comparison of calculated and observed SN values for pregnane-3α,20α-diol with three phases.

|  | Steroid number | | |
|---|---|---|---|
|  | SE-30 | NGS | QF-1 |
| C$_{21}$ steroid, A/B trans | 21.0 | 21.0 | 21.0 |
| 3α(eq.)-ol | 2.4 | 7.5 | 5.1 |
| 20α-ol | 2.2 | 6.9 | 4.4 |
| A/B cis | -0.3 | -0.3 | -0.3 |
| Calc. | 25.3 | 35.1 | 30.2 |
| Observed | 25.3 | 35.1 | 30.2 |

TABLE 6.

|  | SN values | |
|---|---|---|
| Steroid | Calculated | Found |
| Androstane | 19.0 |  |
| 3-one | 2.6 |  |
| 17β-ol | 2.4 |  |
|  | 24.0 | 24.1 |
| Acetate |  |  |
| Androstane | 19.0 |  |
| 3-one | 2.6 |  |
| 17β-acetoxy | 3.7 |  |
|  | 25.3 | 25.4 |
| DMH derivative |  |  |
| Androstane | 19.0 |  |
| 3-DMH | 4.0 |  |
| 17β-ol | 2.4 |  |
|  | 25.4 | 25.4 |
| Acetate-DMH derivative |  |  |
| Androstane | 19.0 |  |
| 3-DMH | 4.0 |  |
| 17β-acetoxy | 3.7 |  |
|  | 26.7 | 26.7 |

If you have a steroid number of 24.1, you can be sure it is not a cholestane. That would have an F value of at least 27. In addition, we possess information that we have at least two functional groups. Therefore the "F" contribution would be at least four. The observed value of 25.4 is too low for a pregnane, so the unknown must be an androstane. A 3-keto group possesses an F value of 2.6. A likely position for the hydroxyl group is at $C_{17}$. Table 6 shows the calculated and experimental values for the unknown (androstane-17β-ol-3-one) and its derivatives.

I hope that what I have said this morning will indicate some of the things you can do with GLC. I mentioned the problem of rearrangements. We obtained a great deal of information about the structure of these products by using retention times of model compounds. But of equal importance was the fact that we had in Dr. Horning's laboratory a combination mass spectrometer and gas chromatograph, thanks to Professors Bergström and Ryhage of the Karolinska Institute. There is perhaps nothing that can equal the power of a method like this. You have a gas chromatograph hooked up to the mass spectrometer. The former effects the separation and also gives you a good deal of information about the sample. With the molecular separator you separate the carrier gas from the steroid molecules, which pass into the ion chamber and then you obtain the mass spectrogram.

This is really a powerful technique, and will be used more and more by biologists. You can get the structure from microgram or smaller quantities of steroid. I am sure Dr. Fales will talk about this later.

One further point, however; in addition to silanizing your column support, it is also a good idea to silanize your glass wool plugs and your column. Don't use too much glass wool in the vaporizer zone. You have plenty of heat to volatilize your sample. We use a small amount of glass wool, mainly to catch the pieces of rubber septum that occasionally fall down into the columns.

## DISCUSSION

DR. TOUCHSTONE: I wonder if Dr. VandenHeuvel has noticed that sometimes, better responses can be obtained with certain types of compounds when combination columns are used? Also we have noticed that by merely changing the percent of phase, for instance, from one percent QF-1 to 5 or 10% QF-1 we can get an increased response with estriol.

With respect to the comments about dehydration of some of the hydroxy sulfates, although this may happen during GLC, it may not be possible to quantitate the product.

DR. VANDEN HEUVEL: The loss of sulfonate groups that one observes is quantitative, so while it may not be the sulfonate which comes through the column, if you have essentially 100% conversion to the olefin you are all right. The same may be true with the sulfates.

As far as the response to a steroid is concerned, one must be careful not to confuse the effect of structure on response with the effect of loss on the column. When you increase the amount of your liquid phase, you may be covering up some active sites which were causing adsorption. With less adsorption you will observe greater response.

DR. SMITH: Suppose you have an unknown mixture which you wish to analyze and the amounts of the unknown constituents present are relatively small. You would like to know how many peaks you have. You are using a column which has some adsorption -- you don't know how much. Should you repeat the same sample several times in succession until the height of each peak on the chromatogram becomes constant?

This question raises a serious problem. It seems there is no adequate method of priming a column because you don't know what mixture to prime it with.

DR. VANDEN HEUVEL: I would say this. Work very, very diligently trying to obtain column packing which does not inhibit tailing, which does not absorb your sterols. This can be done quite effectively by acid washing, base washing, silanizing or by applying other methods of deactivation. The preparation of derivatives which convert the polar hydroxyl groups into nonpolar groups is also necessary.

If you are interested in determining whether loss of sterols is occurring on the column, you can use a test mixture of cholestane and cholesterol. One can assume that the cholestane is not going to be lost to any great extent because it has no functional groups. If no decrease in the cholesterol-cholestane peak area ratio occurs as you apply successively smaller samples of the mixture, there is no selective adsorption. Your plot of micrograms versus peak area ratio will remain constant.

DR. SMITH: But you are approaching the problem backwards. It won't come to you that way normally.

DR. VANDEN HEUVEL: If you are working with 10 or 20 µg per peak and you are losing 0.2 µg it is not going to make much difference. As you reduce the amount of sample, this absolute loss of 0.2 µg of sterol is going to start showing up. At the 1 µg level this is a 20% loss.

If your peak area ratio is constant down to at least less than one µg of sterol you are doing quite well.

DR. FALES: Furthermore, if you lose cholesterol on your column, you are probably going to lose the steroid that you have in mind. You can use the cholesterol cholestane ratio to check losses of more polar steroids.

DR. SMITH: Isn't it true that you would start with a large amount of material on the column? You won't find any loss of small amounts you would inject later because you are starting with a column which is saturated, so to speak, with the compound.

From the information that Dr. Wotiz gave us yesterday it is obviously possible to inject a small quantity of some unknown material and lose it on the column.

DR. VANDEN HEUVEL: The point that Dr. Fales is making is this: cholestane should show no trailing. If you have a trailing cholestane peak, then you are in trouble. This may well be a physical problem. The column may be poorly packed, or perhaps the instrument isn't functioning properly. A good cholestane peak but a bad cholesterol peak means irreversible loss on the column somewhere.

DR. WOTIZ: I would like to show you a rather old slide (Tab. 1). We carried out some experiments with the deliberate idea of overemphasizing the column adsorption of a steroid. We used a very large column, I think a total of 58 g of 3% SE-30 on 30-60 mesh packing was involved.

As you can see, after the first injection of 500 µg we lost a considerable amount of the material. As I said, this is unusually high by comparison since the average column may have only 4 or 5 g of packing. But, there is a loss of about 2 µg of steroid per g of packing.

If you chromatograph a urine or plasma extract, after the first one or two injections on a primed column, the interfering peaks are usually relatively small. After that they may get very much larger. I don't know how you can overcome this difficulty if you have a microgram or less of unknown material. You may not see this little amount even on some very good columns. I haven't experimented with the approach Dr. Fales and Dr. VandenHeuvel took, comparing the cholesterol/cholestane peaks.

We prefer taking a new column and analyzing steroids of specific interest to see how much they are adsorbed. There is considerable difference not only with respect to the adsorption of different steroids, but also various derivatives adsorb differently. In general, we prefer to work with compounds similar to the type to be analyzed. This, however, still does not solve your problem. I do not think there presently is a good answer to how you can find minute quantities of an unknown material.

DR. WILSON: Dr. VandenHeuvel and others talked about using a number of different kinds of columns and trying for instance that column and this column. I understand that if you let a column cool off it loses efficiency. Then if you prepare a new one it may have to be aged several days before you can use it. I would like to ask Dr. VandenHeuvel how he handles his columns when he is changing them.

DR. VANDEN HEUVEL: At the present time I only have one machine, and I use four or five different columns. In Houston I generally worked with two instruments. I would occasionally use five or six different columns on the same day. They can be taken out and cooled off and put back in. It usually takes 45 min to equilibrate a column. You don't have to worry about cooling a column. It doesn't hurt. Dr. Luukkainen said he had used a column off and on for five years in Finland.

TABLE 1. Recovery of injected steroids from a new column.

| Steroid | μg Injected | μg Recovered | % Adsorbed | μg Steroid adsorbed |
|---|---|---|---|---|
| Estrone acetate | 500 | 335 | 33 | 2.85 |
| Estrone acetate | 500 | 415 | 13 | 1.46 |
| Estrone acetate | 500 | 465 | 7 | .60 |

The material was trapped at the exit port in a test tube surrounded by liquid nitrogen.

Total packing 58 g SE-30 (3%) on 30-60 mesh Chromsorb.

The column was cured at 250 C for approximately two weeks.

DR. FALES: If you have a polyester column of any sort and any concentration, and you take it out before it has cooled off, the polyester will hydrolyze. The silicone columns can stand considerably more. If you take the top of the column off before the gas pressure has dropped you will get a backsurge which obviously does not do any great good to the column.

DR. VANDEN HEUVEL: On one occasion I actually lost about a foot of packing from the front end of a column due to pressure backsurge. The performance of the column actually improved. You do observe some strange results. Apparently it doesn't hurt to have a foot or two of empty tube at the front.

DR. WOTIZ: With respect to this business of cooling a column, taking it out and reheating it, I fully agree with Drs. VandenHeuvel and Fales. We have almost always managed to take apart the instrument and put other columns in that we want without losing the columns. In fact, one of my SE-30 columns, handled this way, is now better than four and a half years old and is still a good column.

DR. WILSON: My other question or remark had to do with the TMS ethers of pregnanediol. We have observed different retention times for the pregnanediol derivative. Do you have any explanation of this?

DR. VANDEN HEUVEL: It is my experience, and of most os us, that if one uses a selective or polar stationary phase such as XE-60 or the polyester NGS, coated on silanized support, pregnanediol di-TMS ether comes off before the monosilyls of androsterone, etiocholanolone and dehydroisoandrosterone. As I indicated earlier, one can observe an effect from the surface of the support, since the use of unsilanized support leads to a different elution order. Elution patterns do not change unless there are major changes in the conditions of the separation. Perhaps your di-TMS ether of pregnanediol hydrolyzed to the mono-silyl, or perhaps much of your phase was stripped off, allowing a support surface effect to become operative. As long as you are consistent, your sample peak is going to come at a reproducible time.

DR. EIK-NES: "After all that we have suffered and achieved we find ourselves still confronted with problems and perils not less but more formidable than those to which we have so narrowly made our way." This is not a statement by Eik-Nes, it is Churchill in "The Gathering Storm."

Some of my views on GLC have been challenged. I find this natural and I had hoped that they would be challenged. We are here to disagree and discuss not to agree that GLC is the best or the worst of all techniques for steroid separation and estimation.

With particular frankness and sincerity, I question the following: why go into the technique of GLC without analyzing critically where this technique may take the field of steroid endocrinology? I feel strongly that scholars and shoemakers alike should pause and decide this for themselves. I should be the last one to disregard data on the behavior of synthetic steroids on the GLC column; without such information we would not be here today. We have, however, progressed a great deal since 1960-1961 and I shall admit that I liked Dr. Rosenfeld's publication in Steroids since it is a new approach to GLC of adrenocorticoids. Moreover, in this work it was demonstrated that the published technique could be applied to urinary steroids.

Many of the early reports on GLC of standard steroids led to the impression that all problems in steroid methodology were finally solved. This is not the case, and I adhere strictly to the view that unless you have been through the suffering of many years purifying steroids from biological samples and knowing all the tricks from such experimentation, you may have limited or even no benefit from GLC. We are driving for sample purity and this is also required for adequate work with GLC.

Dr. Engel, for whom I have great respect, suggested yesterday that maybe some of us would do better with a box without a column. This box should contain the sensitive detector systems used in GLC. This is not a ridiculous thought; as a matter of fact, it is a brilliant idea. We need detectors of extreme sensitivity in order to measure some of the plasma steroids. It is quite possible that purification of such steroids by standard technique followed by vaporization and detection of the steroid in the gas phase with sensitive detectors is a solution to many of our problems.

I have challenged the investigators on the "Atlantic Ocean" and I should care to add to those who like to drift on the crest: don't jump into the drink. I would not date to calculate steroid concentration on my dogs from such chromatographic tracings. Some investigators may say: We need the same answers in a hurry and it is good enough for clinical purposes. I should like to challenge such philosophy by something the grand master of steroid methodology, Dr. H. Mason, wrote about 20 years ago: "What is good enough for clinical purposes?" Gentlemen, we are dealing with our fellow man. "All things are moving at the same moment, year by year, month by month, they have all been moving forward together. While we have reached certain positions in thought, others have reached certain positions in fact. What we should not dream of doing in a year ago, what we should not dream of doing even a month ago, we are doing now." Again, not Eik-Nes but Churchill in "The Gathering Storm."

# SOME PRACTICAL PROBLEMS IN GAS-LIQUID CHROMATOGRAPHY

Henry M. Fales, Laboratory of Metabolism, National Heart Institute, Bethesda, Md.

I do agree with Dr. Eik-Nes that most of you have different problems than we usually encounter in injecting pure materials on the column. The wonder to me is that this technique works at all at the levels that I have seen during the last two days. I think at this point I have only a few more points to make concerning some actual problems.

Some of you have been exhibiting peaks which are clearly the result of decomposition on the column. If I see a peak that appears as in Fig. 1, I consider it *a priori* evidence of decomposition. It shouldn't even be called one peak. Rather it is a superimposition of peaks.

Something else we see quite frequently on a paper column is the phenomenon that Dr. Vanden-Heuvel has already described partially. For example, if a chromatogram of cholesterol appears as in Fig. 2, it simply means that the sterol is dehydrating beginning in the flash heater and then undergoing continuous dehydration in the column from this point on. Thus the first peak represents cholestadiene which tails into the main cholesterol peak.

But certainly, unless the peak is perfectly symmetrical we have either decomposition or adsorption. What about adsorption? That is demonstrated by the sort of peak shown in Fig. 3. Consideration of the partition demonstrates clearly that overloading the liquid phase results in this type of peak. Many people refer to an "overloaded column" as exhibiting a peak as in Fig. 3. This is indeed an overload not, however, of the liquid phase but of the gas phase.

It is true that an increase in the concentration of liquid phase on the support results in disappearance of this type of asymmetry. The reason it disappears is, because with a higher concentration of liquid phase, higher temperatures are required to elute the sample in the same retention time and it is the higher temperature which increases the capacity of the compound to exist in the vapor phase rather than in the liquid or solid state.

A second point involves the characterization of derivatives. When one proposes a new derivative it is important to characterize it as thoroughly as possible in terms of classical methods of chemistry. It is especially important to know how many substituent groups one has attached to the molecule. It is not completely clear in my mind from the discussion this week whether the phrase "completely characterized" refers to such complete characterization in the chemical sense. I realize it is annoying when one is not equipped with some of these expensive tools to have to find someone to do infrared spectroscopy, mass spectrometry, or nuclear magnetic resonance. All that I can suggest at the moment is that our laboratory will be willing to help if other assistance can't be obtained.

Concerning internal standards, when one uses internal standards, it is fairly important to use a compound as similar to the one in question as one can get. Cholestane is not really a good internal standard for sterols and steroids. It is very convenient, easy to get, I admit. For example, it is important, if one is studying alcohols and ketones, to use something containing alcohol and ketonic fractions in order to test the adsorption or loss on glass surfaces and hot spots.

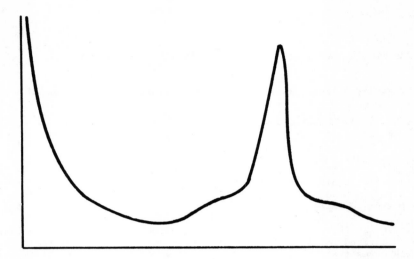

Fig. 1. Evidence of decomposition. Single compound injected.

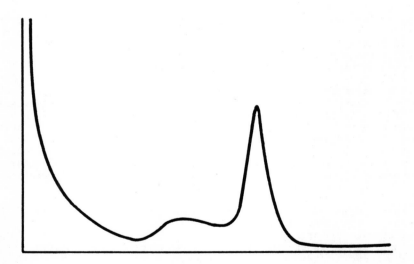

Fig. 2. Dehydration on column.

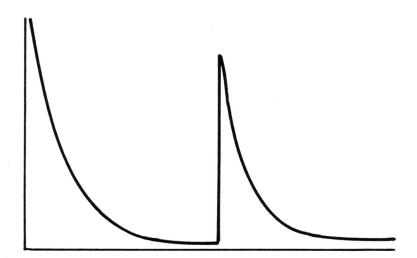

Fig. 3. Adsorption of steroid.

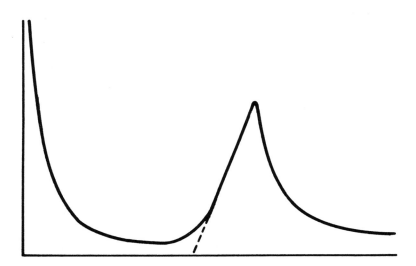

Fig. 4. Measurement of retention time.

Concerning flame ionization and argon ionization detectors, I believe theory dictates clearly that a flame ionization detector would be about one-third as sensitive as the argon detector for the same electronic amplification and recorder sensitivity. Of course, there is considerably less noise in the flame detector so greater amplification can be employed with higher over-all gain. The flame ionization detector and $^{90}$Sr detector have about the same overall sensitivity in terms of signal to noise ratio.

If one achieves better results when you increase the sensitivity of the recorder via a 1 mv slide wire this simply means that your recorder amplifier is better than your electrometer amplifier. Increasing recorder gain is usually achieved at some sacrifice in stability and usually the same amplification is more properly achieved in the electrometer. Finally, I would like to emphasize again Dr. VandenHeuvel's plea not to overlook the chemical information that is contained within these curves. Measure the retention times as accurately as possible. If one has a distorted peak it is usually better to use the initial retention time. This is the point where a tangent drawn on the slope of the leading part of the peak intersects the baseline (Fig. 4). Only this point is sensitive to sample load or peak asymmetry and, in my opinion, it is the most reproducible point.

I think the day will come when we will be able to characterize all the peaks we see in a given chromatogram. Even today with mass spectrometry we could analyze the formula of every single peak in a gas spectrogram, although it would be a formidable job.

These spectra can be fed into a computer as is being done by Prof. Klaus Bieman of MIT and eventually should yield an exact picture of the substance in question. Thus, further sophistication in instrumentation may be expected to increse greatly the usefulness of GLC.

# COMMENTS ON TECHNIQUES ON GAS CHROMATOGRAPHY

H. H. Wotiz, Boston University, School of Medicine, Boston, Mass.

I would like to continue the discussion of some of the factors affecting GLC. It may be reasonably stated that the design of an instrument has much to do with quantitation of steroids. In addition to the deleterious effects of inefficient injection systems on zone spreading, chromatography at high temperature requires an efficient arrangement for the maintenance of uniform temperature throughout the system. Variations in temperature along the column may give rise to so-called 'hot-spots' which in turn cause localized deterioration of the column packing. The same effect is produced when the column packing extends into the flash heating zone. Adsorption of compounds onto the thus altered packing becomes more pronounced when sub-microgram amounts are injected. A temperature drop in the connection between column exit and the detector, or in the detector itself, may lead to a partial condensation of the vapor and thus may result in decreased detector response as well as eventual line clogging. Excessive dead volume in the injector component will lead to column inefficiency and zone diffusion during chromatography. A similar effect, despite high column efficiency can be noted when such dead volume exists between the column outlet and the detector.

Lower limits of detection and versatility of detector characteristics appear to be important practical considerations. Experience with the use of an argon ionization detector for urinary extracts presented a serious problem of detector fouling, resulting in leakage, cracking of the electrode connection and even complete corrosion of the detector cell. In contrast, flame ionization detectors have been shown to have far greater reliability. The latter do not corrode, rarely require shut-down for cleaning, and notably, allow the use of higher column temperatures. This in turn permits the use of a longer and more highly resolving column, limited only by the heat stability of the stationary phase. Such detectors also have a greater range of linearity ($1 \times 10^5$). The sensitivity of flame ionization detectors is adequate to make them suitable for the measurement of estrogens as low as 0.01-0.02 µg per injection. Indeed, smaller quantities may well be sensed by the detector, but may not be capable of passing through the column due to adsorption, thereby limiting overall sensitivity.

A variety of commercially available supporting materials (plain, acid washed, acid and base washed, silanized or un-silanized) were used in a preliminary investigation of optimum conditions for estrogen determinations. It was observed that in many cases adsorption of estrogens, particularly estradiol diacetate, increased markedly after a few days or weeks on columns kept consistently at 250 C. Commercially obtained Diatoport S (F&M Scientific Co., Avondale, Pa.) fortunately permitted the maintenance of a highly efficient column for several weeks or months (depending on operating temperature) under identical conditions. Another solid support Gas Chrom Z (Applied Science Lab. Inc., State College, Pa.) was used in a series of experiments and was found to be adequate for the compounds under study.

Examination of the VanDeemter equation shows that the size distribution of the support particles plays a critical part in the efficiency (i.e., number of theoretical plates) of the column. Uniform mesh size of the packed material, therefore, is of considerable importance. Most commercially obtained diatomaceous earth can be purchased in mesh cuts of 20, in some instances even in cuts of 10. Two problems tend to arise. First, fragmentation of the rather brittle support occurs during transhipment from the factory. Secondly, during the process of coating the stationary phase on the support further fragmentation tends to occur. VandenHeuvel et al. (1) have developed an interesting method of coating stationary phase by filtration. This is sufficiently gentle to prevent much of the fragmentation. The only disadvantage of this particular method rests in the uncertainty of the final concentration of stationary phase. Frequently this is of little significance unless, as occurs with an occasional column, the phase concentration is too low. Another method of preparing the column consists simply of the evaporation of dissolved stationary phase onto the solid support while constantly stirring the mixture. This in general leads to significant fragmentation (one notable exception being the new Gas Chrom Q support). We have generally been able to obtain good results by suspending the dry, coated support in absolute ethanol, and gently floating the fragments (fines) off the top of the liquid. After decantation of the alcohol and careful drying in an oven the stationary phase now is ready to be packed into the column. In either case careful attention to the preparation should result in the making of GLC columns with an efficiency of 500-700 theoretical plates per foot. Indeed, it is this kind of high efficiency which imparts considerably greater specificity to the methods.

Almost all of the liquid phases found suitable for steroid chromatography have also been utilized for estrogen analysis. Among them are the silicone polymers such as Dow-Corning stopcock grease, SE-30 (dimethylsilicone polymer), XE-60 (cyanoethyl-methyl silicone polymer), QF-1 (fluoroalkyl silicone polymer), and polyesters like NGS (neopentyl glycol succinate), EGA (ethylene glycol adipate) and EGIP (ethylene glycol isophthalate). It may be worth repeating here that ideal liquid phase should be non-volatile, thermally stable and chemically inert toward the solutes of interest at the useful column temperatures. SE-30, a non-selective elastometer fulfills these criteria very well. As a results, it has been widely applied not only for work with steroids, but also in a variety of other fields. In fact, for the determination of the three classical estrogens, as well as some of the newer metabolites in urine extracts from pregnant women, this specific stationary phase was shown to offer great versatility (2, 3, 4) particularly for use with the acetates. The other silicone polymers (QF-1; XE-60) and polyester phases (NGS; EGA) show variable degrees of selectivity toward ketones and hydroxyls respectively and have also been applied for the determination of estrogens. The interesting possibility of adjusting the properties of a chromatographic column by application of mixed stationary phases of different characteristics has been explored to some extent (5, 6). In most of the single phase columns, the separation of 16$\alpha$-hydroxy estrone diacetate and 16-ketoestradiol diacetate is poor while in the mixed phase column of EGA (0.5%) and SE-30 (2.5%) complete resolution of these two compounds could be achieved (7).

In general the type of phase to be used is primarily determined by the physical and chemical properties of the compounds to be analyzed. However, the contaminants in the sample to be analyzed, and the relative amounts to be measured occasionally require reevaluation of this selection. It has been stated earlier that the non-selective dimethyl silicone elastomer SE-30 has been shown to be suitable for measurements of estrogens in pregnancy urine. However, use of larger aliquots of urine becomes necessary for nonpregnancy urine. In consequence it was observed that in some such extracts measurement was unreliable because of the presence of interfering compounds having retention times similar to those of some estrogens in the SE-30 column. Furthermore, displacement of the retention time of some estrogens, particularly small amounts of estriol, in the presence of contaminants makes some such analyses doubtful. When these same extracts were gas chromatographed on a QF-1 column no interference or displacement effects have as yet been observed. In consequence it would appear that to obtain more reliable gas chromatograms, the use of this ketone-selective stationary phase should be recommended despite its lower thermal stability. The poor resolution of estrone and estradiol acetates is of little consequence since these two substances are separated prior to GLC.

*Irreversible Adsorption*. The point raised by Whittier et al. (8) about adsorption on exposed surfaces is an extremely important one. It has been our day to day experience that when working quantitatively with relatively small amounts of steroids, either very small peaks, or occasionally, no peaks at all can be seen following the first one or two injections. This phenomenon is caused by a form of 'irreversible adsorption' and requires the analyst to saturate the column each day with a few micrograms of the substances to be analyzed. Saturation can be attained by the injection of high concentrations of steroids, or, for the rarer compounds, through the injection of some well preserved old extracts. When working with relatively high concentrations of steroid the adsorption effect frequently remains unnoticed.

Evidence for the adsorption of acetylated steroids was presented in the discussion following Dr. VandenHeuvel's paper (page 294 ). Collection efficiency using a liquid nitrogen trap was shown to be approximately 90%. A considerable amount of estrogen is lost on such a column following the first injection with further smaller losses following the second injection. Figure 1 presents similar evidence following chromatography of estrone acetate on a well-conditioned 1/8 in SE-30 column. Injections were made on two successive mornings and the peak heights are plotted for the first three injections each morning. On the first morning, the column was not primed, while on the second day 12 µg of estrone acetate was passed through the column in three successive injections, followed by three injections of pure acetone to make certain that no residual material was left in the injector. Then three injections of standard were made.

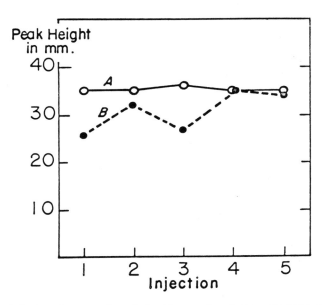

Fig. 1.  O-O-O - Consecutive injections of estrone acetate on primed column. ... - Consecutive injections of estrone acetate on unprimed column.

Solvents. In general, reagent grade solvents, when used for direct injection present only one fast moving solvent peak. However, in the process of extraction or solvent partitioning a fairly large amount (100-500 cc) of solvent is frequently utilized. This material is subsequently evaporated to dryness and it has been our experience that numerous GLC peaks may be obtained from even the purest solvents (i.e., spectrograde) handled in this manner. Figure 2 shows a chromatogram obtained following evaporation of 200 ml acetone (reagent grade) redisolving the residue in 50 μl of acetone and injecting 2 μl of this solution into the gas chromatograph. As can be seen nearly all of the estrogen metabolites would likely be obscured if present in extremely low concentrations. These solvent impurities would not be noticeable in the analysis of hormones present in anything but very small amounts. In Table 1 are shown the retention times relative to estrone acetate and the relative peak heights at the various attenuations for a number of solvents before and after distillation and some before and after treatment with acetic anhydride and pyridine. In all instances, at least two and frequently many more peaks can be discerned. To date, we have found only one solvent which produces no extraneous peaks following simple distillation of commercial grade material and injection into the gas chromatograph. This solvent was dichlomethane.

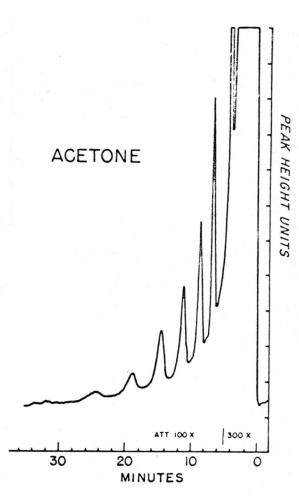

Fig. 2. Chromatogram of residue of reagent grade acetone.

TABLE 1. Gas chromatography of distillation residues from common solvents.

| Solvent | Volume evaporated (ml) | $R_T$ (min) | $\frac{R_T}{R_{E_1A}}$ * | Peak height (mm) |
|---|---|---|---|---|
| Acetone (acetic anhydride and pyridine treated) | 250 | 4.1 | .63 | o.s.** |
| | | 5.1 | .78 | o.s.** |
| | | 6.4 | .99 | 150 |
| | | 8.4 | 1.35 | 85 |
| | | 11.2 | 1.73 | 55 |
| | | 14.4 | 2.23 | 35 |
| | | 19.0 | 2.94 | 13 |
| | | 24.5 | 3.80 | 6 |
| Acetone (singly distilled) (spectroanalyzed) | 500 | 6.35 | .99 | 11 |
| | | 8.50 | 1.32 | 70 |
| Benzene (distilled) | 100 | 4.4 | .64 | 22 |
| | | 5.4 | .82 | 14 |
| | | 7.1 | 1.09 | 14 |
| | | 9.3 | 1.43 | 100 |
| | | 12.2 | 1.89 | 6 |
| | | 15.6 | 2.48 | 4 |
| Benzene (acetic anhydride and pyridine treated) | 100 | 6.3 | .98 | 12 |
| Ethyl ether | 500 | 5.2 | .81 | o.s.** |
| | | 6.2 | .96 | 100 |
| | | 9.1 | 1.41 | 70 |
| | | 11.3 | 1.75 | 37 |
| | | 15.4 | 2.39 | 16 |
| Toluene | 400 | 3.55 | .55 | 28 |
| | | 5.03 | .78 | 15 |
| | | 6.35 | .99 | 5 |
| | | 8.55 | 1.32 | 4 |
| | | 11.25 | 1.74 | 18 |
| Petroleum ether | 200 | 4.4 | .68 | o.s.** |
| | | 5.2 | .80 | o.s.** |
| | | 6.4 | .99 | 30 |
| | | 8.1 | 1.25 | 88 |
| Methelene chloride | 200 | no peaks | | |

\* Relative to estrone acetate, $R_T$ = 6.45 min.
\*\* Off scale.

Column: 6 ft 1/8 in, 3% SE-30 on 80-100 mesh Diatoport S. Total material injected - 1/50 of final solution. Attenuation = 100X.

$E_2$ - acetate ($R_T$ = 9.15 min, $R_t$ = 1.42)

## REFERENCES

1. VandenHeuvel, W. J. A., C. C. Sweeley, ane E. C. Horning, J Am Chem Soc 82: 3481, 1960.

2. Wotiz, H. H., Biochim Biophys Acta 74: 122, 1963.

3. Fishman, J. and J. B. Brown, J Chromatog 8: 21, 1962.

4. Wotiz, H. H. and S. C. Chattoraj, Anal Chem 36: 1466, 1964.

5. Touchstone, J. C., A. Nikolski, and T. Murawec, Steroids 3: 569, 1964.

6. Nair, P. P., I. J. Sarlos, D. Solomon, and D. Turner, Anal Biochem 7: 96, 1964.

7. Wotiz, H. H., unpublished data (1965).

8. Whittier, M. B., L. Mikelsen, and N. Armstrong, Presented at the 14th Ann Mid-Amer Spectroscopy Symposium, May 20-23, 1963, Chicago, Ill.

## DISCUSSION OF DR. WOTIZ'S PAPER

DR. LIPSETT: Do the commercially available phases often meet the criteria that you and others have proposed?

DR. WOTIZ: Yes some of them have lately. Gas Chrom Q shows almost no tailing with cholesterol. But SE-30 on Gas Chrom Q is one combination that usually gives good results. With more selective phases there is less tailing. I think that for the last two years now some of the commercially available, carefully acid washed, silanized, the most expensive that the company has to offer, do meet or almost meet the requirements. I must confess though that we still make most of our own packing.

DR. VANDEN HEUVEL: We still make most of our packings, at least I do at work and the people in Dr. Horning's group do too. But things are improving. My recent experience with a commercially available XE-60 column has shown that a commercial packing can be very good. It has taken our group and other groups quite a while to become familiar with these methods for preparing supports and packings. For someone who is not at all experienced with this, I don't think you can really expect top notch packing starting with the unwashed support.

DR. FALES: I will go along with what has been said. The Gas Chrom Q appears a very good phase. We also have been quite happy with commercially coated Diatoport S with 3% SE-30. We had a very good 3% QF-1 on Gas Chrom P. These have been our best commercially obtained columns although by and large we make our own.

One should remember that preparing a packing still has a good deal of art involved and that following the directions cook book style will not necessarily give a good packing every time. When we get a bad column, we just throw it out and try again.

VOICE: I would like to ask Dr. Karmen to discuss the advantages of temperature programming.

DR. KARMEN: We haven't done very much temperature programming except to explore some of the possibilities involved. In general, if you can get a peak to come out at 10 min or less, you are running the detector at about the sensitivity that you can hope to get out of it. On the other hand, if this is a peak that is supposed to come out in 40 min, you can make it appear like a 10 min peak by temperature programming. So, in a sense temperature programming offers another order of magnitude of increased sensitivity and may offer you a little more resolution.

DR. VANDEN HEUVEL: Temperature programming is most important in work with mixtures that have a wide variety of compounds of varied molecular weight. You can spread out compounds that range in molecular weight from 200 to 500, see each one, and quantitate each. As far as my own work goes, we have observed, with Dr. Luukkainen, that although programming is helpful there was one case where the separation factor for two of the estrogens was actually decreased when we used temperature programming. What we are depending upon in GLC are the relative volatilities, and these to a certain extent are temperature dependent. When you are running a column isothermally at 210 C you may get good separation of two compounds. When the temperature is changing continually, we can't be sure that the same relative volatilities will hold. Furthermore, sometimes it is necessary to change the detector when using temperature programming. For example, we were doing work with fatty acid methylesters with a certain polyester stationary phase. We got poor quantitative results when we used an argon detector with programming, but when we switched to the flame detector and kept everything else constant we got good quantitative results.

# THE UTILITY OF GLC FOR ANALYZING SPECIFIC STEROIDS IN BIOLOGICAL FLUIDS

A. Karmen, The Johns Hopkins Medical Institutions, Baltimore, Maryland.

The utility of GLC for analyzing steroids in biological fluids depends on several factors; how well pure steroid can be chromatographed, how conveniently interfering compounds can be removed from the sample before analysis, and how small a quantity of steroid can be distinguished from the other compounds still present after purification. I will attempt to consider each of these points.

## Quantitative Chromatography of Steroids

In the early years of GLC, analysis of steroids was not achieved because they decomposed at the high temperature necessary to elute them from any of the available columns. Their analysis became feasible after Cooke (1) demonstrated that many compounds could be chromatographed at temperatures far below their boiling points if columns that contained only a small quantity of liquid phase were used. The idea of using lightly coated columns was then extended to steroid analysis by Horning, Sweeley and VandenHeuvel and a technique was evolved for chromatographing steroids and other high boiling compounds that included the use of glass inlet systems, glass columns, and carefully deactivated liquid phases and solid supports. However, even with careful attention to the details of the technique, quantitative elution of steroids injected into a column has not been routinely and reproducibly obtained even in single laboratories. There seems to be even less certainty that the same results will be obtained in different laboratories even when the same column packings are used. It thus would seem advisable, if not mandatory, that each of us recheck the quantitative aspects of our chromatography frequently. Fortunately, this is not difficult. Sternberg et al. (2), Ettre (3), and Perkins et al. (4) described a method for predicting the response of the hydrogen flame ionization detector to a variety of compounds containing many different functional groups. This method is applicable to steroids as well. To determine whether chromatography of a given steroid is quantitative, therefore, all that must be done is to add a known quantity of a known volatile compound to a measured quantity of pure steroid and compare the responses obtained with those predicted. Still another method of determining whether chromatography has been quantitative is to analyze a known quantity of purified radioative steroid and to compare the response of a radiation detector to the steroid with its response to a known quantity of a labeled compound that is known to be volatile. Neither of these techniques has been widely used, however. Instead, most reports of steroid analysis by GLC give as evidence for quantitative or successful chromatography only the fact that symmetrical peaks were obtained. The only studies of the quantitative aspects of steroid GLC were those using the argon ionization detector, with which most of the early work was done. Both Sweeley et al. (5) and Bloomfield (6) found that the responses of the argon ionization detector to different steroids were less in proportion to the amount of oxygen on the molecule. While this was consistent with the known properties of the argon ionization detector, these compounds were also the most polar, and tended most to be adsorbed and retained on the column. It was therefore difficult to determine from the data presented whether the different responses were attributable to the detector alone or also to failure of the compounds to be eluted from the column.

One encounters the same difficulty in attempting to interpret the variation in the sensitivity of the electron affinity detector to different steroid haloacetates, as described by Landowne and Lipsky (7). Although the trichloroacetates were reported to be detected with less sensitivity than the monochloroacetates, they were also least volatile and least easy to chromatograph. No evidence was presented that any of these compounds were eluted from the column quantitatively.

Attempts to recover radioactive steroids from a column have not always been successful even when methods were used that were successful with more easily chromatographed columns.

One must conclude that methods for successful and quantitative GLC of steroids on a routine basis have not yet been developed or described and that its application to biological measurements requires that the methods to be used in each laboratory be calibrated for quantitative accuracy.

## Sample Preparation and Sensitivity

It is possible to use GLC for steroid analysis after subjecting the sample to extensive purification by such techniques as solvent extractions and other forms of chromatography. Compounds in the sample that could interfere with the analyses are thus removed and the material injected into the gas chromatography column consists almost entirely of the steroid dissolved in a pure solvent. Since the function of the GLC column is then primarily to deliver the sample to the detector, very little resolving power is required.

This approach undoubtedly does not fully utilize the capability of GLC since the column can perform many of the prepurification steps. In any analysis, the amount of prepurification necessary before GLC is, of course, an inverse function of the specificity of the detector. As an example, water soluble organic compounds in an aqueous solution can be assayed simply by injecting an aliquot of the solution into the column, if a hydrogen flame ionization detector, which is insensitive to water, is used. It is apparent that this is a real advantage when one considers the difficulty encountered in removing the water from the sample to make it suitable for analysis by the argon ionization detector or the electron affinity detector or any of the other detectors that respond to water. Similar reasoning applies when analyzing steroids in the presence of nonsteroidal organic compounds. Less prepurification is necessary before the analysis if the detector is not sensitive to the other compounds. The minimum concentration of a steroid that can be measured in a biological fluid also depends on the specificity of the detector since that determines how much sample can be injected.

Still another problem sometimes arises in that the responses of certain detectors to one compound may be changed if another is present. The hydrogen flame detector is particularly useful in trace analyses because this effect is not observed. The same cannot be said for the argon ionization detector and the electron affinity detector. The responses of these detectors to given steroids may be profoundly altered by the presence of other compounds in the gas. More prepurification of the sample is therefore required when these detectors are used.

The chief advantage offered by the electron affinity detector is its high sensitivity. Under ideal conditions, smaller quantities of certain compounds can be distinguished from detector noise with the electron affinity detector than with any other detector now available. Although many steroids are not markedly electronegative, many can be detected with very high sensitivity if haloacetates of the steroids are prepared (7). Although the electron affinity detector is considered to be quite specific, it was of interest to note that the papers presented at this conference by people who used this technique described prepurifying the sample considerably before injecting it into the column. It must be assumed that most of this purification is necessary because of the relative non-specificity of the detector. Other, more specific, detectors are available with sensitivity that approaches that of the electron affinity detector which may have even greater merit for this application. For example, Goulden, Goodwin and Davies reported that for analyses of chlorine containing insecticides, a halogen detector, such as one that is commonly used for detecting Freon leaks, offered appreciably greater specificity and higher practical sensitivity than the electron affinity detector (8).

We noted a similar improvement in practical sensitivity for certain compounds in our experiments with a halogen detection method that is even more specific. In our procedure, the effluent of the column was burned in a hydrogen flame ionization detector (9). In addition to the usual electrode arrangement, in this detector the flame impinged upon a metal screen that was pretreated with a salt of one of the alkali metals. The rate of volatilization of the alkali metal from the screen was increased by the presence of halogen or phosphorus in the flame. A second hydrogen flame was placed above the heated area of the screen. This hydrogen flame ionized the alkali metal released from the screen and detected it but was completely unresponsive to organic compounds in the lower flame because those compounds had already been burned. The upper flame therefore responded only to the concentration of the halogen while the lower flame responded to all organic materials in the effluent. When chloroacetyl derivatives of the alcohols or amines in a mixture were analyzed, the upper flame responded in proportion to the number of moles of alcohol or amine present. A similar molar response was obtained to fatty acids and aldehydes by analyzing their chloroethyl derivatives. A striking advantage of this detector was its insensitivity to the compounds present that did not contain halogens and phosphorus. Relatively minute amounts of chlorine and phosphorus containing organic compounds could be analyzed with no interference from the presence of much larger quantities of esters of fatty acids (Fig. 1). The response of the upper flame, shown on the upper record, was proportional to the quantity of chlorine and phosphorus present, while the entire composition of the mixture was detected by the lower flame (lower record).

Successful application of this detector to steroid analysis requires that halogen-containing derivatives suitable for chromatography be prepared. Our attempts to chromatograph dichloroacetyl derivatives of testosterone have so far been unsuccessful. If these can be successfully chromatographed, it should be possible to detect them specifically in quantities less than a tenth of a microgram.

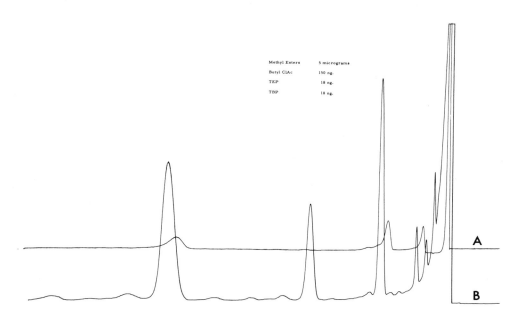

Fig. 1. Analysis of a mixture containing butyl chloroacetate 0.150 µg; triethyl phosphate and tributyl phosphate 0.018 µg each; and methyl esters of fatty acids 5.0 µg. The upper tracing (A) was that of the upper flame; the lower tracing (B) that of the lower flame of the hydrogen flame detector and specific for halogens and phosphorus. Only the halogen and phosphorus were detected in the upper flame. There was no interference from the methyl esters.

Metabolic Studies

One of the potential applications of GLC of which we have heard very little is the study of the metabolism of individual steroids in both tissue homogenates and in vivo. The ability to analyze rather complicated mixtures of steroids rapidly and quantitatively and the ability to measure radioactivity in the effluent of the gas chromatography column makes GLC ideal for this purpose.

How well gas chromatography can compare with alternative methods of analysis of steroids depends, of course, on the interplay of all of the factors we have considered. Many of the shortcomings of GLC, it seems to me, are attributable to the non-specificity of the response of the GLC detector as compared to the highly specific detection methods used for other forms of chromatography. In this respect, it should be pointed out that many of the detection methods ordinarily used with other kinds of chromatography can probably also be applied to gas chromatography and in addition, even more sensitive, highly selective gas detectors are feasible and can be developed. On the other hand, some of the separations that can be achieved by thin layer and liquid chromatography would be difficult by gas chromatography. It seems only reasonable to expect that the highly sensitive GLC detectors and the automatic recording techniques associated with them should be applicable to these other kinds of chromatography as well.

## REFERENCES

1. Cooke, W. D., Abstracts, Am Chem Soc National Meeting, April 1960.

2. Sternberg, J. C., W. S. Galloway, and D. T. L. Jones, In Gas Chromatography. Brenner, N., J. E. Callen, and M. D. Weiss, eds., Academic Press, New York, 1962, p. 231.

3. Ettre, L. S. Ibid. p. 307.

4. Perkins, G., Jr., G. M. Rouayheb, L. D. Lively, and W. E. Hamilton. Ibid. p. 269.

5. Sweeley, C. C. and T. Chang, Anal Chem 33: 1860, 1961.

6. Bloomfield, D. K., Anal Chem 34: 737, 1962.

7. Landowne, R. E. and S. R. Lipsky, Anal Chem 35: 532, 1963.

8. Goulden, R., E. S. Goodwin, and L. Davies, Analyst 88: 951, 1963.

9. Karmen, A., Anal Chem 36: 1416, 1964.

# INDEX

## A

Adsorption, 50, 51, 53, 292, 297, 302
Aldosterone diacetate, behavior on column, 104, 121
Allopregnanediol, separation from pregnanediol, 187
Androstenediol, 182
Atlantic Ocean, 47, 121, 195

## C

Column
    preparation, 125, 302
    priming, 50, 51, 124, 302
    specificity, 81, 265, 266
    support, 124, 301
    testing, 292

Corticoids
    breakdown on column, 123
    comparison of GLC with standard method, 113
    specificity of GLC, 116, 123

Cortisol, suggested method with GLC, 124

## D

Decomposition on column, 297
Derivatives, 195, 279
    chloroacetates, 26,
    halogens, 51, 52
    trimethylsilyl ethers, 67, 89, 93, 94
Displacement phenomena, 77, 302

## E

Electron capture
    detector, 25
    pulsed voltage, 266
    purification necessary, 266
    sensitivity, 30, 157, 159, 270, 271
Epitestosterone, 53
Error of estimate, 181
Estrogens
    fluorometry vs GLC, 264, 265
    paper chromatography vs GLC, 263
    specificity of GLC, 204, 223
Etiocholanolone, separation from 3β-hydroxyandrosterone, 190

## F

Flame ionization detector, sensitivity, 179, 300, 301

## G

Gas-liquid chromatography
    purification necessary, 268, 269
    resolution, 271
    vs conventional techniques, 125, 191

Gas-liquid radiochromatography, 53

## H

Halogen-containing derivatives, 51, 52
16α-hydroxydehydroepiandrosterone, 182
18-hydroxyetiocholanolone, 44
20β-hydroxyprogesterone
    in plasma, 179
    separation from 20α- , 179

## I

Injection techniques, 75, 77, 113, 115
Internal standards, 26, 27, 47, 50, 154
Isotope fractionation, 104

## K

17-Ketosteroids
    comparison of GLC and other methods, 7, 43, 46
    phases for GLC of, 46
    specificity of GLC, 5, 9

## P

Phases, 302, 307
    curing, 272
    mixed, 265, 284, 292
    polar vs non-polar, 278
    preparation, 121, 122
    stability, 121, 182, 272
Pregnanediol
    comparison of GLC with Klopper method, 136, 185, 187
    in plasma, 190
    mono and di-TMS ethers, 294
Pregnenediol, 182
Pregnenetriol, 182
Progesterone
    in plasma, 121
    in rat ovarian vein plasma, 177
    radio gas-liquid chromatography, 177
Progesterone metabolites, GLC of, 188, 189

## R

Retention time, 300

## S

Separation of steroids in biological mixtures, 267
Septum, artefacts produced by, 77, 115
Solid sample techniques, 117, 143, 272
Solvent residues, 247, 304
Steroid number concept, 290
Support
    treatment, 278, 281
    effect on separation, 279, 280, 283

## T

Testosterone, plasma
    double isotope derivative method, 48
    error in measurement in women, 43
    levels, 24

    precision, 47
    specificity of electron capture method, 31
    spermatic vein, 47
Testosterone, urinary
    errors in measurement by GLC, 43
    precision, 44
    purification of sample, 44
    separation from epitestosterone, 53, 55
Transformation on column
    cholesterol, 288
    18-hydroxyetiocholanolone, 44
    steroid sulfates, 289